Interfacial Mechanics of Offshore Engineering Equipment

Yu Zhang

Interfacial Mechanics of Offshore Engineering Equipment

Springer

Science Press
Beijing

Yu Zhang
College of Safety and Ocean Engineering
China University of Petroleum
Beijing, China

ISBN 978-981-95-3199-8 ISBN 978-981-95-3200-1 (eBook)
https://doi.org/10.1007/978-981-95-3200-1

Jointly published with Science Press

The print edition is not for sale in China (Mainland). Customers from China (Mainland) please order the print book from: Science Press.

This Springer imprint is published by the registered company Springer Nature Singapore Pte Ltd.
The registered company address is: 152 Beach Road, #21-01/04 Gateway East, Singapore 189721, Singapore

Competing Interests The author has no competing interests to declare that are relevant to the content of this manuscript.

Contents

Application and Challenges of Carbon Fiber Reinforced Polymers in Marine Structures and Equipments

1 Introduction of Carbon Fiber

Carbon fiber is a high-strength, high-modulus fiber material composed of at least 95% carbon by weight. Its tensile strength exceeds eight times that of conventional carbon steel, while its density is less than one-fourth of steel (Table 1) [1]. Carbon fiber can be classified through multiple approaches. Based on raw material sources, it is categorized into: polyacrylonitrile-based (PAN) carbon fiber, pitch-based carbon fiber, viscose-based carbon fiber, phenolic-based carbon fiber, and gas-phase grown carbon fiber. Classification by mechanical properties yields high-strength (HS), ultra-high-strength (UHS), high-modulus (HM), and ultra-high-modulus (UHM) carbon fibers. Physical forms include continuous filament, short fiber, and chopped fiber. By product specification, carbon fibers are divided into aerospace-grade (small tow) primarily used in defense, high-tech industries, and sports equipment, and industrial-grade (large tow) for civil applications. At present, PAN carbon fiber

Table 1 Comparison of the properties of carbon fiber with other village materials [1]

Name of material	Tensile strength (MPa)	Elastic modulus (GPa)	Percentage of breaking elongation (%)	Densities (g/cm^{-3})
Q235 steel	420	210	25	7.85
2A12 aluminum	410	75	12	2.82
E glass fiber	3430	73	4.8	2.54
Aramid fiber 1414	3600	131	2.8	1.44
T300 carbon fiber	3530	230	1.5	1.76
T700S carbon fiber	4900	230	2.1	1.81
T800H carbon fiber	5490	294	1.5	1.8
T1000G carbon fiber	6370	294	2.2	1.91
M55J carbon fiber	4020	540	0.8	7.85

Y. Zhang, *Interfacial Mechanics of Offshore Engineering Equipment*,
https://doi.org/10.1007/978-981-95-3200-1_1

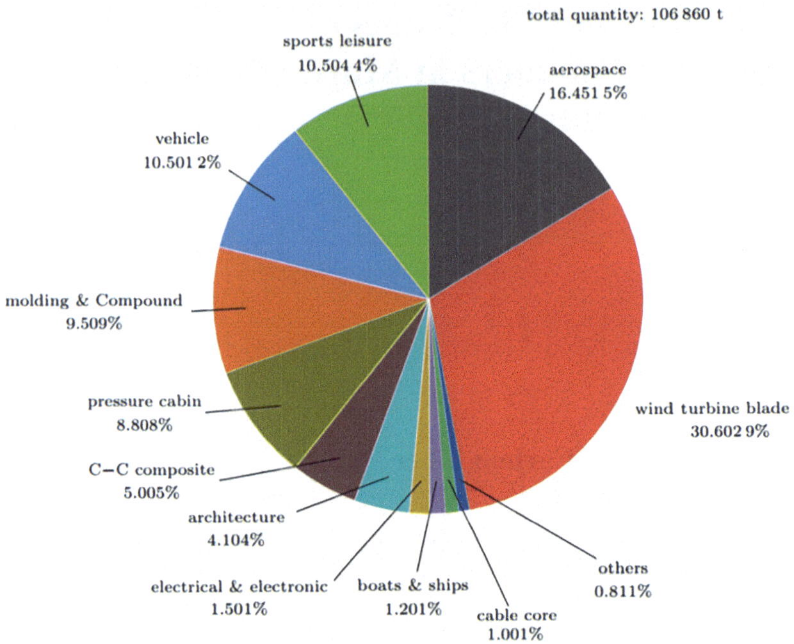

Fig. 1 Global carbon fiber demand-applications in 2020 (kt) [5]

dominates the market, representing roughly 90% of total worldwide carbon fiber manufacturing output [2].

Owing to its outstanding comprehensive performance, carbon fiber has seen progressively expanding applications across multiple industries. In the aerospace sector, carbon fiber reinforced composites are extensively utilized in aircraft structural components, particularly for manufacturing wings, nose cones, forward/center fuselage sections, and fairings [3]. The material also serves critical functions in rocket engine nozzles. For defense applications, carbon fiber composites enable weapon systems with superior miniaturization, reduced weight, enhanced strength, extended service life, and improved stability/safety. The sports equipment domain represents a historically significant market, accounting for one-third of Japan's early carbon fiber consumption [4]. Typical applications range from fishing rods to badminton rackets, golf clubs, and tennis rackets. In automotive engineering, carbon fiber has become a pivotal lightweight solution, with applications spanning exterior panels, battery enclosures, powertrain components, and braking systems. The advent of new energy vehicles is expected to further accelerate its automotive adoption. Civil engineering applications include structural reinforcement for bridges and buildings, where asphalt-modified carbon fiber reinforced cement demonstrates exceptional performance (Figs. 1 and 2).

In recent years, carbon fiber has also gradually emerged in marine structures and equipment. With the development of deep-water oil and gas resources and clean ocean energy, the application of carbon fiber in marine structures and equipment has also attracted much attention.

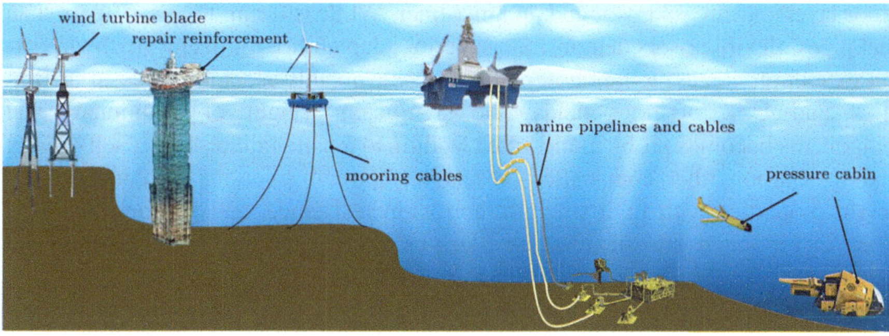

Fig. 2 The application of carbon fiber in ocean engineering

2 Carbon Fiber Applications in Marine Structures and Equipment

High-strength steel has traditionally been the primary material for marine structures and equipment. However, as offshore resource exploration extends from shallow to deep-water environments, there is growing demand for structures with superior lightweight properties, corrosion resistance, and fatigue performance. This emerging requirement has created an urgent need for advanced material solutions. Composite materials, with their unique advantages, offer innovative approaches for marine engineering applications. Currently, carbon fiber composites are seeing increasingly widespread adoption in critical marine components including underwater pipelines and cables, mooring systems, turbine blades, pressure-resistant hulls, and equipment rehabilitation.

2.1 Carbon Fiber in Marine Pipeline Applications

In deepwater and ultra-deepwater offshore oil and gas development, the standard configuration typically consists of a surface floating facility connected to a subsea production system through various pipelines and cables. These connecting systems primarily include production risers and umbilical cables. The production riser serves as the conduit for transporting hydrocarbons and other media, while the umbilical cable provides essential utilities to subsea equipment, including electrical power, hydraulic control signals, and chemical injection fluids. These dynamic systems operate in harsh marine environments where they are continuously subjected to wind, wave, and current forces. Consequently, their design must account for critical mechanical properties such as tensile strength, bending resistance, torsional capacity, and fatigue performance.

Production risers primarily consist of steel pipes and flexible risers. The latter feature a multilayer composite construction, typically comprising an outer sheath, armor layers, a carcass layer, and an inner liner. While offering lower stiffness than steel pipes, flexible risers demonstrate superior adaptability for deepwater hydrocarbon extraction. As exploitation depths increase—notably reaching 3000 m in Brazilian and U.S. offshore fields—the significant weight of traditional steel components becomes problematic, often leading to tensile failures under self-weight loading. Carbon fiber composites present an optimal solution, combining lightweight properties with high strength, corrosion resistance, and fatigue durability. When incorporated into flexible riser designs, these advanced materials maintain structural integrity while substantially reducing weight [6]. Meniconi [7] studied a 1000 m tension leg with production risers, employing ±20° carbon fiber laminated material and ring-direction glass fiber composite as the structural layer. The wall thickness of this composite structural layer was determined based on relevant damage criteria, followed by an analysis of the axial tensile-compressive failure load envelope of the overall structure. Chen et al. [8] proposed a structural design method for carbon fiber composite gas pipelines, incorporating reliability theory and structural optimization methods. In 2003, DNV released the design specifications and acceptance standards for composite risers [9]. Yan et al. [10] established an analytical model to determine the safety factor of composite materials according to the limited strain criterion, accounting for the property characteristics of carbon fiber composites. Amaechi [11] conducted stress analysis on a deep-water carbon fiber composite riser using finite element analysis software, investigated the overall structural safety factor under different loading conditions, and optimized key parameters of the structure.

The umbilical cable is a key component of the underwater control system, characterized by a complex cross-section that integrates multiple functional units for transmitting hydraulic pressure, electrical power, control signals, and chemicals between surface facilities and subsea production systems. Deepwater umbilical cables typically incorporate a steel tensile armor layer (composed of an even number of strands with opposing helical angles). However, as water depth increases, the excessive self-weight of the steel armor layer may lead to tensile failure of the umbilical cable. Replacing the steel armor layer with carbon fiber reinforced rods can provide sufficient tensile strength while significantly reducing weight. Yi [12] established a mechanical model of carbon fiber helically wound reinforcement rods for umbilical cables, accounting for the material's anisotropic properties, and analyzed both its stiffness and the dynamic response of the umbilical cables. Zhang and Xia [13, 14] investigated failure mechanisms including tensile and torsional failure, subsequently evaluating system reliability. The application of carbon fiber composites to marine pipes and cables achieves both weight reduction and enhanced structural strength. Nevertheless, the long-term durability of carbon fiber pipes and cables in complex marine environments requires further validation through extended engineering practice. Although the manufacturing costs of carbon fiber pipes and cables are higher compared to steel counterparts, the total lifecycle cost—considering installation, operation, and maintenance—may ultimately be lower.

At present, hoses used in hydrate and mineral resource mining, LNG cryogenic transfer, CO_2 injection, and oil and gas offloading operations have stringent

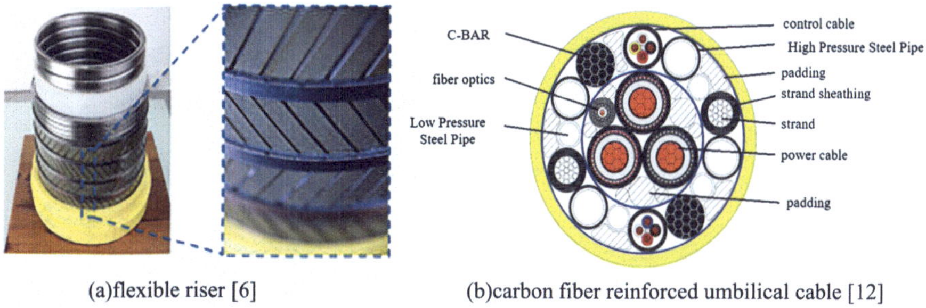

Fig. 3 Marine pipe cables. (**a**) Flexible riser [6]. (**b**) Carbon fiber reinforced umbilical cable [12]

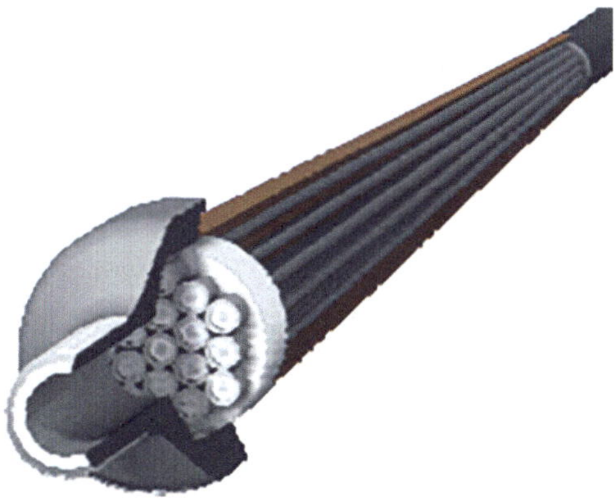

Fig. 4 Carbon fiber composite bollard [16]

requirements for stiffness, corrosion resistance, and lightweight properties. Therefore, carbon fiber composites hold significant potential for application in these types of hoses (Fig. 3).

2.2 *Application of Carbon Fiber in Mooring Systems*

In deep-sea oilfield exploitation, production platforms are prone to instability on the sea surface due to wind, waves, and currents. The primary function of the mooring system is to maintain the platform's position. Traditional mooring systems primarily use steel structures, which are susceptible to corrosion in seawater, resulting in relatively short service lives and high maintenance costs. Furthermore, the weight of steel mooring systems increases significantly with water depth. For instance, a drilling platform operating at 1500 m requires steel mooring cables weighing approximately 6500 tons [15] (Figs. 4 and 5).

Composite materials exhibit excellent mechanical properties and superior corrosion resistance. Replacing traditional steel mooring systems with composites can

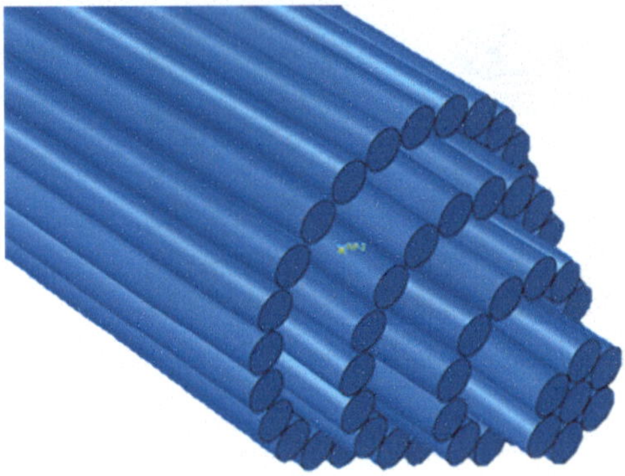

Fig. 5 Carbon fiber helical rod mooring cable model [19]

significantly reduce weight while effectively mitigating seawater corrosion, thereby extending service life. When platform operating depths exceed 1600 m, the total cost of composite mooring cables becomes lower than that of steel alternatives [16]. Jackson [17] developed a carbon fiber mooring cable specifically for mobile offshore drilling units. Zhou et al. [18] invented a novel composite anchor chain structure consisting of interlocking ring units, where the outer shell and inner core materials are carbon fiber and glass fiber, respectively. Felipe [19] conducted a mechanical analysis of carbon fiber rod mooring systems under tensile and bending loads, demonstrating that carbon fiber mooring cables possess greater tensile strength and smaller minimum bending radii compared to traditional structures. However, as a viscoelastic material, carbon fiber requires special consideration of dynamic viscoelastic behavior and creep effects under tension and cyclic loading conditions. The instantaneous and long-term tensile characteristics of carbon fiber mooring systems differ fundamentally from steel cables, necessitating specialized design and analysis approaches.

2.3 Carbon Fiber Applications in Pressure-Resistant Modules

Pressure-resistant hulls serve as core components for underwater equipment including deep-sea submersibles and gliders. Their primary design objectives require achieving both sufficient structural integrity and the lowest possible weight-to-buoyancy ratio [20]. The application of carbon fiber composites to pressure-resistant hulls provides significant advantages, including greater operational depth capability, reduced weight, and improved weight-to-buoyancy performance.

In the 1990s, the United States employed carbon fiber composites to optimize deep-sea submersible structures, achieving significant upgrades to underwater search systems. Field applications demonstrated that while meeting buoyancy requirements, these modifications substantially reduced the weight-to-buoyancy ratio, enabling operational depths up to 6096 m. In 2013, U.S. company OceanGate initiated the Cyclops deep-submergence vehicle development program, which included two variants: the Cyclops I with a carbon fiber hull (500 m depth rating) and Cyclops II featuring a carbon fiber-titanium alloy composite structure (4000 m operational depth) [21]. The United Kingdom's Southampton Oceanography Centre developed the AUTOSUB autonomous underwater vehicle (AUV), which incorporates a carbon fiber cylindrical pressure hull with titanium alloy end caps. This design maintains positive buoyancy while reducing structural weight, achieving a 6000 m diving capability. In March 2017, the Shenyang Institute of Automation launched the "Sea-Wing 7000" underwater glider for scientific missions in the Mariana Trench. The glider's pressure-resistant structure extensively utilized carbon fiber components, reaching a record operating depth of 6239 m during the mission and setting multiple world records [22]. In 2021, Professor Pan's research team at Northwestern Polytechnical University investigated carbon fiber applications for underwater gliders. Their work focused on the buckling failure mechanisms of carbon fiber pressure-resistant columns, including comprehensive analysis of buckling resistance across various wall thicknesses and elliptical ratios [23, 24] (Figs. 6 and 7).

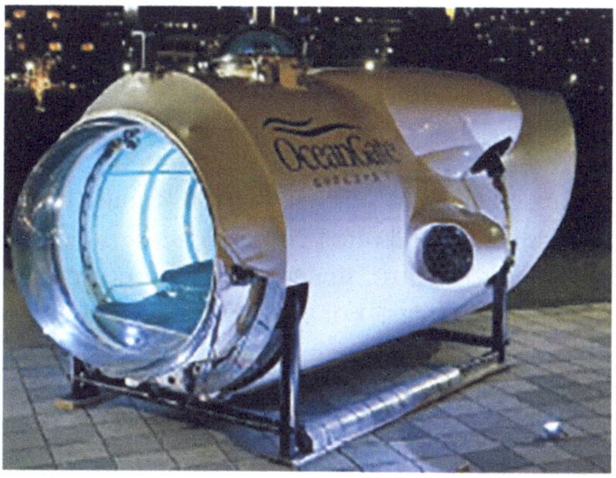

Fig. 6 Cyclops I manned deep submersible [21]

Fig. 7 China's "Sea Wing-7000" underwater glider [22]

2.4 *Carbon Fiber Applications in Marine Clean Energy Structures and Equipment*

As shown in Fig. 1, wind turbine blades accounted for the largest share of global carbon fiber demand in 2020. Benefiting from China's abundant offshore wind resources and the "carbon neutrality" and "carbon peaking" development goals proposed in 2020, the offshore wind power industry has experienced rapid growth as a green energy solution. In 2020, carbon fiber consumption for wind turbine blades reached 30,600 tons, representing 29% of global carbon fiber consumption. During the 14th Five-Year Plan period (2021–2025), the demand for carbon fiber in wind blades remains strong and is projected to exceed 93,000 tons by 2025 [25]. With advancements in wind turbine manufacturing and supporting technologies, both turbine capacity and blade size continue to increase. Research indicates that applying carbon fiber composites to key structural components (such as spar caps and main beams) in blades longer than 40 m can reduce blade weight by 38% and lower costs by 14%. As wind turbine capacities progressively increase and blade lengths continue to grow, carbon fiber has become an essential material choice for the industry.

The load distribution varies across different positions of wind turbine blades. Based on the distinct stress conditions and failure modes at various locations, different materials can be strategically employed in different blade sections. Carbon fiber composites, with a stiffness two to three times greater than glass fiber composites and a lower density, are typically suitable for manufacturing blade spars. As illustrated in Fig. 8, carbon fiber composites are primarily used for the main beam, leading edge, and trailing edge, while glass fiber composites are generally applied to the blade shell and other components [27, 28]. Hybrid composites are engineered by reinforcing a matrix material with two or more distinct materials, combining their optimal properties. For instance, one of the world's longest wind turbine blades utilizes glass/carbon hybrid reinforcement [29]. Furthermore, nanoengineered composites represent the latest advancement in composite fabrication. This approach incorporates nanoscale reinforcements into the matrix to enhance material properties. Merugula [30] demonstrated that adding 1–5 wt% carbon nanofibers to glass

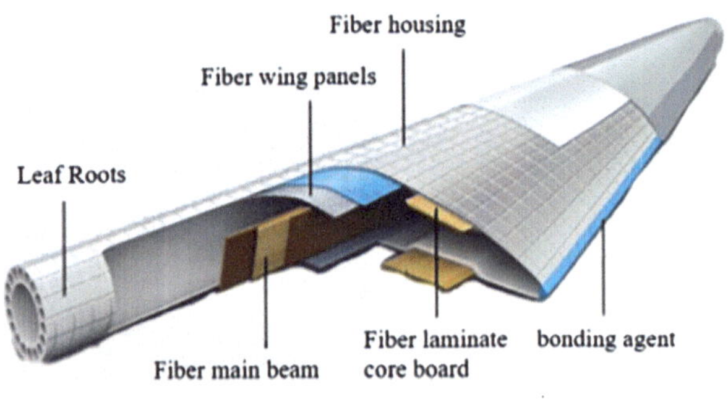

Fig. 8 Structural composition of wind turbine blades [26]

Fig. 9 Hydrogen storage cylinders [33]

fiber composite blades significantly improves their tensile strength and service life (Fig. 9).

Offshore hydrogen production technology represents a crucial component of clean energy development, where hydrogen storage technology becomes particularly critical due to hydrogen's low volumetric energy density at ambient conditions. Unlike metallic materials that suffer from hydrogen embrittlement, carbon fiber composites demonstrate excellent resistance to such damage, as evidenced by their successful application in hydrogen storage cylinders for new energy vehicles—a development that provides important references for offshore hydrogen storage solutions. Significant progress has been made in high-pressure hydrogen storage technology, beginning with Germany's Lincoln Composites achieving 95 MPa working pressure in 2002 using high-density polyethylene-lined carbon fiber fully-wound cylinders [31], followed by Canada's Dynetek developing aluminum-lined carbon fiber fully-wound vessels rated at 70 MPa [32], with China having now reached comparable 70 MPa production capabilities [33]. The field continues to advance through innovations like the $2.7 million U.S. DOE-funded project awarded to IACMI in 2020 to develop cost-reduced carbon fiber technology that could lower production costs by 25% for composite fuel tanks [34]. Recent research by Guo (2021) [35] has further contributed through simulation-based optimization of carbon fiber gas storage tanks, systematically evaluating key parameters including carbon/glass fiber winding configurations (layer counts, angles) and individual layer thicknesses to enhance structural performance.

In addition to hydrogen storage cylinders, carbon fiber is also widely used in oxygen tanks and other pressure vessels. While China's pressure vessel industry accounts for 17% of global carbon fiber demand, domestic production capacity remains limited due to current technological constraints. As a result, China primarily relies on imported carbon fiber for pressure vessel applications.

2.5 Application of Carbon Fiber in Repair and Reinforcement Operations of Marine Structures

Marine structures operating in complex and harsh ocean environments endure long-term working loads and environmental stresses, inevitably developing structural defects such as fatigue cracks and corrosion damage during service. Carbon fiber composites have gained widespread application in marine structure repair and reinforcement due to their advantageous properties, including low density, high strength, corrosion resistance, ease of construction, excellent fatigue performance, and minimal impact on structural integrity. Chen et al. [36] developed a mechanical model for carbon fiber-repaired cracked conductor rack platforms. The QinetiQ Company successfully addressed a 60 mm crack in the bulkhead of an FPSO (Floating Production Storage and Offloading vessel) operating in Norwegian waters using carbon fiber materials [37]. Zhang et al. [38] investigated the mechanical behavior of carbon fiber-repaired cracked steel plates, establishing theoretical expressions for interface stress and developing an interface failure prediction model. In subsequent research, Zhang et al. [39] examined the durability of carbon fiber-repaired corroded marine pipelines under combined load-temperature-seawater immersion conditions, analyzing the relationship between repair durability, carbon fiber material properties, and the degradation behavior of carbon fiber-metal interfaces. Alrsai et al. [40] evaluated the feasibility of carbon fiber-based anti-crush devices for marine pipelines, conducting comprehensive comparative analyses between carbon fiber solutions and traditional sliding or integral anti-crush devices. Elchalakani et al. [41] studied deformation patterns in carbon fiber cloth-repaired pipelines under three-point bending and direct indentation, investigating the influence of parameters including wall thickness corrosion penetration, longitudinal corrosion extent, and carbon fiber cloth type. While carbon fiber repair effectively enhances the ultimate load-bearing capacity of marine structures containing corrosion or cracks, inhibits defect propagation, and improves residual fatigue life, the multi-material interfaces in repaired structures remain vulnerable to damage under complex marine environmental conditions (temperature cycling, static/dynamic loading). These interface integrity issues can be monitored using acoustic emission, image processing, and other non-destructive evaluation techniques (Fig. 10 and 11).

3 Challenges of Carbon Fiber in Marine Structures and Equipment Applications

As marine resource exploration advances into deeper waters and marine clean energy development progresses, the performance requirements for materials used in marine structures and equipment continue to increase. Carbon fiber composites, with their excellent properties including lightweight characteristics, high strength, corrosion resistance, and fatigue resistance, are finding increasingly extensive

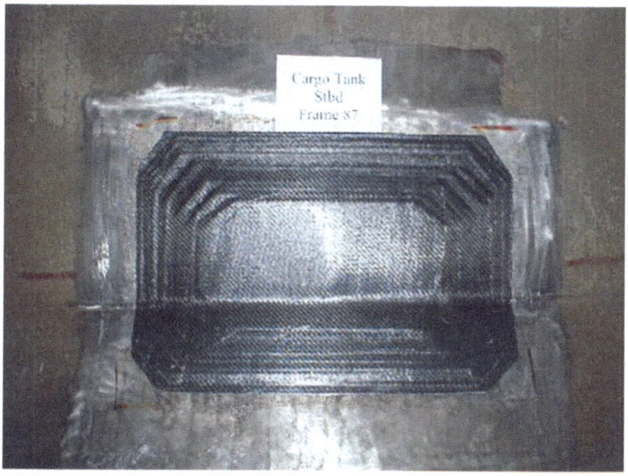

Fig. 10 CFRP repair of FPSO bulkheads [37]

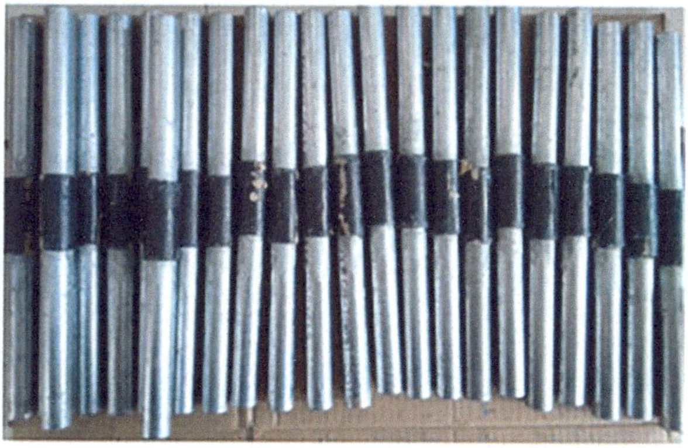

Fig. 11 CFRP repaired steel pipe [39]

applications in marine structures and equipment. However, the complex marine environment and loading conditions present significant challenges throughout the entire lifecycle of carbon fiber applications, from manufacturing to engineering implementation.

1. Carbon fiber composite structures feature complex heterogeneous material interfaces with multiple bonding surfaces: the intrinsic fiber-matrix interface within the composite material itself; interlayer bonding interfaces between multiple carbon fiber fabric layers in typical shell structures; and metal-composite interfaces when carbon fiber components are integrated with metallic structures in marine applications. Under prolonged exposure to harsh marine environments and complex loading conditions, these dissimilar material interfaces are particularly susceptible to damage due to mismatched material properties and incompatible deformation behavior. Typical failure modes include interfacial delamination and cracking, necessitating design and manufacturing process optimization to enhance interface performance.

2. Carbon fiber-based marine structures and equipment typically employ composite configurations. These systems must withstand complex multi-axial loading conditions including tension, compression, bending, and torsion, while simultaneously meeting stringent sealing and fatigue resistance requirements. The anisotropic nature of carbon fiber composites presents unique design challenges, as their superior mechanical properties are primarily manifested along the fiber orientation. This characteristic necessitates careful consideration during structural design, often requiring material hybridization to achieve optimal performance. Furthermore, the relatively inferior permeability resistance of resin-based carbon fiber composites compared to metallic materials imposes additional constraints on sealing system design, frequently demanding composite solutions incorporating other materials. The structural complexity is further compounded when carbon fiber is integrated with dissimilar materials. The manufacturing processes, failure mechanisms under harsh marine environments, and mechanical response analyses of such hybrid structures are significantly more complex than those of conventional monolithic metal structures.
3. Carbon fiber composites often face dispersion challenges during manufacturing. These materials typically incorporate carbon fibers or fiber fabrics as reinforcement within a carbonaceous or graphitized resin matrix. However, achieving uniform fiber dispersion in the matrix remains a significant technical hurdle. To address this issue, it is essential to conduct systematic research on processing technologies and optimize the matrix composition ratios. Such optimization ensures improved dispersion of either discrete carbon fibers or woven fabrics throughout the composite material, thereby enhancing overall material performance.
4. The long-term durability of carbon fiber composites in marine environments requires careful consideration. Marine structures and equipment must withstand prolonged exposure to complex environmental loads, including wind, waves, currents, and ice, while simultaneously resisting corrosion and temperature variations. Under these demanding conditions, the interfacial performance between carbon fiber and other materials tends to degrade over time, ultimately compromising the overall structural integrity. To address these challenges, optimization efforts should focus on both enhancing the intrinsic properties of carbon fiber materials and improving manufacturing and processing technologies.

4 Conclusion

The development of marine structures and equipment for deepwater oil and gas resources represents a critical "bottleneck" technology in China's marine energy sector. From the 11th to the 13th Five-Year Plan periods, China has achieved remarkable progress in this field, with several marine equipment successfully localized and implemented in engineering applications. However, complete technological independence and self-sufficiency remain to be accomplished. In 2020, China established its "carbon neutrality" and "carbon peak" development goals, positioning marine clean energy (including offshore wind power and offshore hydrogen production) as key components of this initiative. The advancement of these technologies fundamentally depends on the parallel development of supporting marine structures and equipment. This paper reviews current applications and research progress of carbon fiber composites in marine structures and equipment. Owing to their exceptional properties—including lightweight characteristics, high strength, and corrosion resistance—carbon fiber materials demonstrate significant potential for deepwater and clean energy applications, particularly in weight-sensitive designs that can substantially reduce lifecycle costs. Nevertheless, certain challenges persist in marine applications due to inherent material characteristics and manufacturing limitations. Addressing these issues requires multidisciplinary collaborative research spanning material development, production processes, and structural design optimization. With continued advancements, carbon fiber composites are expected to play an increasingly vital role in future deepwater resource exploitation and clean energy marine infrastructure development.

References

1. Yang C, Peng X. Nonferrous metals industry in-depth research: carbon fiber, technology strengthening + domestic substitution, black gold ushering in industry windfall. Beijing: Tianfeng Securities; 2020.
2. Qi Y. The development of carbon fiber and composites. Adv Mater Ind. 2017;(12):2–6. (in Chinese).
3. Zhang F. Reguirements of aerospace structural composites for carbon Firbers. Mater Rep. 2000;14(11):5–7. (in Chinese).
4. Kataoka K, Harada A, Nagasaki Y. Block copolymer micelles for drug delivery: characterization and biological significance. Adv Drug Deliv Rev. 2012;64(1):37–48.
5. Lin G. Global carbon fiber composite market report in 2020. Text Sci Res. 2021;(05):27–49. (in Chinese).
6. Vidigal da Silva J, Damiens A. 3000 m water depth flexible pipe configuration portfolio. In: Offshore technology conference, Houston, USA; 2016.
7. Meniconi LCM, Reid SR, Soden PD. Preliminary design of composite riser stress joints. Compos A: Appl Sci Manuf. 2001;32(5):597–605.
8. Chen J, Song D. Structural design for composite pipes. Acta Mater Compos Sin. 1996;13(2): 110–6. (in Chinese).
9. DNV offshore standard DNV-RP-F202: composite risers. In: Det-Norske-Veritas, editor. 2003.
10. Yan F, Yao W, Li H. Statistical analysis on safety factor of composite materials. J Nanjing Univ Aeronaut Astronaut. 2009;(05):105–9. (in Chinese).
11. Amaechi CV, Nathaniel G, Agbomerie CO, et al. Composite risers for deep waters using a numerical modelling approach. Compos Struct. 2018;210:486–99.
12. Yi Z. Analysis of mechanical properties of umbilical spiral carbon fiber reinforced rod. Master thesis, China University of Petroleum, Beijing; 2016. (in Chinese).
13. Zhang Y, Guo ZS, Zhao HK, et al. Analysis of torsional instability and reliability of marine flexible pipelines. Ocean Eng. 2021;228:108947.
14. Xia R, Zhang Y, Zheng LJ, et al. Reliability analysis of an umbilical under ultimate tensile load based on response surface approach. Underw Technol. 2020;37(3):87–93.

15. Dai Z, Song P, Gao Z, et al. Application of fiber reinforced composites in offshore oil and gas exploration. Ocean Eng Equip Technol. 2014;1(03):249–53. (in Chinese).
16. Sparks C, Zivanovic I, Sparks F, et al. Carbon fiber composite tendons for Deepwater tension leg platforms. In: Offshore technology conference, Houston, USA; 2003.
17. Jackson D, Shepheard B, Kebadze E, et al. CFRP mooring lines for Modu applications. In: Offshore technology conference, Houston, USA; 2005.
18. Lifeng Z, Zhengwen L. A composite material-based anchor chain and its preparation method: China. CN209365540U; 2019.
19. Luz FF, De Menezes EW, Cimini CAJ, et al. Numerical analysis of carbon fiber cables for mooring lines under tensile and bending loading. Int J Offshore Polar Eng. 2018;28(4):387–92.
20. Zhang J, Zuo XL, Wang WB, et al. Overviews of investigation on submersible pressure hulls. Adv Nat Sci. 2014;7(4):54–61.
21. Luo S, Li Y, Wang W. Development and prospects of non-metallic submersible pressure hull. Chin J Ship Res. 2020;15(4):9–18. (in Chinese).
22. Lai, S. Design and experiments of buoyancy-driven system of Full Ocean depth autonomous underwater glider. Master thesis, Tianjin University, Tianjin; 2018. (in Chinese).
23. Shen KC, Pan G. Buckling and strain response of filament winding composite cylindrical shell subjected to hydrostatic pressure: numerical solution and experiment. Compos Struct. 2021;276:114534.
24. Li Z, Shen KC, Zhang XH, et al. Buckling of composite cylindrical shells with ovality and thickness variation subjected to hydrostatic pressure. Def Technol. 2021:(1).
25. Peng G, Li G, Shi F, et al. Development status and analysis of domestic PAN—based large tow carbon fiber. Hi-Tech Fiber Appl. 2021;46(06):11–6. (in Chinese).
26. Arias F. Assessment of present/future decommissioned wind blade fiber-reinforced composite material in the United States. Independent Study, Dept. of Civil Engineering, City College of New York, United States; 2016.
27. Luo Y, Li W, Hu H, et al. Application of carbon fiber reinforced composites in wind turbine blade. Power Syst Clean Energy. 2008;(11):53–7. (in Chinese).
28. Wang Z, Sun J, Liu L. High module glass fiber single axis fabric used in 2MW wind turbine blade. Fiber Compos. 2012;29(04):8–10. (in Chinese).
29. Wind Power Monthly Webpage. https://www.windpowermonthly.com/article/1419306/turbines-year-rotor-blades.
30. Merugula L, Khanna V, Bakshi BR. Reinforced wind turbine blades—an environmental life cycle evaluation. Environ Sci Technol. 2012;46(17):9785–92.
31. Kai F, Liu Z, Fu Q, et al. State-of-the-art of fiber-reinforced high-pressure hydrogen storage tanks. J Press Equip Syst. 2004;2:59–63.
32. Dynetek Luxer Dynecell lightweight high-pressure hydrogen-storage cylinders. http://www.uxfercylinders.com/products/alternative-fuel-solutions/688-luxfr-dynec.
33. Kesheng O. Research on fully-wrapped carbon fiber reinforced composite high-pressure hydrogen storage Clinder Subiected to localized fire. PhD thesis, Zhejiang University, Hangzhou; 2014. (in Chinese).
34. Iacmi—the composites institute; new IACMI collaboration aims to develop advanced carbon fiber for hydrogen, natural gas storage tanks. Energy Weekly News. 2020.
35. Guo W, Wang X, Wu M, et al. Structural parameters optimization of vehicle gas storage tank made of fiber reinforced polymer based on kriging model. Plastics. 2021;50(03):115–20. (in Chinese).
36. Chen T, Chen G, Lin H, et al. Simulation study of CFRP-based repair effect for cracked tubular joint of offshore platforms. China Pet Mach. 2008;356(10):1–4, 50, 96. (in Chinese).
37. Turton TJ, Dalzel-Job J, Livingstone F. Oil platforms, destroyers and frigates-case studies of QinetiQ's marine composite patch repairs. Compos Part A Appl Sci Manuf. 2005;36:1066–72.
38. Zhang Y, Zhang K, Zhao HK, et al. Stress analysis of adhesive in a cracked steel plate repaired with CFRP. J Constr Steel Res. 2018;145:210–7.

39. Zhang Y, Liu ZZ, Xin JH, et al. The attenuation mechanism of CFRP repaired corroded marine pipelines based on experiments and FEM. Thin-Walled Struct. 2021;16:108469.
40. Alrsai M, Karampour H, Hall W, et al. Carbon fibre buckle arrestors for offshore pipelines. Appl Ocean Res. 2021;111:102633.
41. Elchalakani M, Karrech A, Basarir H, et al. CFRP strengthening and rehabilitation of corroded steel pipelines under direct indentation. Thin-Walled Struct. 2017;119:510–21.

Key Technologies and Challenges in Marine Composite Pipe and Cable Design

1 Introduction to Marine Composite Cables

Marine pipes and cables serve as critical lifelines for offshore engineering structures and represent essential components in marine resource development. Marine composite pipes and cables primarily consist of several types: composite flexible pipes (including flexible risers, floating hoses, cryogenic hoses, and thermoplastic pipes), umbilical cables, and submarine cables. Compared to traditional steel pipes, these composite systems integrate metallic and non-metallic materials to achieve multifunctional performance while offering superior mechanical properties. They combine optimal stiffness-strength characteristics with enhanced flexibility, along with significant advantages in weight reduction, high strength, and corrosion resistance. These attributes make composite pipes and cables particularly suitable for demanding applications in deepwater and ultra-deepwater environments, as well as for projects requiring specialized functionalities and extreme environmental resilience.

Composite flexible pipes and cables can be classified into non-bonded and bonded types based on their manufacturing processes, with flexible risers representing a typical non-bonded configuration whose structure is shown in Fig. 1. These risers are primarily used for connecting offshore floating production platforms to subsea production systems in deepwater environments, featuring a multilayered construction comprising: (1) a skeleton layer that resists external pressure and supports the liner; (2) an inner liner layer ensuring fluid containment; (3) a compressive armor layer resisting radial pressures; (4) a tensile armor layer bearing axial loads; (5) an anti-friction layer reducing intermetallic wear; and (6) an outer jacket layer preventing seawater intrusion and fluid leakage. Flexible riser design follows international standards including API SPEC 17J, API RP 17B, and ISO 13628-2.

The umbilical cable serves as a critical component for subsea production systems, providing power transmission, electrical signal communication, hydraulic power

Y. Zhang, *Interfacial Mechanics of Offshore Engineering Equipment*,
https://doi.org/10.1007/978-981-95-3200-1_2

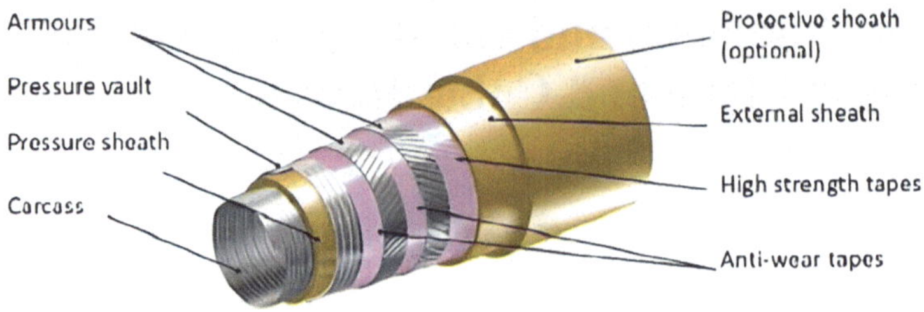

Fig. 1 Typical structure of a flexible riser [1]

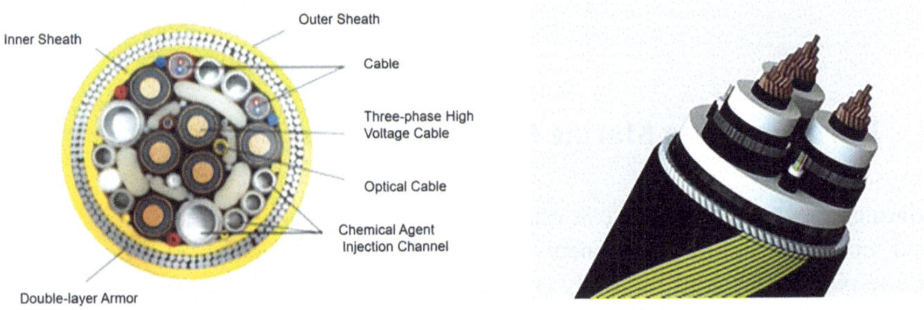

(a) Umbilical cable (b) dynamic cable

Fig. 2 Typical structure of umbilical cord cables and cables [2]. (**a**) Umbilical cable. (**b**) Dynamic cable

supply, and chemical reagent transportation. As illustrated in Fig. 2a, its cross-sectional configuration typically incorporates multiple functional units including electrical cables, optical fibers, hydraulic tubes, and chemical injection lines. The design and manufacturing of umbilical cables generally comply with international standards such as ISO 13628-5 and API 17E. Marine cables (shown in Fig. 2b) are commonly designed according to IEC 60287 standards. However, for dynamic marine cables used in wind turbine applications, there currently exists a lack of specific standards, necessitating reference to both umbilical cable standards and relevant marine cable specifications during design processes.

The floating hose represents a typical bonded composite hose (as illustrated in Fig. 3), primarily employed for crude oil transfer operations between FPSOs, buoys and shuttle tankers, coastal tanker loading/unloading, and naval vessel replenishment at sea. Its multilayered structure consists of: (1) inner and outer rubber layers providing corrosion resistance and protective functions; (2) a cord reinforcement layer composed of rubber-impregnated cords for tensile load bearing; (3) a spiral steel wire layer embedded within the inner rubber layer to withstand combined tensile, bending, and pressure loads; and (4) a buoyancy layer of foam material positioned between the cord reinforcement and outer rubber layers for flotation capability. Floating hose design and manufacturing generally adhere to API 17K, API 17B, and OCIMF 2009 standards.

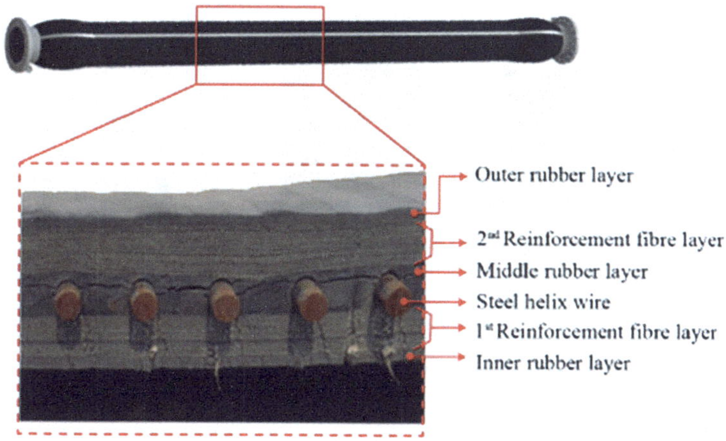

Fig. 3 Typical structure of bonded composite flexible hose [3]

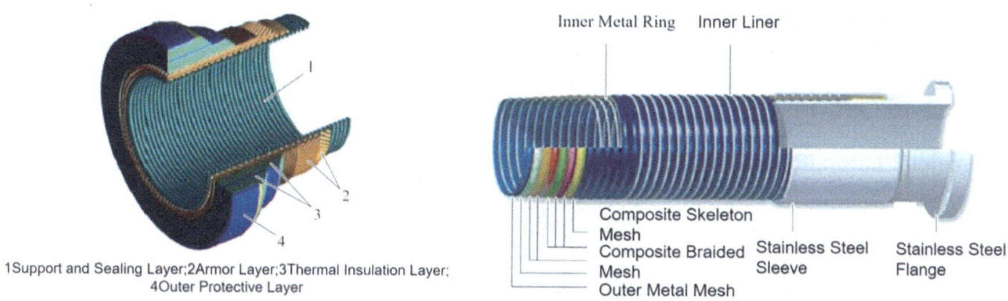

(a) Metal bellows (b) Composite cryogenic hose

Fig. 4 Typical structure of a cryogenic hose [4]. (**a**) Metal bellows. (**b**) Composite cryogenic hose

Cryogenic hoses are primarily designed for transferring cryogenic media such as liquefied natural gas (LNG), offering significant advantages over traditional rigid loading arms, particularly in harsh marine conditions. Based on their structural configuration, cryogenic hoses are classified into two types: metal bellows hoses and non-bonded composite cryogenic hoses. As illustrated in Fig. 4a, the metal bellows hose structure consists of four key components: (1) an inner bellows layer, (2) a tensile armor layer, (3) an insulation layer, and (4) an outer protective layer. The composite cryogenic hose (Fig. 4b) features a different construction comprising: (1) a spiral metal strip, (2) a composite skeleton mesh, (3) an inner liner, (4) a composite braided mesh, and (5) an outer metal mesh. The design of LNG cryogenic hoses typically follows the EN 1474-2 standard.

Thermoplastic Composite Pipe (TCP) offers significant advantages including spoolability, corrosion resistance, fatigue resistance, and lightweight properties, making it particularly suitable for marine engineering applications where traditional steel risers and flexible risers often fail due to corrosion. As shown in Fig. 5, the TCP

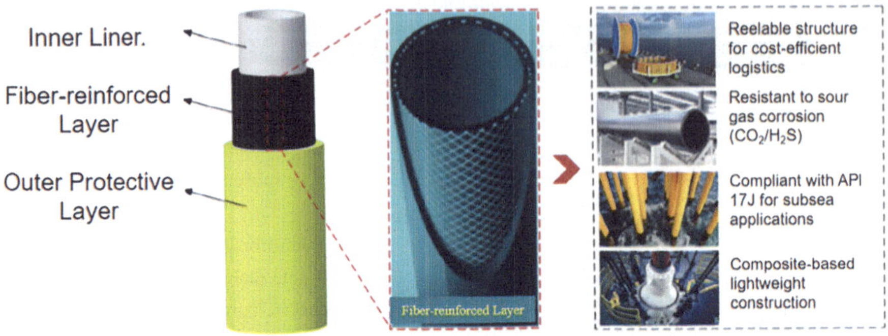

Fig. 5 Typical structure of a thermoplastic pipe [5]

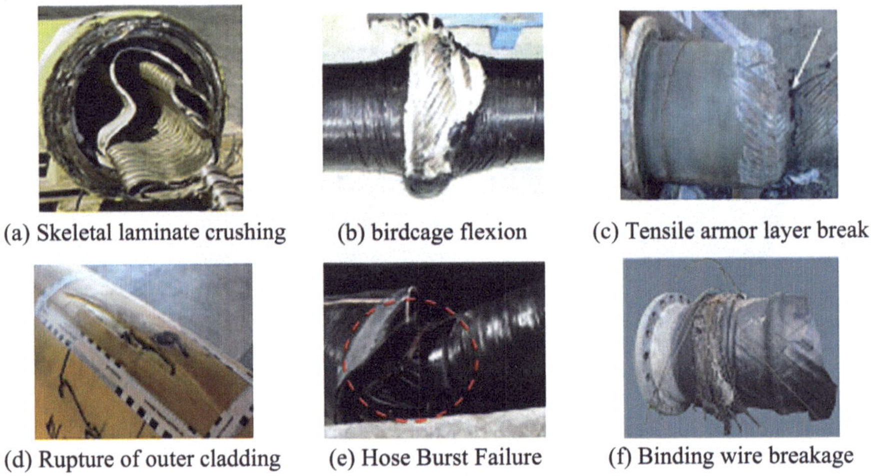

Fig. 6 Failure modes of composite tubular cables [6, 7]. (**a**) Skeletal laminate crushing. (**b**) Birdcage flexion. (**c**) Tensile armor layer break. (**d**) Rupture of outer cladding. (**e**) Hose Burst Failure. (**f**) Binding wire breakage

structure consists of three functional layers: (1) an inner liner providing fluid containment and corrosion protection; (2) an outer protective layer maintaining pipeline integrity during transportation, installation, and service; and (3) a fiber reinforcement layer that delivers the required structural rigidity and strength. The design and qualification of thermoplastic pipes typically comply with DNVGL-RP-F119 and ISO 23936-1 standards.

Marine composite pipe-cable systems play a vital role in offshore oil and gas development, offshore wind power generation, and deep-sea mining operations, though their structural complexity in harsh operating environments leads to various damage mechanisms as illustrated in Fig. 6. Common failure modes include skeleton layer corrosion, buckling deformation, and pressure collapse; liner layer aging and degradation; tensile/compressive armor layer fatigue fracture, cracking, and deformation; outer sheath rupture, cracking, and abrasion; bonded hose bursting failure and spiral steel wire fracture; as well as joint debonding and bundling wire rupture. These potential failure mechanisms highlight the critical importance of breakthroughs in marine composite pipe-cable design technologies to reduce failure probabilities and advance marine resource development.

2 Design Specification for Composite Cables

The basic design of marine composite pipe-cable systems, as outlined in relevant design specifications, comprises three fundamental components: cross-sectional design, line configuration design, and accessory design. According to API 17B specification [8], flexible pipe-cable designs are typically classified as either static or dynamic configurations. Existing design standards [8–10] define the static pipe-cable design methodology as a systematic process involving: (1) material selection, (2) cross-sectional dimensioning and functional design, (3) line configuration planning, (4) accessory design with service life evaluation, and (5) installation engineering. In contrast to static designs, dynamic pipe-cable design incorporates additional dynamic analysis and optimization following the initial line configuration phase.

The material selection for marine composite cables is primarily determined by operational environmental conditions and functional requirements, particularly the need for lightweight, high-strength, and corrosion-resistant properties in marine applications. Following material selection, the cross-sectional dimensions and functional design are developed based on existing engineering experience and case studies, with preliminary cross-section design adhering to principles of symmetry and compactness. The cable routing configuration must be designed according to environmental conditions, with three typical patterns employed: steep wave type, gradual wave type, and steep S-type configurations. Upon completion of accessory design, the preliminary cable design is finalized. However, subsequent structural verification and optimization are essential, typically involving numerical analysis to evaluate mechanical performance and service life under various operational and installation conditions. When initial designs fail to meet performance requirements, multiple iterative calculations are necessary. For detailed design specifications, API 17K and API 17J provide distinct guidelines for bonded and unbonded composite cable systems respectively.

Furthermore, as marine composite pipe-cable technology continues to advance, specialized design requirements have emerged for different application scenarios. The EN 1474-2 standard [11] specifies that cryogenic hose designs must incorporate buoyancy layer and thermal insulation layer considerations alongside conventional requirements including material selection, cross-sectional dimensioning, and load calibration. Similarly, DNVGL-RP-F119 [12] mandates specific design provisions for thermoplastic composite pipes regarding corrosion resistance and layered structure parameters, while the OCIMF 2009 guidelines [13] emphasize the need to account for installation impacts when designing floating hoses intended for long-distance fluid transportation.

3 Technical Challenges in Composite Cable Design

Current design specifications present several methodological limitations in composite pipe-cable engineering. First, the segregated approach to design verification and calibration analysis necessitates multiple iterative calculations when initial designs fail to meet performance requirements, resulting in excessive computational workload. Second, the decoupled methodology between global line design and local cross-section design often creates suboptimal solutions where mechanically favorable line configurations may compromise sectional performance, making it challenging to simultaneously achieve cost efficiency and optimal performance. Furthermore, beyond compliance with existing specifications, the design process requires dedicated research into: (1) integrated optimization of composite pipe-cables and accessories, and (2) comprehensive evaluation of material nonlinearities, geometric nonlinearities, and interfacial nonlinearities on mechanical behavior.

3.1 Composite Pipe Cable Design Structure Optimization Technology

3.1.1 Key Technologies for Umbilical Cable Cross-Section Design

Umbilical cable design necessitates comprehensive cross-sectional layout optimization due to its integration of multiple functional units. While current design criteria primarily follow compactness and symmetry principles [14], relying solely on geometric constraints proves insufficient for achieving optimal mechanical performance in tensile strength, bending resistance, and fatigue life under operational conditions. Practical umbilical cable design must prioritize the rational arrangement of functional units—including hydraulic, electric power, control signal transmission, and chemical injection systems—considering their thermal and electromagnetic interactions. During operation, fluid transport and electrical energy conduction generate significant heat, adversely affecting mechanical properties. Simultaneously, electromagnetic fields induced by power transmission and signal conduction may cause performance interference. Consequently, umbilical cable design represents a complex multidisciplinary optimization challenge requiring balanced consideration of mechanical, thermal, and electromagnetic performance objectives.

Ye et al. [15] developed a BEAMSHELL finite element type capable of accounting for filler materials in complex umbilical cable cross-section designs, demonstrating accurate simulation of filler effects on fatigue life. Zhang et al. [16] systematically analyzed mechanical property influencers in umbilical cable cross-sections through ANSYS simulations, validating results against theoretical models to establish finite element references for cross-sectional design. To address multidisciplinary optimization requirements in multi-physical field designs, advanced methodologies including full sensitivity equation analysis, subspace optimization, and collaborative optimization have been implemented. Traditional umbilical cable cross-section design processes suffer from inefficient over-reliance on repetitive iterative analysis and calibration. The advent of intelligent algorithms, such as approximation models, neural network training, and genetic algorithms, has significantly advanced integrated optimal design and layout optimization capabilities. Yang [17] proposed a geometric two-scale optimization method for umbilical cables, modeling both cross-sectional and global line parameters with NSGA-II algorithm implementation, resulting in structurally optimized designs with enhanced fatigue life. Ma [18] established theoretical and finite element analysis frameworks for electro-thermal-mechanical coupling in umbilical cables. Luo [19] conducted comprehensive comparative analyses of temperature field distributions using equivalent thermal circuit methods, ABAQUS electro-thermal coupling, and COMSOL electromagnetic-thermal coupling approaches, correlating results with experimental data to develop robust multi-field coupling analysis methodologies.

3.1.2 Key Technologies for the Optimal Design of Other Composite Pipeline and Cable Structures

Beyond umbilical cable cross-section design, other marine pipe-cable systems with multilayer structures and multiple functional units—including offshore wind dynamic cables, flexible risers, and thermoplastic composite pipes—also require structural design optimization. Okasha [20] developed a cross-section optimization method for multilayer composite pipe-cable structures, introducing an enhanced MDO-MOPSO algorithm (structural cross-section normalization algorithm) to address multi-objective optimization challenges in composite pipe-cable design. Li and Murilo [21] investigated lateral instability in flexible pipeline armor wire structures, establishing both an analytical model for wire instability and a stability limit prediction model to optimize armor wire design. Zhao et al. [22] focused on tensile armor layer optimization in flexible risers, proposing a combined finite element and evolutionary strategy approach that simultaneously considers mechanical responses to structural parameters like helix angle, cross-section dimensions, and friction coefficients. For thermoplastic composite pipes, Wang et al. [23] conducted ABAQUS-based finite element analyses comparing single-angle and multi-angle ply designs, demonstrating the mechanical superiority of multi-angle layups while reducing manufacturing costs, thus providing valuable engineering references. Li et al. [24] formulated anti-torsion structural design principles for offshore wind dynamic cables, with comparative program results yielding an optimized structure validated through low-temperature torsion resistance testing.

The outcomes of structural design optimization for marine composite cables directly influence both their mechanical performance and production costs. As offshore resource extraction advances to greater depths and harsher environments, the performance demands on these cables continue to escalate. This necessitates ongoing refinement of optimization methodologies, specifically focusing on: (1) the correlation between internal structural configurations and cable mechanical properties, (2) mechanical behavior under electro-thermal-mechanical coupling conditions, and (3) the development of intelligent algorithm-assisted approaches for efficient cross-sectional design iteration. Such comprehensive optimization strategies are essential to achieve marine composite cable systems that simultaneously satisfy three critical objectives: optimal mechanical performance, minimized manufacturing costs, and maximized design efficiency.

3.2 Nonlinear Characteristics of Composite Tube Cable

The mechanical analysis of composite pipe-cable systems must account for various nonlinearities arising from interactions between their multilayer structures and multiple functional units. These nonlinear effects can be categorized into three primary types: material nonlinearity, geometric nonlinearity, and interfacial nonlinearity.

3.2.1 Material Nonlinearity

Material nonlinearity in composite pipe-cable systems stems from the unique characteristics of their constituent nonmetallic materials, particularly the hyperelastic behavior of rubber components and viscoelastic properties of polymer matrices. These material properties exhibit nonlinear responses under varying environmental conditions. Fu et al. [25] developed theoretical and analytical methods for nonlinear dynamics of viscoelastic materials, formulating transverse pipeline control equations based on Euler-Bernoulli beam theory. Their work provides fundamental insights into how viscoelastic materials affect pipeline nonlinear dynamic performance. Sun et al. [26] investigated polyurethane material nonlinearity effects on mechanical behavior of offshore flexible pipe bend restrictors. Through comparative analysis of numerical simulations and theoretical models with varying material properties, they demonstrated the necessity of accounting for material nonlinearity in bend curvature analysis, offering critical safety assessment references for flexible riser bend restrictor design. Wang and Huang [27] systematically analyzed bend restrictor material parameter influences on flexible riser performance. Their finite element studies evaluated how different design and material parameters affect riser bending characteristics, establishing important correlations between material properties and structural performance.

Composite pipe-cable systems in marine environments experience creep deformation in their non-metallic components [28, 29] due to prolonged exposure to both harsh external conditions and internal transported media, particularly under high-

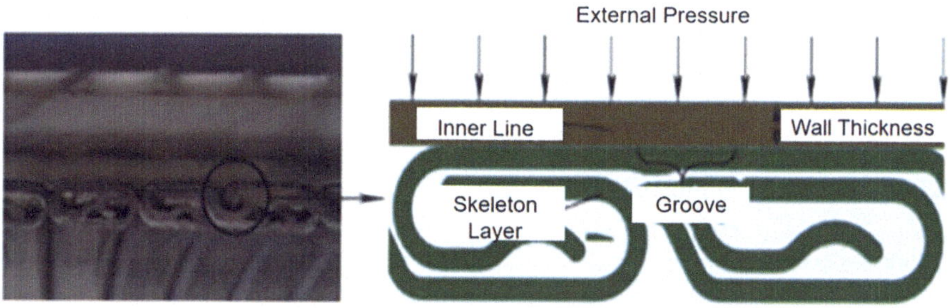

Fig. 7 Creep of flexible riser [30]

temperature service environments where polymer materials exhibit accelerated creep behavior (Fig. 7). The API 17J standard explicitly limits liner thickness variation to 30% for such systems. Recent research has advanced creep analysis methodologies: Hocine et al. [31] developed a life prediction model for multilayered structures under creep loading; Shen et al. [32] proposed a gap-span creep analysis method for unbonded flexible risers that evaluates remaining life through thickness reduction measurements; while Liu et al. [30] implemented both time-hardened and multiple-integral nonlinear constitutive models to characterize flexible riser liner creep behavior, establishing critical design and evaluation benchmarks.

3.2.2 Geometric Nonlinearity

Marine composite cables exhibit significant geometric nonlinearities—including large displacements, rotations, axial deformations, and bending deformations—due to boundary constraints imposed by environmental loads, floating structure motions, and seabed interactions [33–36]. Wagne et al. [37] developed a mathematical model for flexible beams accounting for large displacements and rotations, providing theoretical foundations for analyzing geometric nonlinear effects on flexible riser modal states. Yang et al. [38] established a stress decomposition-based methodology to assess lifespan reduction at flexible riser touchdown points caused by geometric nonlinearities. Through combined Lump-Mass method and ANSYS simulations, Li [39] systematically evaluated parameter effects on deep-sea riser nonlinear vibration responses. Zhang et al. [40] formulated a nonlinear dynamic riser model using absolute nodal coordinate formulation that accurately characterizes riser nonlinear behavior. Koh et al. [41] derived buoyancy load equations specifically for geometric nonlinear analysis of flexible risers, enhancing beam theory with continuum mechanics formulations while considering boundary effects. Neto et al. [42] investigated installation-phase nonlinearities in marine pipe-cable systems, proposing a geometrically nonlinear beam model to predict loop formation during deployment. Zhang et al. [43] analyzed mechanical properties of flexible pipelines after wave-induced looping/knotting at touchdown zones, developing bending failure response surface functions to evaluate installation-phase reliability, thereby establishing critical references for nonlinear behavior analysis during offshore installation operations.

3.2.3 Interface Nonlinearity

1. Interface strength
   Flexible hoses are composite structures incorporating both metallic and non-metallic materials. In bonded flexible hoses, components such as rubber matrices, cord reinforcements, and spiral reinforcements are integrated through vulcanization processes, while flanges are bonded to rubber fittings using adhesive agents. For non-bonded flexible hose connectors, epoxy resin bonds the tensile armor layer to the inner sleeve. These configurations inherently create interfacial layers between dissimilar materials, where mismatched deformation can induce interfacial damage and subsequent debonding failure [44].

   Interfacial strength analysis typically employs cohesive zone models to characterize interface mechanical behavior. Theoretical studies have advanced this methodology through several key contributions: Ouyang et al. [45] developed a torsional load theoretical model for flexible pipe joints using cohesive modeling, while Zhang et al. [46] established an interfacial pipe mechanical model incorporating residual strength factors to correct tensile and bending stresses. Numerical modeling approaches have evolved significantly, with Gao et al. [47] simplifying joint interface interactions through bound constraints in floating hose finite element models, and Zhang et al. [48] analyzing riser joint strength via solid interface finite element modeling. Wei et al. [49] implemented Rebar units for cord layer simulation and Embedded techniques for helical reinforcement interactions, though both methods neglected interfacial strength considerations. Recent research has expanded to environmental effects, with Zhang et al. [50, 51] investigating temperature and material degradation impacts on interfacial structures' ultimate capacity, particularly examining metal-nonmetal interface strength degradation in seawater corrosion. Advanced simulations by Pirondi et al. [52], Xu [53], and Zhang et al. [54] have utilized Cohesive elements to accurately model bonded flexible pipe joint interfaces, including flange-rubber layers and metal-nonmetal boundaries (Fig. 8).
2. Interface fatigue
   Marine composite pipe-cable systems experience significant cyclic loading under complex operating conditions, with interfacial debonding identified as the predominant failure mode. Jones et al. [56] demonstrated this through fatigue tension tests on interfacial steel plates, where all specimens failed at the material interfaces, confirming these zones as structural weak points. Experimental investigations by Boccoarelli et al. [57] and Chia-Hsuan et al. [58] have quantitatively characterized metal-nonmetal and CFRP-concrete interface fatigue respectively, establishing fundamental load-cycle life relationships. Beyond conventional S-N curve approaches, advanced numerical methods using fatigue cohesive zone

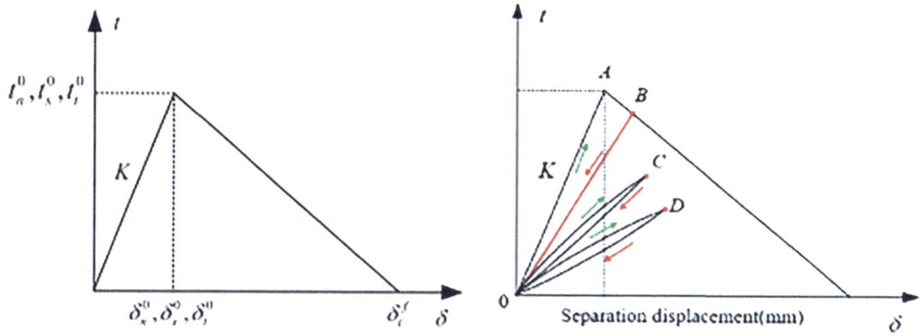

Fig. 8 Bilinear cohesion model [55]

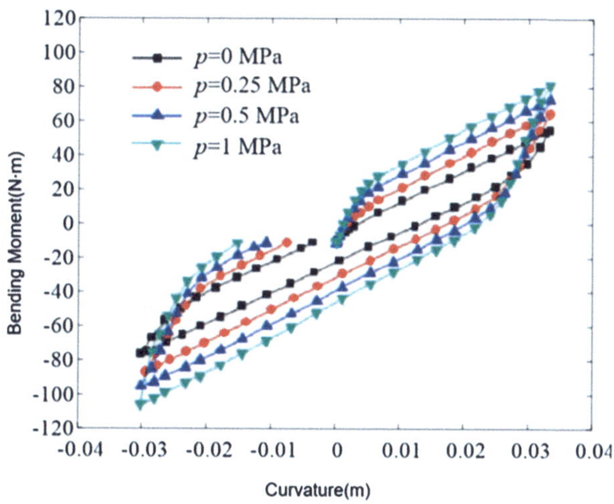

Fig. 9 Bending moment-curvature curves under different external pressures [62]

models have been developed by Dávila et al. [59], Huang et al. [60], and He et al. [61] to analyze composite interface failure mechanisms, while Zhang et al. [55] specifically investigated epoxy-bonded tensile armor layer fatigue in unbonded flexible pipe joints.

3. Interface friction

Unbonded flexible pipe-cable systems exhibit layer independence with interlayer gaps, enabling dynamic contact-separation-slip adjustments under varying loads and boundary conditions. The interfacial interactions transition between static and dynamic friction states, resulting in pronounced hysteresis nonlinearity under cyclic bending loads (Fig. 9). Significant theoretical advancements include: Witz et al. [63] developed analytical models for axial and bending stresses in armor layers under rotational bending, deriving complete pipe-cable bending stiffness; Kraincanic et al. [64] applied virtual work principles to analyze helical unit sliding during bending; Ramos et al. [65] formulated full-sliding state bending stiffness equations incorporating interlayer radial contacts. Leroy et al.

[66] enhanced bending stiffness formulations by considering radial contact effects, improving stress-slip expressions for double-layer helical armor, and establishing friction-inclusive bending behavior models. Zhang et al. [67, 68] investigated helical strip mechanics through equilibrium equations, energy conservation principles, deformation compatibility conditions, and Coulomb friction laws, deriving slip-stress relationships and moment-curvature hysteresis. Lu et al. [69] combined penalty function methods with Kebadze bending theory to model flexible pipe response under combined axisymmetric and bending loads.

The nonlinear analysis of marine composite cable design is essential, requiring comprehensive consideration of material nonlinearity, geometric nonlinearity, and interfacial nonlinearity. Establishing appropriate mechanical and numerical models enables accurate evaluation of critical mechanical characteristics, including ultimate load capacity, dynamic response, and fatigue life.

3.3 Composite Pipe and Cable Accessories

Composite cable accessories primarily consist of connecting joints, bend restrictors (including anti-benders, bend limiters, and flares), buoyancy modules, and J-tubes. These components serve distinct functions: joints provide critical interconnections between cable segments; bend restrictors prevent excessive curvature that could compromise flexible composite cables; buoyancy modules are integrated into cable system designs to achieve optimal submerged weight; while J-tubes offer essential protection for dynamic cables by minimizing frictional wear and abrasion during operation.

3.3.1 Composite Pipe and Cable Joints

The bonded flexible hose coupling system comprises several key components: a flange, rubber matrix, bundled steel wires, cord reinforcements, and spiral steel elements (structural configuration shown in Fig. 10). This design presents significant engineering challenges due to the substantial stiffness transition between the rigid flange connection and flexible hose body. The flange-pipe interface, which serves as the primary load transfer mechanism, is particularly susceptible to debonding failures and complete joint disconnection [70]. China has experienced multiple bonded hose leakage incidents, with statistical analysis indicating that approximately 76% of failures originate at joint interfaces [71]. Notably, recurrent leakage events in Maoming's single-point mooring system subsea oil transfer hoses have been attributed to interfacial debonding at flange-pipe connections [72]. These operational failures underscore the critical importance of investigating both interfacial strength and fatigue characteristics in bonded flexible hose joint design, particularly for ensuring long-term structural integrity in offshore applications.

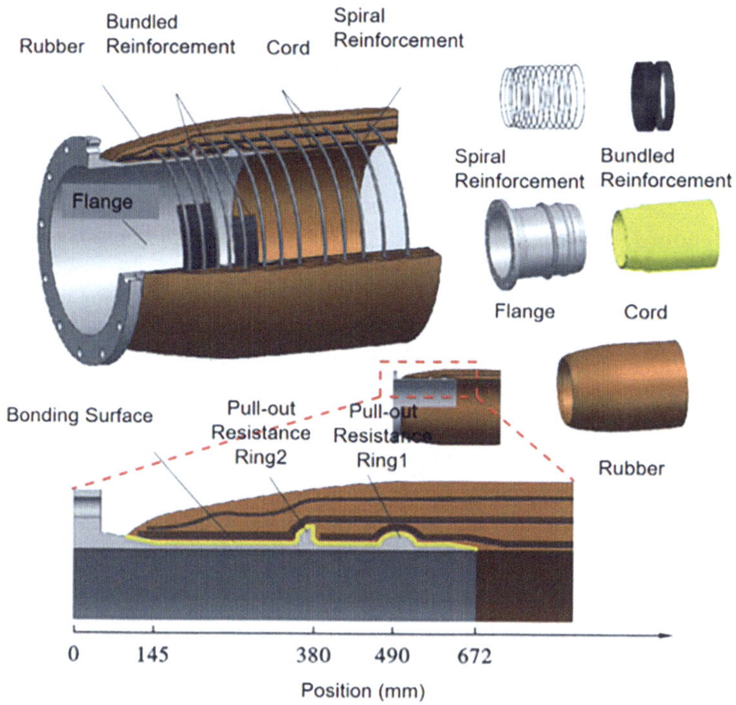

Fig. 10 Schematic diagram of the structure of bonded flexible hose couplings [7]

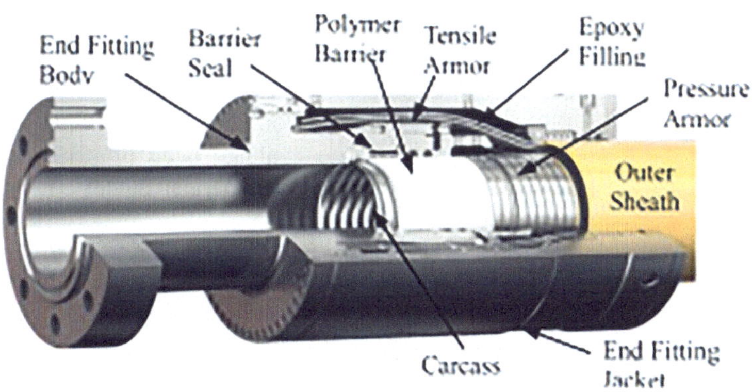

Fig. 11 Cross-section of non-bonded flexible hose coupling [73]

Figure 11 illustrates the joint configuration of non-bonded flexible pipe-cable systems, featuring a tensile armor layer connected through combined welding and epoxy resin injection into the cavity between the joint jacket and inner sleeve, ensuring robust connection strength. The sealing mechanism employs a ring positioned between the liner layer and joint inner sleeve, where bolt-induced axial displacement generates interfacial contact stresses to achieve effective sealing

[74]. Primary failure modes in non-bonded flexible hose couplings involve connection integrity loss and sealing performance degradation. Consequently, investigating epoxy resin degradation mechanisms and metal-to-metal sealing behavior under operational conditions is critical for optimizing joint design and ensuring long-term reliability.

3.3.2 Bending Limiter

Bend restrictors (shown in Fig. 12) are typically constructed from polyurethane elastomers and serve to mitigate excessive bending risks in dynamic cables, flexible risers, and floating connections, thereby enhancing cable fatigue life. Current mechanical analysis approaches include: Deruntz et al. [75] developed a slender beam analytical model, Boet et al. [76] formulated a variable cross-section beam model, and Drobyshevski et al. [77] established a profiled beam model. While these beam models provide valuable insights, their accuracy is inherently limited by the restrictor's slenderness ratio, making three-dimensional solid finite element modeling essential for accurate cross-sectional response simulation and precise mechanical property calculation. Structural optimization research has advanced through multi-objective approaches: Yang et al. [78] and Tang et al. [79] simultaneously minimized mass and maximized fatigue life in restrictor dimension optimization, while Fan et al. [80] implemented topology optimization techniques that outperform traditional dimensional optimization by achieving superior material distribution patterns.

The bend limiter (shown in Fig. 13) comprises multiple identical cylindrical structural units, typically manufactured from high-carbon steel and polyurethane elastomers. Adjacent subunits maintain movable gaps until the flexible pipe-cable system reaches the limiter's locking radius, at which point interlocking occurs to prevent further bending [82]. Current research predominantly employs finite element analysis incorporating contact nonlinearity to evaluate bend limiter mechanical performance. Li et al. [83] developed a coupled numerical model of composite

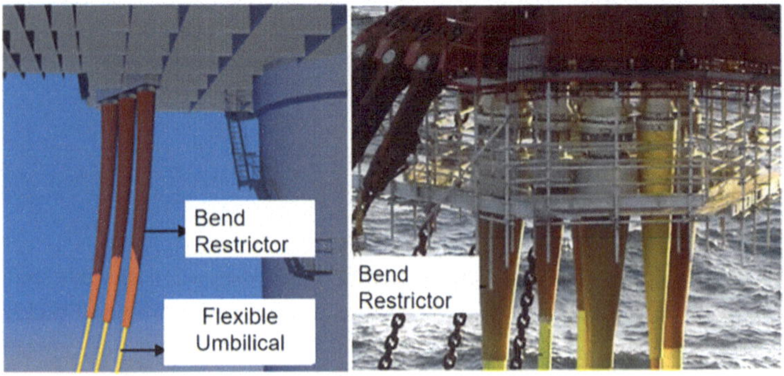

Fig. 12 Schematic diagram of anti-bending device structure [81]

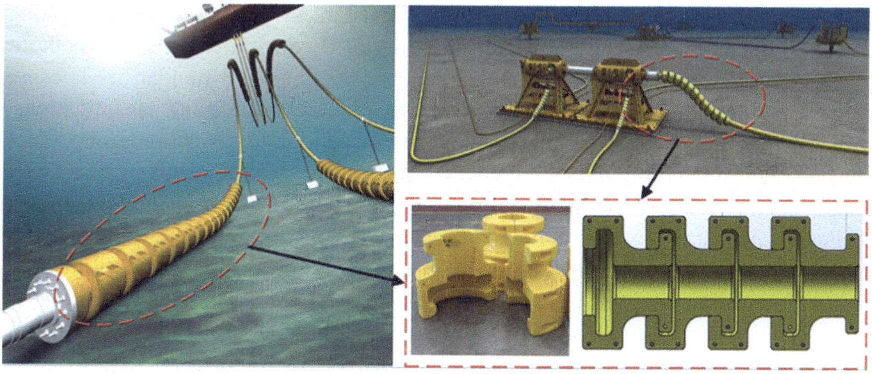

Fig. 13 Schematic diagram of the structure of the bending limiter

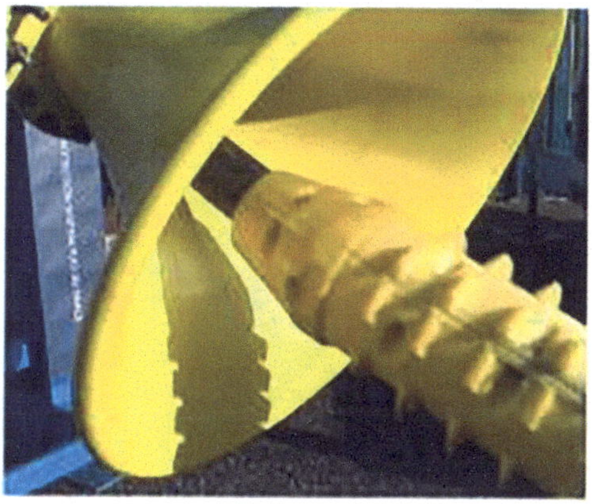

Fig. 14 Schematic diagram of the horn structure [85]

pipe-cable systems with bend limiters, examining polyurethane constitutive behavior and temperature effects on curvature radius. Noh et al. [84] conducted comprehensive finite element analysis of bend limiter mechanics followed by structural topology optimization.

The flare (illustrated in Fig. 14) is a hollow, horn-shaped structural component typically fabricated from either metallic or composite materials. Its design must specifically address frictional wear considerations at the interface with flexible pipe-cable systems [85].

4 Conclusion

Marine composite pipe-cable systems represent critical components in offshore engineering with significant application potential for extreme environments and specialized functional requirements. This paper systematically reviews: (1) typical structural configurations (flexible risers, umbilical cables, submarine cables, floating hoses, cryogenic hoses, and thermoplastic pipes); (2) conventional design methodologies; and (3) prevailing technical challenges in the field. The standardized design process for marine composite cables, as prescribed by industry specifications, sequentially comprises: material selection, cross-sectional dimensioning and functional design, line configuration, dynamic analysis (for dynamic cables), accessory design, service life assessment, installation engineering. However, several technical challenges persist, including: coupled cross-section/line optimization, multi-physics field analysis, advanced modeling techniques accounting for material nonlinearity, geometric nonlinearity, and interfacial behavior (strength, friction, fatigue), along with dynamic performance prediction of material degradation, interface evolution, and damage progression. Furthermore, the engineering of critical accessories—including joints, bend restrictors, buoyancy modules, and flares—demands rigorous attention to structural design, material selection, mechanical modeling, sealing mechanisms, and lifespan prediction. Beyond design considerations, the complete lifecycle of composite pipe-cable systems (encompassing manufacturing, testing, installation, operation, maintenance, and repair) presents additional technological challenges that must be addressed to ensure both operational safety and cost-effectiveness.

References

1. Hanonge D, Luppi A. Special session: advances in flexible riser technology: challenges of flexible riser systems in shallow waters. In: Offshore technology conference. OTC; 2010. OTC-20578-MS.
2. Kumar N, Paul R, Kamal R, et al. Challenges faced in design & selection of umbilical for offshore facilities. In: Abu Dhabi international petroleum exhibition & conference. OnePetro; 2019.
3. Wei D, An C, Zhang J, et al. Effect of winding steel wire on the collapse pressure of submarine hose. J Mar Sci Eng. 2022;10(10):1365.
4. Abbasi T, Hults J E, Mesnage O, et al. LNG flexible integrated with fiber optic distributed leak system enhancing safety, integrity monitoring and life assessments. In: Offshore technology conference, USA; 2016.
5. He Y, et al. Ultimate tension prediction of marine fiber-reinforced thermoplastic composite riser. China Ocean Eng. 2023:1–19.
6. Simonsen A. Inspection and monitoring techniques for un-bonded flexible risers and pipelines. Norway: University of Stavanger; 2014.
7. Wei D, An C, et al. Study on damage mechanism and debonding of adhesive surface of floating hose joint. China Offshore Oil Gas. 2023;35(4):173–80.
8. API 17B. Recommended Practice for Flexible Pipe; 2002.
9. API 17J. Specification for Unbonded Flexible Pipe; 2014.
10. API 17K. Specification for Bonded Flexible Pipe; 2005.
11. EN 1474–2. Installation and equipment for liquefied natural gas-Design and testing of marine transfer systems; 2008.
12. DNVGL-RP-F110. Global buckling of submarine pipelines; 2019.
13. OCIMF2009. Guide to manufacturing and purchasing hoses for offshore moorings; 2009.
14. ISO 13628-5. Specification for subsea umbilicals; 2009.

15. Ye N, Gjosteen JKO, Sævik S. On choice of finite element type for the filled bodies in umbilicals for deep water application. Int Conf Offshore Mechanics Arctic Eng. 2010;49132: 679–85.
16. Zhang K, Guo H, et al. Sectional mechanical properties analysis and numerical simulation of marine umbilical cable. Period Ocean Univ China. 2019;1
17. Yang Z. Study on geometric bi-scale analysis and optimization design of the structure of marine umbilical. Dalian: Dalian University of Technology; 2019.
18. Ma G. Electro-thermo-mechanical coupling analysis of deep-water strong-current composite umbilical cable. Beijin: China University of Petroleum, Beijing; 2019.
19. Luo Q. Study on mechanical response of deepwater strong-current composite umbilical cable based on electro-thermo-mechanical analysis. Beijin: China University of Petroleum, Beijing; 2021.
20. Okasha NM, Alzo'ubi AK, Mughieda O, et al. A near-optimum multi-objective optimization approach for structural design. Ain Shams Eng J. 2023:102388.
21. Li X, Vaz MA. Analytical model for lateral instability of quasi-rectangular or circular armor wires in flexible pipes and umbilicals. Ocean Eng. 2019;190:106423.
22. Zhao Y, Jiang D, Xu X. Optimization design of flexible riser based on nonlinear finite element and evolutionary strategy. Fly Ash Compr Utiliz. 2020;
23. Wang H, Shi C, et al. Study on single angle and multi angle stacking section design of thermoplastic composite pipe. China Pet Mach. 2022;50(11):50–7.
24. Li J, Su J, Liu Y, et al. Optimization of structural design of multi-core control cables for offshore wind power systems. Opt Fiber Electr Cable Their Appl. 2021;(5):39–40,46.
25. Fu G, Tuo Y, Zhang H, et al. Effects of material characteristics on nonlinear dynamics of viscoelastic axially functionally graded material pipe conveying pulsating fluid. J Mar Sci Appl. 2023;22(2):247–59.
26. Sun K, Yue Q, Yan J, et al. Finite element analysis on bending stiffener of flexible marine riser based on material non-linearity. Comput Aided Eng. 2014;23(6):66–9.
27. Wang H, Huang D. Nonlinear modeling and parameter sensitivity analysis of ocean flexible riser bending restrictors. China Offshore Platform. 2020;35(1):1–5.
28. Ma Q, Chen X, Gao C, et al. Research and application of dynamic submarine wet storage technology for submarine cables. Chin New Technol Prod. 2020;
29. Shen Y, Zhao J, Tan Z, et al. Analysis of the creep behaviour of the polymer barrier layer in Unbonded flexible pipes under different fluid temperatures. In: International conference on offshore mechanics and Arctic engineering, vol. 44908. American Society of Mechanical Engineers; 2012. p. 153–8.
30. Liu J, et al. Analysis of nonlinear mechanical behavior of flexible riser inner liner. J Mech Eng. 2021;57(20):89–97.
31. Hocine A, Maizia A, Ghouaoula A, et al. Deterministic analysis for creep behavior of damaged composite tubular structure. J Fail Anal Prev. 2020;20:1858–63.
32. Shen Y, Burton A, Birkinshaw P, et al. Influence of gap span creep behavior of polymer barrier layer on service life in the HPHT Deepwater Unbonded flexible risers. In: ISOPE international ocean and polar engineering conference. ISOPE; 2018. ISOPE-I-18-642.
33. Song R. Submarine pipelines and ocean risers in deepwater development. Ships Yachts. 2003;(6):31–42.
34. Huang W, Bai X, Li H. State of the art of Research and Development of overseas Deepwater steel catenary risers. Period Ocean Univ China. 2009;39(2):290–4.
35. Liang H. Review of research on interactions between deepwater steel catenary risers and soft clay seabeds. J Mar Sci Appl. 2009;2(8):163–7.
36. Gu J, Wang Y, Zhang Y, et al. Analytical solution of mean top tension of long flexible riser in modeling vortex-induced vibrations. Appl Ocean Res. 2013;41:1–8.
37. Wagner G, Lima R, Sampaio R. A versatile strategy to compute nonlinear normal modes of flexible beams. Nonlinear Dyn. 2023;111(11):9815–37.

38. Yang H, Jiang H, Yang Q. Enhanced multi-layer fatigue-analysis approach for unbonded flexible risers. China Ocean Eng. 2014;28:363–79.
39. Lin L. Research on nonlinear vibration characteristic of large dimension deep-sea risers. Tianjin University; 2004.
40. Zhang C, Lu L, Cao Q, et al. Nonlinear motion regimes and phase dynamics of a free standing hybrid riser system subjected to ocean current and vessel motion. Ocean Eng. 2022;252: 111197.
41. Koh KJ, Yassin AYM, Latheef M. On the buoyancy load formulation for geometrically nonlinear analysis of flexible marine risers. Ocean Eng. 2018;157:313–24.
42. Gay Neto A, Pimenta PM, Martins CA. Loop formation in catenary risers on installation conditions: a comparison of statics and dynamics. In: International conference on offshore mechanics and arctic engineering, vol. 55379. American Society of Mechanical Engineers; 2013. V04BT04A046.
43. Yu Z, Zhansheng G, Haikun Z, et al. Analysis of torsional instability and reliability of marine flexible pipelines. Ocean Eng. 2021;228
44. Chen B, Zhang Y, Wang Q, et al. The signal analysis of lamb wave in steel plates strengthened by CFRP with interface debonding. Ocean Eng. 2023;283:115061.
45. Ouyang Z, Li G. Cohesive zone model based analytical solutions for adhesively bonded pipe joints under torsional loading. Int J Solids Struct. 2009;46(5):1205–17.
46. Zhang Y, Cheng N, Cheng Z, Luo Q. The debonding failure analysis of corroded pipes repaired with CFRP under tension and the bending moment. Int J Adhes Adhes. 2020;103:102714–24.
47. Gao Q, Zhang P, Duan M, et al. Investigation on structural behavior of ring-stiffened composite offshore rubber hose under internal pressure. Appl Ocean Res. 2018;79:7–19.
48. Zhang Y, Qin TY, Noda NA, et al. Strength analysis of adhesive joints of riser pipes in deep sea environment loadings. Appl Adhes Sci. 2013;1:1–9.
49. Wei D, An C, Wu C, et al. Torsional structural behavior of composite rubber hose for offshore applications. Appl Ocean Res. 2022;128:103333.
50. Xin J, Zhang Y, Zhong C, et al. Analysis of ultimate pressure of corroded subsea pipe repaired with composite considering temperature and material degradation. J Mar Sci Technol. 2020;25: 285–97.
51. Zhang Y, Liu Z, Xin J, et al. The attenuation mechanism of CFRP repaired corroded marine pipelines based on experiments and FEM. Thin-Walled Struct. 2021;169:108469.
52. Pirondi A, Moroni F. Clinch-bonded and rivet-bonded hybrid joints: application of damage models for simulation of forming and failure. J Adhes Sci Technol. 2009;23(10):1547–74.
53. Zejin X. Study on tensile property of the joint of floating hose. Beijin: China University of Petroleum, Beijing; 2021.
54. Zhang P, Zhang Y, Xin J, et al. The failure mechanism analysis of corroded pipelines repaired with CFRP under bending loads using FEM. Ships Offshore Struct. 2023:1–8.
55. Zhang Y, Zhang P, Yang P, et al. The analysis of debonding damage between steel tension wire and epoxy resin in flexible end fitting under cyclic loading. Appl Ocean Res. 2023;137:103606.
56. Jones S, Civjan S. Application of fiber reinforced polymer overlays to extend steel fatigue life. J Composit Constr. 2003;7(4):331–8.
57. Bocciarelli M, Colombi P, Fava G, et al. Fatigue performance of tensile steel members strengthened with CFRP plates. Compos Struct. 2009;87(4):334–43.
58. Chou JX, Zhang Z, Zhang J, et al. Study on fatigue bond behavior of NSM CFRP-concrete interface. China J Highw Transp. 2022;35(02):234–46.
59. Dávila CG. From SN to the Paris law with a new mixed-mode cohesive fatigue model for delamination in composites. Theor Appl Fract Mech. 2020;106:102499.
60. Qiang H. Research on durability law of CFRP repaired pipeline under fatigue load. Beijing: China University of Petroleum; 2021. https://doi.org/10.27643/d.cnki.gsybu.2021.001182.
61. He L, Wang J, Wu X, et al. A novel fatigue cohesive model for interface between ballastless track slab and self-compacting concrete. Constr Build Mater. 2023;377:130962.

62. Yu Jinjin, Guo Haiyan, Liu Zhen, et al. Parametric sensitivity analysis of unbonded steel cable's bending hysteresis. Period Ocean Univ China,2021,51(11):110–116.
63. Witz JA, Tan Z. Rotary bending of marine cables and umbilicals. Eng Struct. 1995;17(4): 267–75.
64. Kraincanic I, Kebadze E. Slip initiation and progression in helical armouring layers of unbonded flexible pipes and its effect on pipe bending behavior. J Strain Anal Eng Des. 2001;36(3):265–75.
65. Ramos R, Pesce CP, Martins Martins CA. A new analytical expression to estimate the bending stiffness of flexible risers. In: Proceedings of 22nd international conference on offshore mechanics and arctic engineering, Cancun, Mexico; 2003. p. 1–10.
66. Leroy JM, Estrier P. Calculation of stresses and slips in helical layers of dynamically bent flexible pipes. Oil Gas Sci Technol. 2006;56(6):545–54.
67. Zhang Y, Qiu L. Numerical model to simulate tensile wire behavior in unbonded flexible pipe during bending. In: Proceedings of the 26th international conference on offshore mechanics & arctic engineering. San Diego, USA; 2007. p. 17–29.
68. Love AEH. A treatise on the mathematical theory of elasticity. Cambridge University Press; 2013.
69. YuTian L. Research on cross-sectional mechanical model and fatigue life analysis of Deepwater flexible risers. Zhejiang: Zhejiang University; 2017.
70. Amaechi CV, Chesterton C, Butler HO, et al. Review of composite marine risers for deep-water applications: design, development and mechanics. J Compos Sci. 2022;6(3):96.
71. Dandan Y. Dynamic analysis of floating flexible hose arranged in wave and current. Heilongjiang: Harbin Engineering University; 2009.
72. Yifeng D. Damage research and extreme response of submarine hoses subjected to wave loads at Maoming single point mooring. Guangzhou: South China University of Technology; 2011.
73. Fernando US, Karabelas G. Analysis of end fitting barrier seal performance in high pressure unbonded flexible pipes. In: International conference on offshore mechanics and Arctic engineering, vol. 45462. American Society of Mechanical Engineers; 2014. V06AT04A038.
74. Fernando US, Nott P, Graham G, et al. Experimental evaluation of the metal-to-metal seal design for high-pressure flexible pipes. In: Offshore technology conference. OTC; 2012. OTC-23110-MS.
75. DeRuntz JA Jr. End effect bending stresses in cables. J Appl Mech. 1969;36(4):750–6.
76. Boet WJC, Out JMM. Analysis of a flexible-riser top connection with bend restrictor. In: Offshore technology conference. OTC; 1990. OTC-6436-MS.
77. Drobyshevski Y. Investigation into non-linear bending of elastic bars with application to design of bend stiffeners. Mar Struct. 2013;31:102–30.
78. Yang H, Wang A. Fatigue reliability based design optimization of bending stiffener. J Ship Res. 2012;56(02):120–8.
79. Tang M, Yan J, Chen J, et al. Nonlinear analysis and multi-objective optimization for bend stiffeners of flexible riser. J Mar Sci Technol. 2015;20:591–603.
80. Fan Z, Yang Z, Xu Q, et al. Nonlinear stiffness topology optimization for the bend stiffener of flexible riser. Chin J Theor Appl Mech. 2022;54(04):929–38.
81. Liu J, Qi H, Luo X, et al. Fatigue life of flexible riser bend stiffener based on finite element method. Ocean Eng. 2021;39(03):141–7.
82. Kang Y, Xiao W, Zhang D, et al. Performance validation and dynamic response analysis of a deepwater cable bending restrictor. Underw Technol. 2020;37(3):95–101.
83. Li B, Li W, Guo J, Zhang Y. Nonlinear mechanical analysis of subsea bend restrictor under different constitutive models. China Pet Mach. 2019;47(07):48–53.
84. Non J, Ha Y. Multi-body contact analysis and structural design optimization of bend restrictors for subsea pipelines. J Soc Nav Archit Korea. 2018;55(4):289–96.
85. Zongbo X. Research on design and manufacturing for bend limiter of marine flexible risers and umbilicals. Dalian: Dalian University of Technology. p. 201.

Stress Intensity Factors of a Central Interface Crack in a Bonded Finite Plate and Periodic Interface Cracks Under Arbitrary Material Combinations

Abstract Although a lot of interface crack problems were previously treated, few solutions are available under arbitrary material combinations. This chapter deals with a central interface crack in a bonded finite plate and periodic interface cracks. Then, the effects of material combination and relative crack length on the stress intensity factors are discussed. A useful method to calculate the stress intensity factor of interface crack is presented with focusing on the stress at the crack tip calculated by the finite element method.

1 Introduction

Although a lot of papers were published, several fundamental problems have not been solved yet for interface cracks. For stress intensity factors of the interface crack, the most fundamentally known is the results for the bonded infinite plate subjected to the internal stress σ (see Fig. 1a), which is expressed in the following equations.

$$K_I + iK_{II} = (F_I + iF_{II})(1 + 2i\varepsilon)\sigma\sqrt{\pi a}, F_I = 1, F_{II} = 0 \tag{1}$$

$$\varepsilon = \frac{1}{2\pi} \ln\left[\left(\frac{k_1}{G_1} + \frac{1}{G_2}\right) / \left(\frac{k_2}{G_2} + \frac{1}{G_1}\right)\right] \tag{2}$$

$$k_m = \begin{cases} (3 - \nu_m)/(1 + \nu_m) & \text{(Plane stress)} \\ 3 - 4\nu_m & \text{(Plane strain)} \end{cases}$$

$$\nu_m : \text{(Poisson's ratio)} \quad (m = 1, 2)$$

$$G_m : \text{(Shear modulus)} \quad (m = 1, 2)$$

$$\sigma_{x2}^{\infty} = \frac{1}{1 + k_2}\left[\frac{G_2}{G_1}(1 + k_1)\sigma_{x1}^{\infty} + \left\{3 - k_2 - \frac{G_2}{G_1}(3 - k_1)\sigma_y^{\infty}\right\}\right]$$

It should be noted that Fig. 1a is equivalent to Fig. 1b, and the stress intensity factors in Fig. 1c is different from the ones in Fig. 1a, b. In other words, an interface crack subjected to internal pressure in bonded infinite plate is equivalent to an

Y. Zhang, *Interfacial Mechanics of Offshore Engineering Equipment*,
https://doi.org/10.1007/978-981-95-3200-1_3

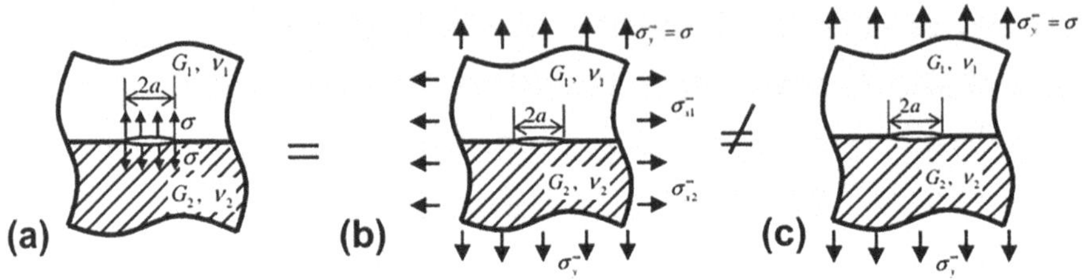

Fig. 1 Infinite plate subjected to (**a**) internal pressure, (**b**) σ_y^∞, σ_{x1}^∞, σ_{x2}^∞ and (**c**) σ_y^∞

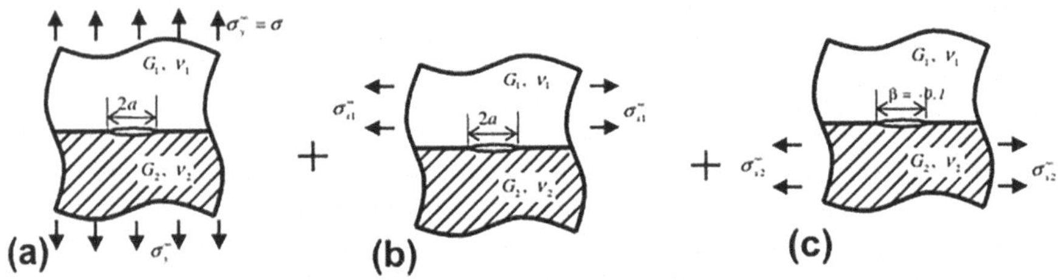

Fig. 2 Infinite plate subjected to (**a**) σ_y^∞, (**b**) σ_{x1}^∞ and (**c**) σ_{x2}^∞

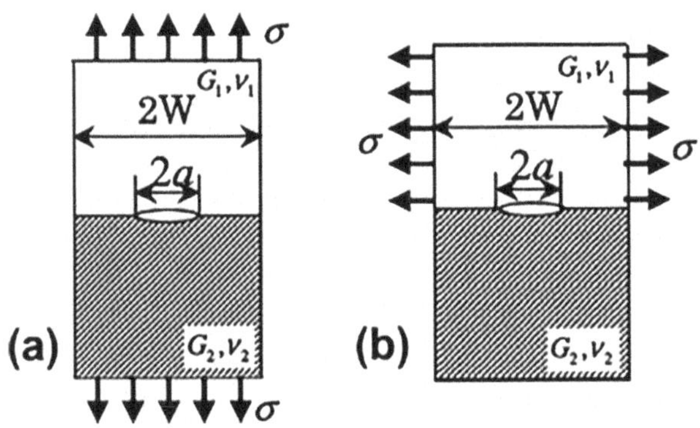

Fig. 3 Bonded plate with a central interface crack with $a/W \to 0$

interface crack subjected to $\sigma_y^\infty = \sigma$ and σ_{x1}^∞, σ_{x2}^∞ which produce $\varepsilon_{x1}^\infty = \varepsilon_{x2}^\infty$ on the interface [1, 2]. Therefore, the stress intensity factors for Fig. 1a, b are expressed as the sum of the ones of Fig. 2a–c, Since the fundamental solutions for Fig. 2a–c were not available, the authors have shown the stress intensity factors for those problems under arbitrary material combination (see Fig. 3a, b) [2].

However, an interface crack in a bonded finite plate has not been discussed vet under arbitrary material combination. Although a lot of related studies were published previously, few solutions are available under arbitrary material combinations. In this paper, therefore, periodic interface cracks as shown in Fig. 4a will be treated in comparison with a central interface crack in bonded finite plates as shown in Fig. 4b. Then, the effects of relative of crack length on the stress intensity factors

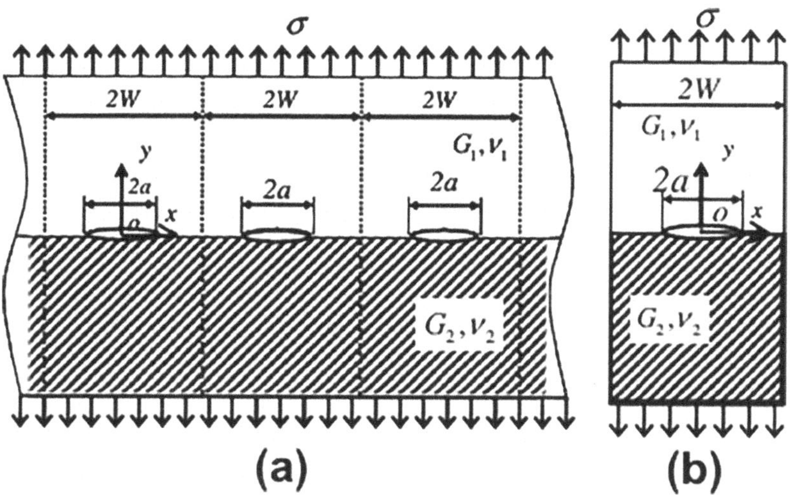

Fig. 4 (**a**) Periodic interface cracks in an infinite bonded plate and (**b**) a central interface crack in a bonded plate

will be analyzed explicitly under arbitrary material combination. In Fig. 4a, along $x = (1 + 2n)W$ (n is the integer) the boundary conditions are $u_x = 0$, $\tau_{xy} = 0$, but $\sigma_x \neq 0$. On the other hand, in Fig. 4b along $x = \pm W$ the boundary conditions are $\sigma_x = 0$, $\tau_{xy} = 0$.

In this paper, a useful method to calculate the stress intensity factor of interface crack is presented with focusing on the stress at the crack tip calculated by the finite element method [3].

2 Analysis Method

The analysis method used in this research is based on the stresses at the crack tip calculated by FEM. By using the proportional stress fields for the reference and given problems, stress intensity factors can be obtained with a good accuracy [4]. For example, for model 1 crack in homogenous plates, the stress distribution near the crack tip ($\theta = 0$) can be expressed by Eq. (3).

$$\sigma_y = K_1 / \sqrt{2\pi r} \tag{3}$$

It is confirmed that the error of FEM mainly comes from the mesh around the crack tip. Therefore if the same mesh size and pattern are applied to the reference and given unknown problems, stress intensity factors K can be obtained from the stresses o, calculated by FEM. At a given distance r the following relationship can be obtained.

$$K_I/\sigma_y = \text{const} \tag{4}$$

If different crack problems A and B are analyzed by applying the same FEM mesh, the following equation can be given at the same distance from the crack tip.

$$\left[K_I^*/\sigma_y^*\right]_A = [K_I/\sigma_y]_B \tag{5}$$

Here, an asterisk (*) means the values of the reference problem, By using Eq. (5) with stress values at crack tip calculated by FEM, accurate stress intensity factors in homogenous plates were successfully obtained by Nisitani et al. [4, 5].

Although this method cannot be applied to interface crack problems without difficulty, an effective method was recently proposed by Oda et al. [3] successfully to analyze interface crack problems. it is well known that there exists oscillation singularity at the interface crack tip, From the stresses σ_y, τ_{xy} along the interface crack tip, stress intensity factors are defined as shown in Eq. (6).

$$\sigma_y + i\tau_{xy} = \frac{K_I + iK_{II}}{\sqrt{2\pi r}}\left(\frac{r}{2a}\right)^{i\varepsilon}, r \to 0, \tag{6}$$

From Eq. (6), it is known that because stress intensity factors for interface crack and the crack in homogenous material are different, it is difficult to separate modes absolutely. So, it is necessary to obtain the following equation from Eq. (7).

$$K_I = \lim_{x \to \infty} \sqrt{2\pi r}\sigma_y\left(\cos Q + \frac{\tau_{xy}}{\sigma_y}\sin Q\right), \tag{7}$$

$$K_{II} = \lim_{x \to \infty} \sqrt{2\pi r}\tau_{xy}\left(\cos Q + \frac{\sigma_y}{\tau_{xy}}\sin Q\right), \tag{8}$$

$$Q = \varepsilon \ln\left(\frac{r}{2a}\right). \tag{9}$$

If the distance r is given as a constant, the following equation can be obtained.

$$Q^* = Q, \frac{\tau_{xy}^*}{\sigma_y^*} = \frac{\tau_{xy}}{\sigma_y} \tag{10}$$

Therefore if Eq. (10) is satisfied, Eq. (11) may be derived from Eqs. (7) and (8). In such case, oscillatory items of the reference and unknown problems are changed into the same.

$$\frac{K_I^*}{\sigma_y^*} = \frac{K_I}{\sigma_y}, \frac{K_{II}^*}{\tau_{xy}^*} = \frac{K_{II}}{\tau_{xy}} \tag{11}$$

Here, $\sigma^*_{y0,FEM}$, $\tau^*_{xy0,FEM}$ are stresses of reference problem calculated by FEM, and $\sigma^*_{y0,FEM}$, $\tau^*_{xy0,FEM}$ are stresses of given unknown problem. Stress intensity factors of the given unknown problem can be obtained by:

$$K_I = \frac{\sigma_{y0,FEM}}{\sigma^*_{y0,FEM}} K^*_I \tag{12}$$

$$K_{II} = \frac{\tau_{y0,FEM}}{\tau^*_{y0,FEM}} K^*_{II} \tag{13}$$

Stress intensity factors of the reference problem are defined by Eq. (14).

$$K^*_I + iK^*_I = (T + iS)\sqrt{\pi a}(1 + 2i\varepsilon) \tag{14}$$

Regarding the reference problem in Fig. 5, denote $\sigma^{T=1,S=0*}_{y0,FEM}$, $\tau^{T=1,S=0*}_{xy0,FEM}$ are values of stresses for $(T, S) = (1, 0)$ and $\sigma^{T=0,S=1*}_{y0,FEM}$, $\tau^{T=0,S=1*}_{xy0,FEM}$ are ones for $(T, S) = (0, 1)$. In order to satisfy Eq. (10), stresses at the crack tip of the reference problem are expressed as

$$\begin{aligned} \sigma^*_{y0,FEM} &= \sigma^{T-1,S-0*}_{y0,FEM} \times T + \sigma^{T-0,S-1*}_{y0,FEM} \times S, \\ \tau^*_{xy0,FEM} &= \tau^{T-1,S-0*}_{xy0,FEM} \times T + \tau^{T-0,S-1*}_{xy0,FEM} \times S, \end{aligned} \tag{15}$$

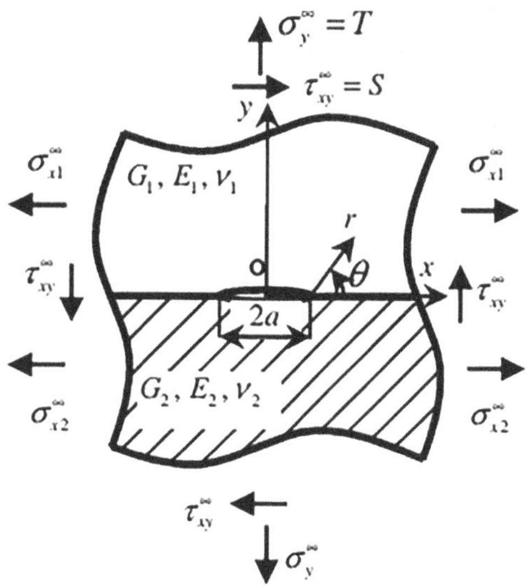

Fig. 5 Reference problem ($\varepsilon_{x1} = \varepsilon_{x2}$ at $y = 0$)

By substituting Eq. (10) into Eq. (15) with $T = 1$, the value of S is obtained as

$$S = \frac{\sigma_{y0,FEM} \times \tau_{xy0,FEM}^{T-1,S-0*} - \tau_{xy0,FEM} \times \sigma_{y0,FEM}^{T-1,S-0*}}{\tau_{xy0,FEM} \times \sigma_{y0,FEM}^{T-0,S-1*} - \sigma_{y0,FEM} \times \tau_{xy0,FEM}^{T-0,S-1*}} \tag{16}$$

The problem that is subjected to $T = 1$ and S expressed by Eq. (16) is considered as the reference problem. Because the exact solution is known, the error of unknown problem can be evaluated by using the same mesh. In the following of the paper, results are shown using dimensionless stress intensity factors F_I, F_{II} defined in Eq. (1). Dundurs' bi-material parameters a, b are defined in Eq. (17).

$$\alpha = \frac{G_1(k_2+1) - G_2(k_1+1)}{G_1(k_2+1) + G_2(k_1+1)}, \beta = \frac{G_1(k_2-1) - G_2(k_1-1)}{G_1(k_2+1) + G_2(k_1+1)} \tag{17}$$

3 Stress Intensity Factors for a Central Interface Crack in a Bonded Infinite Plate

In this paper, the solutions of infinite periodic cracks (see Fig. 4a) and a central interface crack in a bonded finite plate (see Fig. 4b) will be discussed. It should be noted that the limiting case $a/W \to 0$ in Fig. 4a coincides with the solution in Fig. 1a. On the other hand, the limiting case $a/W \to 0$ in Fig. 4b coincides with the solution in Fig. 1c. Since the solution in Fig. 1a is expressed as Eq. (1), another fundamental solution of a central interface crack in bonded plate when $a/W \to 0$ will be indicated in the following [2].

3.1 Effect of Plate Dimensions on the Stress Intensity Factors

In order to discuss bonded infinite plates, it is necessary to consider the effect of the plate dimensions on the stress intensity factors because the finite element method cannot treat the infinite plates directly. The results of central interface crack in Fig. 3a are therefore investigated in Table 1 with varying a/W = 1/1620, 1/3240, 1/6480, 1/6480 and $\alpha = 0.75, \beta = 0, \alpha = 0.9, \beta = 0, \alpha = 0.75, \beta = 0.2$. It is seen that results of $a/W <$ 1/1620 coincide each other and may have more than three digit accuracy. In other words, Table 1 shows that the results for a/W = 1/1620 can be used as the infinite plate $a/W \to 0$ with less than 0.09% error. It is also seen that $F_{II} \to 0$ as $a/W \to 0$ under arbitrary material combination. In the following sections, the results for the bonded infinite plate obtained as shown in Table 1 will be discussed.

Table 1 Dimensionless stress intensity factors of crack in Fig. 3a with different a/w

	a/W	$\alpha = 0.75, \beta = 0$	$\alpha = 0.75, \beta = 0$	$\alpha = 0.75, \beta = 0$
F_I	1/1620	0.93955	0.90859	0.95516
	1/3240	0.93955	0.90883	0.95515
	1/6480	0.93982	0.90943	0.95514
	→0	0.94002	0.91003	0.95513
F_{II}	1/1620	2.21×10^{-4}	2.59 × 10–4	1.11 × 10–4
	1/3240	1.10×10^{-4}	1.28×10^{-4}	5.53×10^{-5}
	1/6480	5.51×10^{-5}	6.42×10^{-5}	2.76×10^{-5}
	→0	0	0	0

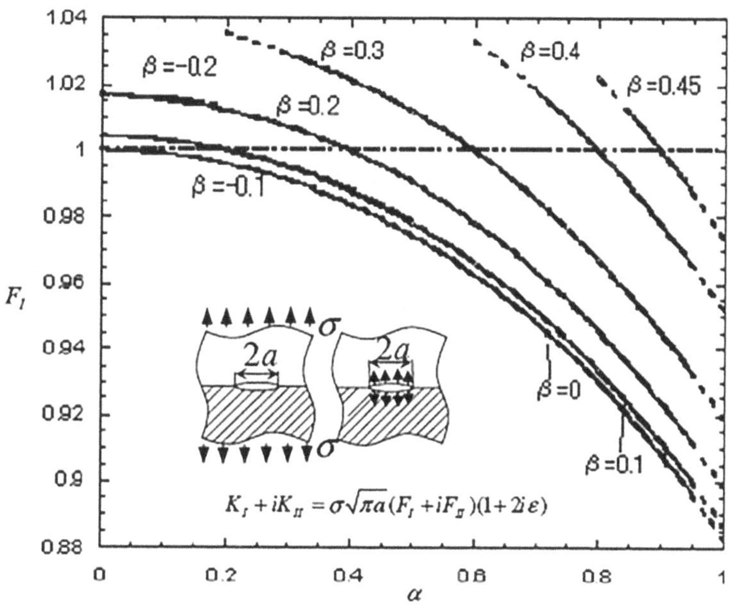

Fig. 6 F_I of a central interface crack in a bonded infinite plate under uni-axal tension which is corresponding to Fig. 3a with $a/W \to 0$

3.2 *Central Interface Crack in a Bonded Infinite Plate Under Uniaxial Tension*

Figure 6 shows the results of a central interface crack in a bonded infinite plate under uniaxial tension in the y-direction as shown in Fig. 3a. In Fig. 6, Dundur's parameter b is fixed, and the variations of F_I are depicted with varying parameter a. From Eq. (19), it is known that when material 1 and material 2 are exchanged, Dundur's parameters (α, β) become $(-\alpha, -\beta)$. Then the stress intensity factors (F_I, F_{II}) become $(F_I, -F_{II})$. Therefore, all material combinations are considered in the range $a > 0$ in Fig. 7.

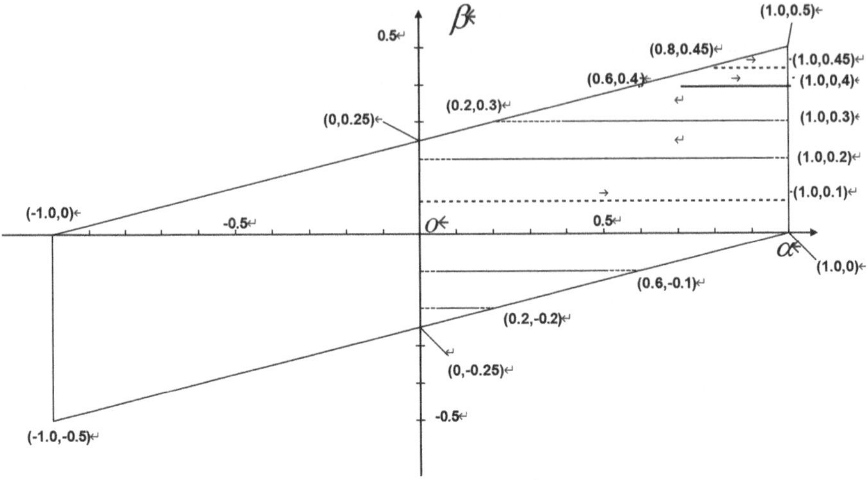

Fig. 7 The map of α and β

In Fig. 6, the solid curves show the results of a central interface crack under remote tension $\sigma_y = \sigma$. The dashed lines are extended from solid lines because some cases of material combination are difficult to be obtained by the FEM. The dash-dotted line shows the results of that under internal pressure r whose solution is known as $F_I = 1$ and $F_{II} = 0$. Figure 6 shows the variation of $F_I = 0.882 \sim 1.036$, which has the minimum value $F_I = 0.882$ when $\alpha = 1.0$, $\beta = 0$, and the maximum value $F_I = 1.036$ when $\alpha = 0.2, \beta = 0.3$. It is also found that $F_{II} = 0$ for the full range of α, β. Therefore, it may be concluded that central interface crack in a bonded infinite plate under remote tension of $r = 1$ is equivalent to that under internal pressure of $\sigma = 0.882{\sim}1.036$. All the values in Fig. 6 are given in Table 2 with three decimal. From Fig. 6 and Table 2, we can conclude that $F_I > 1.0$ when $(a + 2b)(a - 2b) < 0$, $F_I = 1.0$ when $(a + 2b)(a - 2b) = 0$ and $F_I < 1.0$ when $(a + 2b)(a - 2b) > 0$. In Table 3, values for $F_I = 1.0$ are marked by underlines.

3.3 Central Interface Crack in a Bonded Infinite Plate with Material 1 Under Tension in the x-Direction

Table 3 shows stress intensity factors of central interface crack shown in Fig. 3b with different relative crack size $a/W = 1/1620$, $1/13240$, $1/6480$ under different material combinations $a = -0.3, \beta = -0.2$, $a = -0.75, \beta = 0$, $a = -0.8, \beta = 0.4$. It is seen that all the results coincide each other more than three digit when $a/W < 1/1620$.

Figure 8 shows the results of bonded infinite plate $a/W \to 0$ with material 1 under tension in the x-direction. The dashed lines are extended from solid lines because some cases of material combination are difficult to be obtained by the FEM. In

Table 2 Dimensionless stress intensity factor F_I in Fig. 3a with $a/W \to 0$, $F_I = 1.0$ are marked by underlines

α	β —0.2	—0.1	0	0.1	0.2	0.3	0.4	0.45
0.00	1.017	1.004	1.000	1.004	1.017			
0.05	1.016	1.004	1.000	1.004	1.016	–	–	–
0.10	1.016	1.003	0.999	1.003	1.015	–	–	–
0.15	1.015	1.002	0.998	1.002	1.014	–	–	–
0.20	(1.013)	1.000	0.996	1.000	1.012	(1.036)	–	–
0.30	–	0.995	0.991	0.995	1.007	1.029	–	–
0.40	–	0.988	0.984	0.988	1.000	1.022	–	–
0.50	–	0.979	0.975	0.978	0.990	1.012	–	–
0.60	–	(0.966)	0.963	0.966	0.979	1.000	(1.033)	–
0.70	–	–	0.948	0.952	0.964	0.985	1.018	–
0.75	–	–	0.940	0.943	0.955	0.977	1.010	–
0.80	–	–	0.930	0.934	0.946	0.967	1.000	(1.022)
0.85	–	–	0.920	0.924	0.935	0.957	0.990	1.011
0.90	–	–	0.910	0.912	0.924	0.945	0.978	1.000
0.95	–	–	0.896	0.900	0.912	0.933	0.965	0.987
1.00	–	–	(0.882)	(0.886)	(0.898)	(0.919)	(0.952)	(0.974)

Table 3 Dimensionless stress intensity factors of crack in Fig. 3b with different a/W

	a/W	$\alpha = 0.3, \beta = 0.2$	$\alpha = -0.75, \beta = 0$	$\alpha = -0.8, \beta = -0.4$
F_I	1/1620	−0.02540	0.07051	0.17962
	1/3240	−0.02539	0.07056	0.17960
	1/6480	−0.02539	0.07055	0.17959
	→0	−0.02539	0.07054	0.17958
F_{II}	1/1620	1.11×10^4	2.59×10^4	3.28×10^4
	1/3240	5.53×10^4	1.29×10^4	1.64×10^4
	1/6480	2.77×10^5	6.46×10^5	8.19×10^5
	→0	0	0	0

Fig. 8, β in each curve is fixed, and the variations of F_I are depicted with varying parameter a. Previously, it has been thought that tension in the x-direction does not contribute to the stress intensity factors [1]. However, as can be seen from Fig. 8, F_I is not 0 in current research. It should be noted that the stress intensity factor FI is not zero under x-directional tension, except the case when $\varepsilon_{x1} = \varepsilon_{x2}$ is produced along the interface, and it has the minimum value $F_I = -0.034$ when $\alpha = 0.2, \beta = 0.3$, and the maximum value $F_I = 0.267$ when $\alpha = 1.0, \beta = 0.45$. It is also found that $F_{II} = 0$ for the central interface crack under x-directional tension. Therefore, it may be concluded that central interface crack in a bonded infinite plate with material 1 under x-directional remote tension is equivalent to that of under internal pressure of $\sigma = 0.034 \sim 0.267$. All the values in Fig. 8 are provided in Table 4.

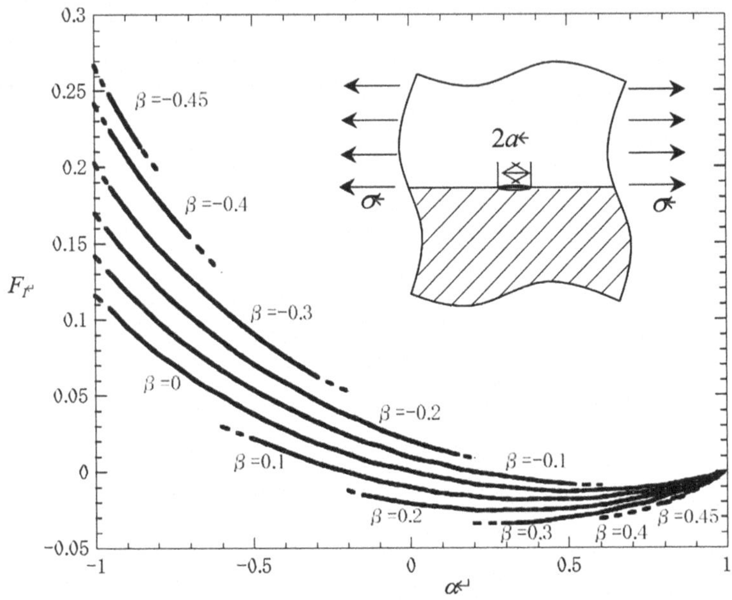

Fig. 8 F_I of a central interface crack in a bonded infinite plate under $\sigma_{x1}^{\infty} - \sigma$ (see Fig. 3b with $a/W \to 0$)

4 Stress Intensity Factors for Periodic Interface Cracks

Periodic interface cracks are one of the most fundamental problems. However, so far as the authors known, the solution is not available under arbitrary material combinations. Similarly to the problem of bonded infinite plate subjected to the internal pressure, the stress intensity factors only depend on the parameter *e*. There, the parameters ε and β have the following relation. That is to say, stress intensity factors for periodic interface cracks only depend on ε or β.

$$\varepsilon = \frac{1}{2\pi} \ln\left(\frac{1-\beta}{1+\beta}\right) \tag{18}$$

First, the problem for homogenous material when $G_1 = G_2$ and $\nu_1 = \nu_2$ is treated. Exact solutions of stress intensity factors for Fig. 4a are expressed in the following [6–8].

$$F_{Ihomo}^{(a)} = \sqrt{\left(\frac{2W}{\pi a}\right) \tan\left(\frac{\pi a}{2W}\right)} \tag{19}$$

Table 5 shows the results of periodic interface cracks $F_I^{(a)}$ for homogenous material. Comparing the present results with the exact solution in Eq. (19), it is found that they

Table 4 Dimensionless stress intensity factor F_I in Fig. 3b with $a/W \to 0$

α	β										
	−0.45	−0.4	−0.3	−0.2	−0.1	0	0.1	0.2	0.3	0.4	0.5
−1.00	(0.267)	(0.242)	(0.203)	(0.170)	(0.142)	(0.116)	–	–	–	–	–
−0.95	0.247	0.224	0.187	0.157	0.130	0.107	–	–	–	–	–
−0.90	0.229	0.208	0.173	0.144	0.119	0.096	–	–	–	–	–
−0.85	0.213	0.193	0.160	0.133	0.109	0.087	–	–	–	–	–
−0.80	(0.199)	0.180	0.148	0.122	0.099	0.078	–	–	–	–	–
−0.75	–	0.167	0.137	0.112	0.090	0.071	–	–	–	–	–
−0.70	–	0.155	0.127	0.103	0.082	0.063	–	–	–	–	–
−0.60	–	(0.131)	0.108	0.086	0.067	0.05	(0.030)	–	–	–	–
−0.50	–	–	0.091	0.071	0.054	0.038	0.022	–	–	–	–
−0.40	–	–	0.076	0.059	0.043	0.028	0.014	–	–	–	–
−0.30	–	–	0.063	0.047	0.033	0.019	0.006	–	–	–	–
−0.20	–	–	(0.053)	0.037	0.024	0.012	0.000	(−0.012)	–	–	–
−0.15	–	–	–	0.033	0.020	0.008	−0.003	−0.015	–	–	–
−0.10	–	–	–	0.028	0.017	0.005	−0.006	−0.017	–	–	–
−0.05	–	–	–	0.024	0.013	0.003	−0.008	−0.019	–	–	–
0.05	–	–	–	0.017	0.007	−0.002	−0.012	−0.022	–	–	–
0.1	–	–	–	0.014	0.005	−0.004	−0.014	−0.023	–	–	–
0.15	–	–	–	0.011	0.002	−0.006	−0.015	−0.024	–	–	–
0.2	–	–	–	(0.009)	0.000	−0.008	−0.016	−0.025	(−0.034)	–	–
0.3	–	–	–	–	0.003	−0.010	−0.018	−0.025	−0.034	–	–
0.4	–	–	–	–	0.006	−0.012	−0.018	−0.025	−0.033	–	–
0.5	–	–	–	–	0.008	−0.013	−0.018	−0.024	−0.030	–	–
0.6	–	–	–	–	(0.009)	−0.012	−0.017	−0.022	−0.027	(−0.031)	–
0.7	–	–	–	–	–	−0.011	−0.014	−0.018	−0.022	−0.027	–

(continued)

Table 4 (continued)

α	β										
	−0.45	−0.4	−0.3	−0.2	−0.1	0	0.1	0.2	0.3	0.4	0.5
0.75	–	–	–	–	–	−0.010	−0.013	−0.016	−0.020	−0.024	–
0.8	–	–	–	–	–	−0.009	−0.011	−0.014	−0.016	−0.020	(−0.021)
0.85	–	–	–	–	–	−0.007	−0.009	−0.011	−0.013	−0.016	−0.017
0.9	–	–	–	–	–	−0.005	−0.006	−0.008	−0.009	−0.011	−0.012
0.95	–	–	–	–	–	−0.003	−0.003	−0.004	−0.004	−0.006	−0.006
1	–	–	–	–	–	(−0.002)	(0.001)	(0.001)	(0.001)	(−0.001)	(0.001)

Table 5 The results of homogeneous plates

	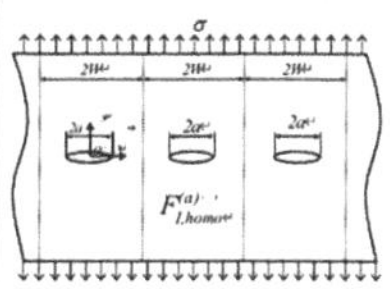		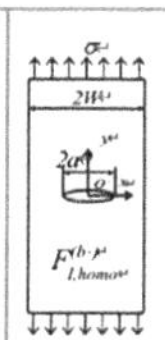	
a/W	Present analysis	Equation (19)	Isida [9]	Equation (20)
→0	1	1	1	1
0.1	1.0042	1.0042	1.006	1.006
0.2	1.017	1.017	1.0246	1.025
0.3	1.0399	1.04	1.0577	1.058
0.4	1.0754	1.0753	1.1094	1.109
0.5	1.1282	1.1284	1.1867	1.186
0.6	1.208	1.2085	1.3033	1.303
0.7	1.335	1.336	1.4882	1.487
0.8	1.561	1.565	1.816	1.814
0.9	2.105	2.1133	2.5776	2.577

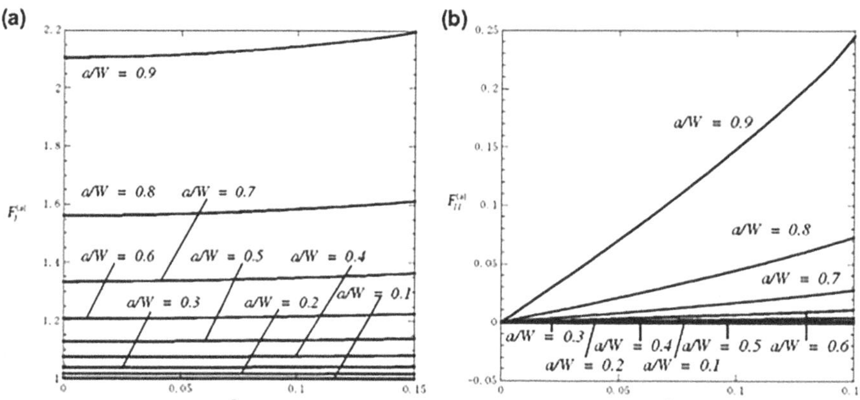

Fig. 9 The relationship between (**a**) $F_I^{(a)}$ vs. ε (**b**) $F_{II}^{(a)}$ vs. e with different a/W in Fig. 4a

are coincide within the error 0.4%. Table 5 also shows the results of a central crack obtained by Isida [9] and the approximate solutions expressed in Eq. (20) [10].

$$F_{Ihomo}^{(b)} = \left\{ 1 - 0.025\left(\frac{a}{W}\right)^2 + 0.06\left(\frac{a}{W}\right)^4 \right\} \sqrt{sec\left(\frac{\pi a}{2W}\right)} \tag{20}$$

Comparing the Eqs. (19) and (20), it is seen that $F_I^{(b)}$ in Fig. 4b is about 20% larger than $F_I^{(a)}$ in Fig. 4a when the relative crack length is large.

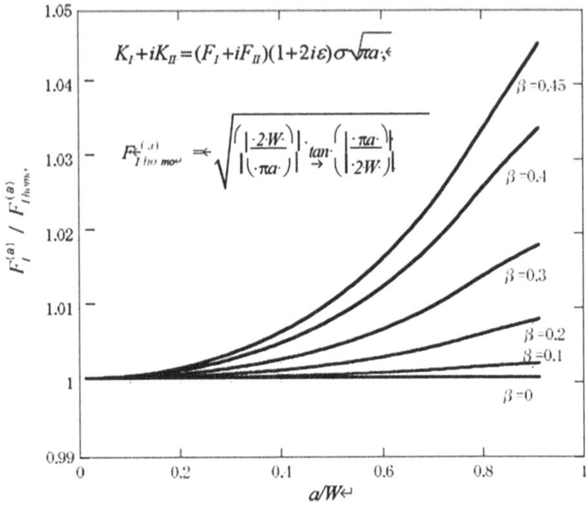

Fig. 10 The relationship between $F_I^{(a)}/F_{I homo}^{(a)}$ vs. a/W in Fig. 4a

Figure 9a shows the relation between $F_I^{(a)}$ and ε with different a/W. From the figure, it is known that $F_I^{(a)}$ slightly increases with increasing ε when a/W is fixed, the increment is also larger when a/W is larger. Figure 9b shows the relation between $F_I^{(a)}$ and ε with different a/W. Similarly to $F_I^{(a)}$, the values of $F_{II}^{(a)}$ also increases with increasing ε when a/W is fixed, and the increment becomes larger when a/W is larger.

Because parameters e and b have the relationship in Eq. (20), the relation F_I and b also can be obtained. Figure 10 shows the relation between F_I divided by $F_{I,\ homo}$ and b. From Fig. 10, it is known that $F_I/F_{I,\ homo}$ is nearly about 1 when a/W is small, and $F_I/F_{I,\ homo}$ becomes larger when a/W is larger.

5 Stress Intensity Factors for the Interface Crack in a Bonded Finite Plate

Regarding another fundamental problem in Fig. 4b, effects of relative crack length a/W on stress intensity factors will be discussed under arbitrary combinations. When material 1 and material 2 are exchanged, Dundur's parameters (α, β) become $(-\alpha, -\beta)$; then the stress intensity factors (F_I, F_{II}) become $(F_I, -F_{II})$. Therefore, all material combinations are considered in the range $\alpha > 0$ as shown in Fig. 7. For special material combinations indicated by the dashed lines in Fig. 7, calculations cannot be executed by the current finite element method code, and the results for the region are obtained by extrapolation using the results that can be calculated.

The interface stress intensity factor $F_I^{(b)}$ is close to the solution $F_{I,\ homo}$ of Isida [9] for homogenous materials as shown in Table 5. To discuss the variation $F_I^{(b)}$ clearly under different material combinations, the ration $F_I^{(b)}$ / $F_{I\text{homo}}^{(b)}$ is indicated in Figs. 11a, 12, 13, 14, 15, 16, 17, and 18a. Those figures also show the maximum value $F_{\text{I}max}^{(b)}$ and minimum value $F_{\text{I}min}^{(b)}$ for all material combinations when β is fixed. Here, $F_{II}^{(b)}$ is directly shown from Fig. 11b, 12, 13, 14, 15, 16, 17, and 18b with the maximum value $F_{\text{I}max}^{(b)}$ and minimum value $F_{\text{I}min}^{(b)}$ for all regions of material combination when b is fixed. When $a/W \to 0$, $F_{II}^{(b)}$ goes to 0, and the absolute value of $F_{II}^{(b)}$ monotonously increases with increasing a/W.

Figure 11 shows the results for $\beta = -0.2$. It is found that $F_I^{(b)}$ / $F_{I\text{homo}}^{(b)}$ is in a small region 1.017–0.976, and it is seen that F_I is close to $F_{\text{I}max}^{(b)}$, and F_{II} is close to $F_{\text{I}max}^{(b)}$ individually. Figure 12 shows the results for $\beta = -0.1$. Similarly to the case

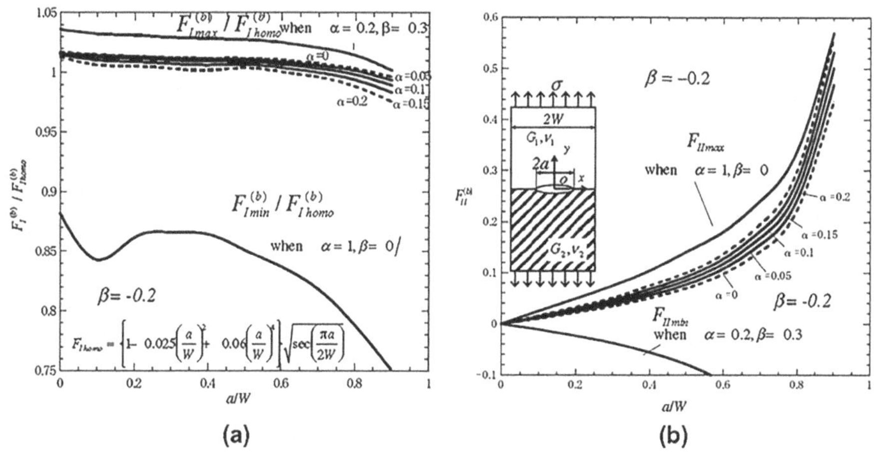

Fig. 11 (**a**) $F_I^{(b)}/F_{\text{homo}}^{(b)}$ vs. a/W and (**b**) $F_H^{(b)}$ vs. a/W when $\beta = -0.2$

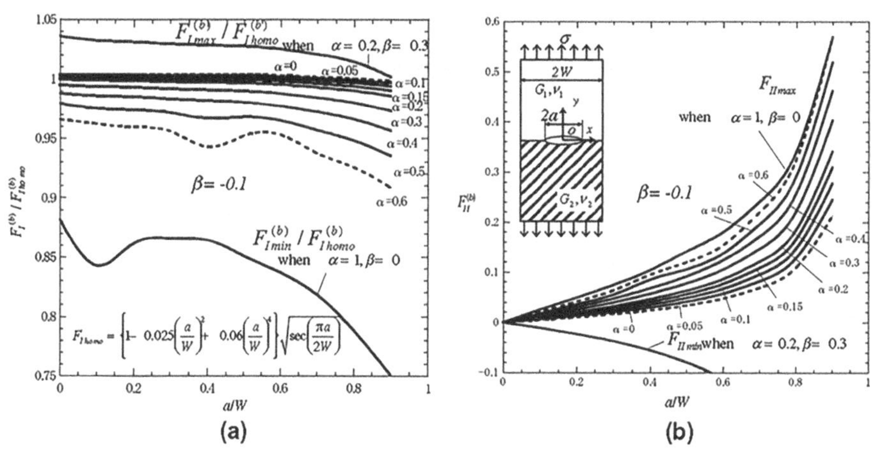

Fig. 12 (**a**) $F_I^{(b)}/F_{\text{homo}}^{(b)}$ vs. a/W and (**b**) $F_H^{(b)}$ vs. a/W when $\beta = -0.1$

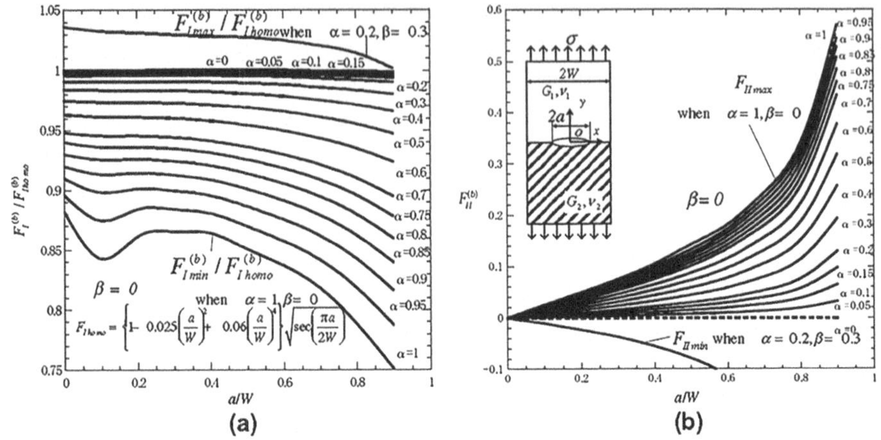

Fig. 13 (**a**) $F_I^{(b)}/F_{\text{homo}}^{(b)}$ vs. a/W and (**b**) $F_H^{(b)}$ vs a/W when $\beta = 0$

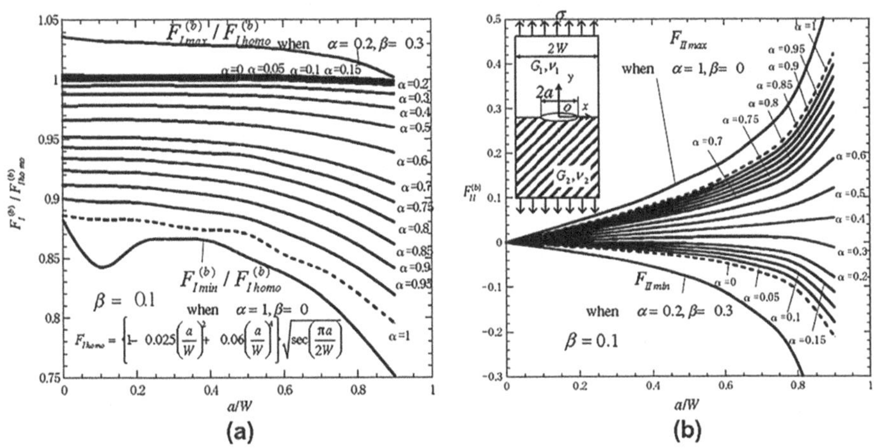

Fig. 14 (**a**) $F_I^{(b)}/F_{\text{homo}}^{(b)}$ vs. a/W and (**b**) $F_H^{(b)}$ vs. a/W when $\beta = 0.1$

$\beta = -0.2$, it is found that $F_I^{(b)} / F_{I\text{homo}}^{(b)}$ is in the small region 1.004–0.909, and $F_I^{(b)}$ is close to $F_{\text{Imax}}^{(b)}$.

On the other hand, when $\beta = 0$, the ratio $F_I^{(b)}/F_{I\text{homo}}^{(b)}$ largely depends on a and distributes in a wide region as 1.000–0.751. The ratio $F_I^{(b)}/F_{I\text{homo}}^{(b)}$ is close to 1 when a is small, and becomes smaller when a is large. When $\alpha = 1$, $F_I^{(b)}/F_{I\text{homo}}^{(b)}$ takes the minimum value for all regions of a/W. On the other hand, $F_{II}^{(b)}$ increases with increasing a, and the increment becomes larger when crack length is larger. For almost all regions of crack length, $F_{II}^{(b)}$ takes the maximum when $\alpha = 1$ except the case when crack length is extremely large.

Similar trend of Fig. 13 can be seen in Fig. 14 for $\beta = 0.1$ and in Fig. 15 for $\beta = 0.2$ although the distribution region is a little smaller in Figs. 14 and 15. On the

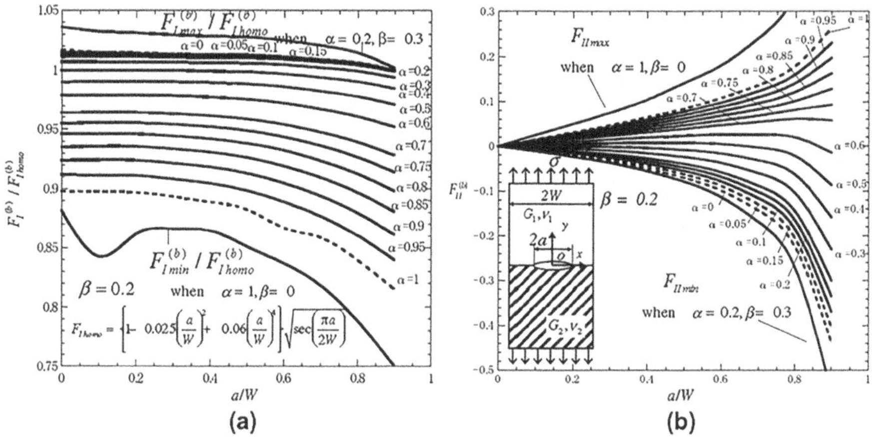

Fig. 15 (**a**) $F_I^{(b)}/F_{\mathrm{homo}}^{(b)}$ vs. a/W and (**b**) $F_H^{(b)}$ vs. a/W when $\beta = 0.2$

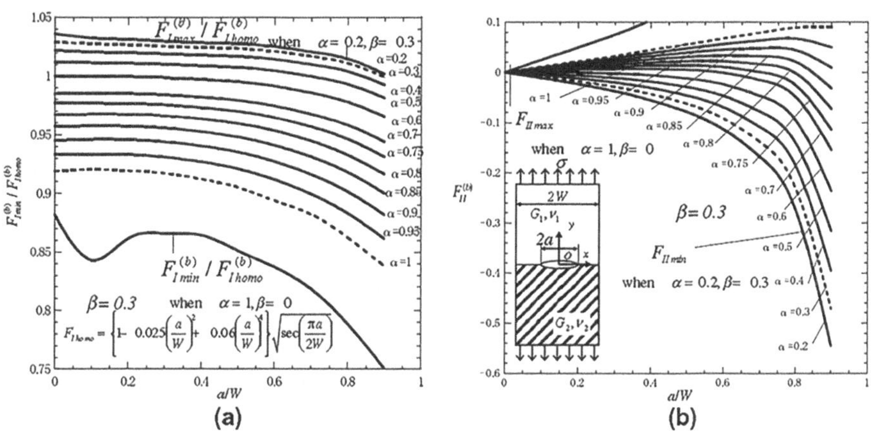

Fig. 16 (**a**) $F_I^{(b)}/F_{\mathrm{homo}}^{(b)}$ vs. a/W and (**b**) $F_H^{(b)}$ vs a/W when $\beta = 0.3$

other hand, $F_{II}^{(b)}$ is always positive when $\beta = 0$, but for $\beta = 0.1$ and $\beta = 0.2$. $F_{II}^{(b)}$ is positive value when a is large and negative when a is small.

For $\beta = 0.3$ in Fig. 16, it is seen that $F_I^{(b)}/F_{I\mathrm{homo}}^{(b)}$ always takes the maximum value for a/W when $\alpha = 0.2$, and $F_{II}^{(b)}$ always takes the minimum value when $\alpha = 0.2$.

Figure 17 shows the case $\beta = 0.4$ and Fig. 18 shows the case $\beta = 0.45$. It is found that $F_I^{(b)}/F_{I\mathrm{homo}}^{(b)}$ decreases with increasing crack length, and on the other hand, $F_{II}^{(b)}$ always has negative values for all a/W.

Considering $F_I^{(b)}/F_{I\mathrm{homo}}^{(b)}$ when $a/W \le 0.9$, it is known that $0.751 \le 0.751 \le F_I^{(b)}/F_{I\mathrm{homo}}^{(b)} < 1.036$, and mostly it distributes in the region which is a little smaller than 1.

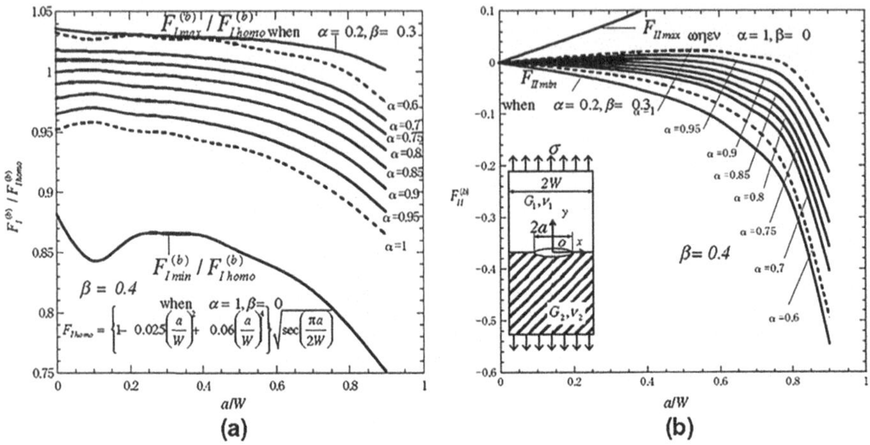

Fig. 17 (a) $F_I^{(b)}/F_{I\text{homo}}^{(b)}$ vs. a/W and (b) $F_{II}^{(b)}$ vs a/W when $\beta = 0.4$

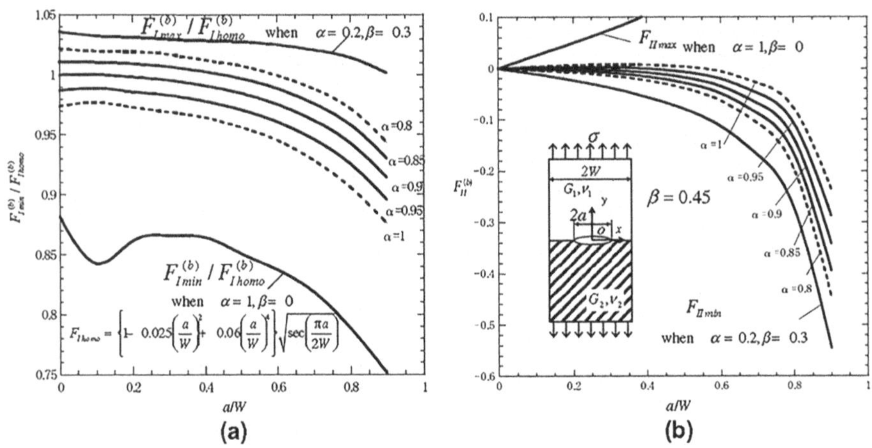

Fig. 18 (a) $F_I^{(b)}/F_{I\text{homo}}^{(b)}$ vs. a/W and (b) $F_{II}^{(b)}$ vs. a/W when $\beta = 0.45$

6 Comparison Between Periodic Interface Cracks and a Center Interface Crack

In order to compare the results of Fig. 4a, b, the results of Fig. 4a are shown as $F_I^{(a)}/F_{I\text{homo}}^{(b)}$ in Fig. 19 when $\beta = 0$, and the results for the case when $\beta = 0.4$ are shown in Fig. 20. For periodic interface cracks, $F_I^{(a)}/F_{I\text{homo}}^{(b)}$ distributes within the region of $F_I^{(b)}/F_{I\text{homo}}^{(b)}$ except the case when the crack length is very large. For a fixed β, the results of periodic interface cracks are close to the results of a central interface crack in a bonded finite plate when a is small and the crack length is small. On the other hand, when, the results of periodic interface cracks are close to the results of a central interface crack in a bonded finite plate when α is large and the crack length is large.

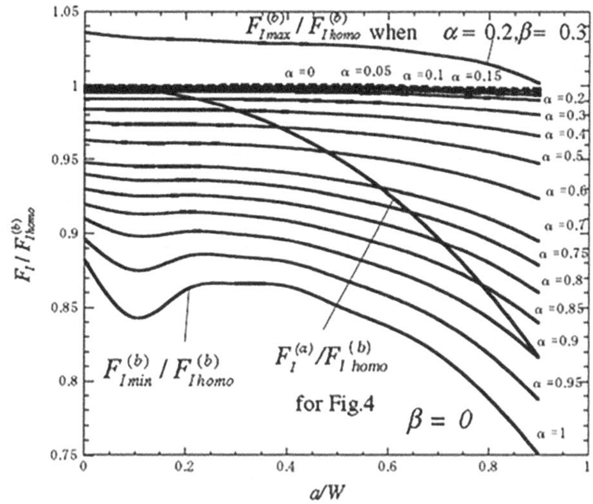

Fig. 19 Comparison between the results in Fig. 4a, b when $\beta = 0$

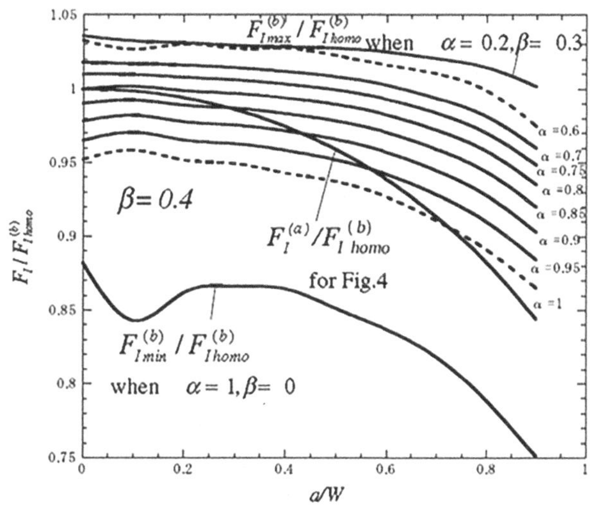

Fig. 20 Comparison between the results in Fig. 4a, b when $\beta = 0.4$

7 Conclusion

In this paper, the stress values at the crack tip calculated by FEM are used and the stress intensity factors of interface cracks are evaluated from the ratio of stress values between a reference problem and a given problem. Then the stress intensity factors are discussed with the following conclusions.

1. For periodic interface cracks in a bonded plate shown in Fig. 4a, the effects of relative crack length and material combinations on the stress intensity factors

have been discussed. Stress intensity factors $F_I^{(a)}$, $F_{II}^{(a)}$ increase with increasing ε (Fig. 9 and Table 5).

2. For a central interface crack in a bonded finite plate shown in Fig. 4b, the effects of relative crack length and material combinations on the stress intensity factors have been discussed. The ratio to the results to the homogeneous material $F_I^{(b)}/F_{I\mathrm{homo}}^{(b)}$ is in the region $0.751 < F_I^{(b)}/F_{I\mathrm{homo}}^{(b)} < 1.036$ when $a/W = 0.9$, and mostly it distributes in the region which is a little smaller than 1.

   Generally, F_I always takes the maximum value when $\alpha = 0.2$, $\beta = 0.3$ and minimum value when $\alpha = 1.0$, $\beta = 0$. On the other hand, F_{II} always takes the maximum value when $\alpha = 1.0, \beta = 0$ and minimum value when $\alpha = 0.2, \beta = 0.3$ except the case when the crack length is extremely large.
3. From the comparison between the results for periodic interface cracks and a central interface crack in a finite bonded plate, it is seen that the results of periodic interface cracks are close to the results of a central interface crack in a bonded finite plate when a is small and the crack length is small. The results of periodic interface cracks are close to the results of a central interface crack in a bonded finite plate when a is large and the crack length is large.

References

1. Yuki Y. Mech interface. Baifuukan; 1992. (in Japanese).
2. Noda N-A, Zhang Y, Lan X, Takase Y, Oda K. Stress intensity factor of an interface crack in a bonded plate under uni-axial tension. J Solid Mech Mater Eng. 2009;4(7):974–87.
3. Oda K, Kamisugi K, Noda NA. Stress intensity factor analysis of interface cracks based on proportional method. Trans Jpn Soc Mech Eng. 2009;A-75:467–82. (in Japanese).
4. Teranisi T, Nisitani H. Determination of highly accurate values of stress intensity factor in a plate of arbitrary form by FEM. Trans Jpn Soc Mech Eng. 1999;A-65:16–21. (in Japanese).
5. Nisitani H, Teranisi T, Fukuyama K. Stress intensity factor analysis of a biomaterial plate based on the crack tip stress method. Trans Jpn Soc Mech Eng. 2003;A-69:1203–8. (in Japanese).
6. Westergaard HM. Bearing pressures and cracks. J Appl Mech. 1939;61:49–53.
7. Irwin GR. Fracture. In: Handbuch der Physik, vol. VI; 1958. p. 551–90.
8. Koiter WT. An infinite row of collinear cracks in an infinite elastic sheet. Ing Arch. 1959;28: 168–72.
9. Isida M. Elasticity analysis and stress intensity factor of crack. Baifuukan 1976. (in Japanese).
10. Okamura H. Introduction linear fracture mechanics. Baifuukan; 1990. (in Japanese).

Effect of Adhesive Thickness on the Intensity of Singular Stress at the Adhesive Dissimilar Joint

1 Introduction

Adhesive joints are most frequently used in numerous industrial sectors such as automobile, shipbuilding, aeronautical, etc., replacing or supplementing traditional joining technologies, such as welding or riveting. Moreover, the adhesive joints have also been used for bonding composite restorations to the dental substrate [1, 2]. The micro-tensile bond test is a laboratory procedure frequently employed today in an attempt to predict the clinical effectiveness of adhesive used for bonding composite restorations to the dental substrate [3], see Fig. 1. This test can be considered as a miniaturized version of the conventional engineering tensile adhesion test using butt-joint specimens, and the specimens can be rectangular in cross-section, see Fig. 2.

However, a mismatch of different materials properties may cause stress singularity at the edge of an interface between different materials, which leads to failure of bonding part in structures, and the singularity is expressed by the intensity of singular stress.

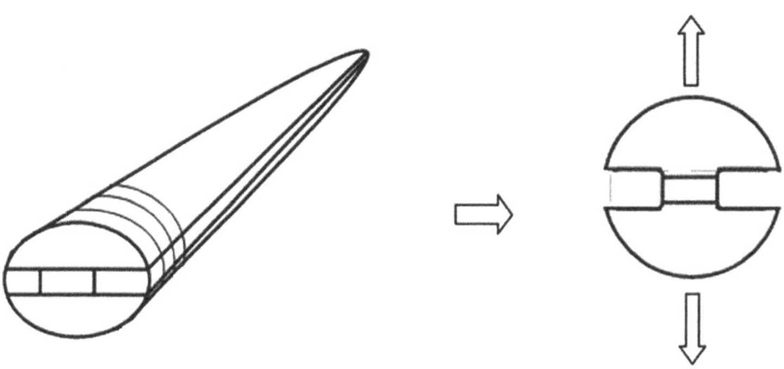

Fig. 1 Micro-tensile bond test

Y. Zhang, *Interfacial Mechanics of Offshore Engineering Equipment*,
https://doi.org/10.1007/978-981-95-3200-1_4

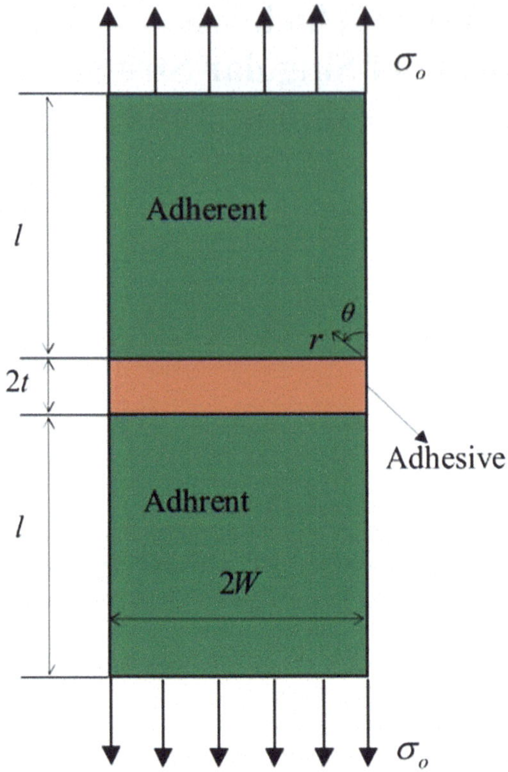

Fig. 2 Adhesive joint

So far, many studies have been done to evaluate the strength of adhesive [4, 5] and few studies has been conducted to describe the stress distribution on the interface between the adhesive and adherents used in micro-tensile bond tests [6]. Besides, until now, no study has obtained the intensity of singular stress at the edge of adhesive joint in micro-tensile bond tests.

From previous experimental results, it is found that the joint strength decreases with increasing of adhesive thickness [7]. However, the reason why the joint strength decreases with increasing of adhesive thickness has not been explained explicitly.

Therefore, in this paper, to explain the reason by simulation, the effect of adhesive thickness on the intensity of singular stress will be analyzed by the finite element method, also the effect of material combination on the intensity of singular stress will be analyzed.

2 Method of Analysis of the Intensity of Singular Stress

For the adhesive joint as shown in Fig. 2, it is known that the interface stress $\sigma_{ij}(ij = rr, \theta\theta, r\theta)$ goes to infinity at the edge of joint and has singularity of $\sigma_{ij} \propto \frac{1}{r^{1-\lambda}}$ when $\alpha(\alpha - 2\beta) > 0$; here, α, β are Dunder's parameters which are expressed by

Passion's ratio v and shear modulus G. The singularity index λ at the joint of interface can be expressed by the following equation [8, 9]. Table 1 shows the values λ obtained by solving Eq. (1). It can be found that $\lambda < 1$ when $\alpha(\alpha - 2\beta) > 0$; $\lambda = 1$ when $\alpha(\alpha - 2\beta) = 0$; $\lambda > 1$ when $\alpha(\alpha - 2\beta) < 0$.

$$\left[\sin^2\left(\frac{\pi}{2}\lambda\right) - \lambda^2\right]^2 \beta^2 + 2\lambda^2\left[\sin^2\left(\frac{\pi}{2}\lambda\right) - \lambda^2\right]\alpha\beta + \lambda^2\left(\lambda^2 - 1\right)\alpha^2 + \frac{\sin^2(\lambda\pi)}{4} = 0 \tag{1}$$

$$\alpha = \frac{G_1(k_2+1) - G_2(k_1+1)}{G_1(k_2+1) + G_2(k_1+1)}, \quad \beta = \frac{G_1(k_2-1) - G_2(k_1-1)}{G_1(k_2+1) + G_2(k_1+1)} \tag{2}$$

$$k_j = \begin{cases} \dfrac{3 - v_j}{1 + v_j} & \text{(plane stress)} \\ 3 - 4v_j & \text{(plane strain)} \end{cases}, \quad k_j = 1, 2 \tag{3}$$

The intensity of singular stress K_σ at the adhesive dissimilar joint is expressed as

$$K_\sigma = \lim_{r \to 0}\left[r^{1-\lambda} \times \sigma_{\theta|\theta=\frac{\pi}{2}}(r)\right] \tag{4}$$

and the dimensionless of intensity of singular stress F_σ is defined by the following equation.

$$F_\sigma = \frac{K_\sigma}{\sigma(2W)^{1-\lambda}} = \frac{\lim_{r \to 0}\left[r^{1-\lambda}\sigma_{\theta|\theta=\frac{\pi}{2}}(r)\right]}{\sigma(2W)^{1-\lambda}} \tag{5}$$

Here, σ is the stress applying to the y direction.

In this paper, the finite element method is used to obtain the stress at the joint of interface, and the software is MSC.MARC 2007. Because of symmetry, one-fourth portion of Fig. 2 is analyzed as a model for analysis. Here, E_1, v_1 are the Young's modulus and the Poisson's ratio of the adherent, and E_2, v_2 are the Young's modulus and the Poisson's ratio of the adhesive. The width of the model $2W = 2000$ mm, and the length $l = 2W$ because it is demonstrated that when; $l \geq 2W$ the interface stresses are the same. The adhesive thickness is changed as $t/W = 0.001, 0.01, 0.1, 0.5, 1, 2, 4$.

We will propose the method of calculating the intensity of singular stress from the results of FEM. In this paper, the ratio of intensity of singular stress; K_σ^1/K_σ^2 will be considered. Here, the superscripts 1, 2 mean specific problems whose t/W are distinct. As shown in Eqs. (4) and (5), the dimensionless intensity of singular stress is related to the distance r, singular index λ, and stress σ_0, width W and limiting stress $\lim_{r \to 0}\sigma_{\theta(\theta=\pi/2)}$. Consider different adhesive thicknesses t_1, t_2, as problem 1 and

Table 1 Values of singular index λ (Red figures indicate $\lambda < 1$, blue figures indicate $\lambda > 1$, black figures indicate $\lambda = 1$)

α	β=-0.45	β=-0.4	β=-0.3	β=-0.2	β=-0.1	β=0	β=0.1	β=0.2	β=0.3	β=0.4	β=0.45
-1.00	0.87624	0.8073	0.7205	0.6646	0.6247	0.5946					
-0.95	0.9349	0.8536	0.7576	0.6975	0.6550	0.6232					
-0.90	1.00000	0.9008	0.7941	0.7295	0.6845	0.6511					
-0.80		1.0000	0.8655	0.7916	0.7415	0.7048					
-0.70		1.1174	0.9348	0.8510	0.7961	0.7564					
-0.60			1.0000	0.9071	0.8480	0.8060	0.7746				
-0.50			1.0558	0.9580	0.8966	0.8532	0.8210				
-0.40			1.0913	1.0000	0.9403	0.8974	0.8655				
-0.30			1.0964	1.0276	0.9761	0.9371	0.9075				
-0.20			1.0756	1.0360	1.0000	0.9699	0.9457	0.9269			
-0.10				1.0251	1.0083	0.9921	0.9777	0.9659			
0.00				1.0000	1.0000	1.0000	1.0000	1.0000			
0.10				0.9659	0.9777	0.9921	1.0083	1.0251			
0.20				0.9269	0.9457	0.9699	1.0000	1.0360	1.0756		
0.30					0.9075	0.9371	0.9761	1.0276	1.0964		
0.40					0.8655	0.8974	0.9403	1.0000	1.0913		
0.50					0.8210	0.8532	0.8966	0.9580	1.0558		
0.60					0.7746	0.8060	0.8480	0.9071	1.0000		
0.70						0.7564	0.7961	0.8510	0.9348	1.1174	
0.80						0.7048	0.7415	0.7916	0.8655	1.0000	
0.90						0.6511	0.6845	0.7295	0.7941	0.9008	1.0000
0.95						0.6232	0.6550	0.6975	0.7576	0.8536	0.9349
1.00						0.5946	0.6247	0.6646	0.7205	0.8073	0.8762

problem 2, both of which have the same stress at infinity σ and material combinations. Therefore, it should be noted that the singular index $\lambda_1 = \lambda_2$. As shown in Eq. (6), the ratio of intensity of singular stress K_σ^1/K_σ^2 is controlled by the ratio of stress $\lim_{r\to 0}\left(\sigma_{\theta|\theta=\pi/2}^1/\sigma_{\theta|\theta=\pi/2}^2\right)$.

Fig. 3 Bonded strip

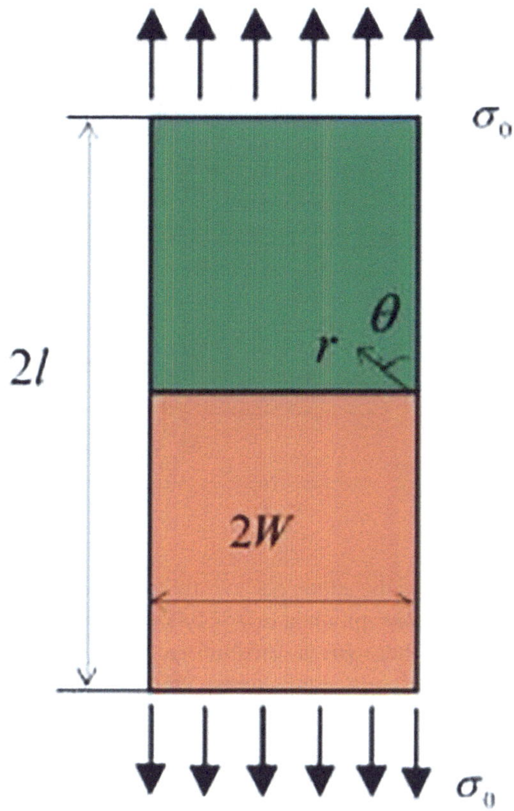

$$\frac{K_\sigma^1}{K_\sigma^2} = \frac{\sigma^1 (2W)^{1-\lambda_1} F_\sigma^1}{\sigma^2 (2W)^{1-\lambda_1} F_\sigma^2} = \frac{F_\sigma^1}{F_\sigma^2} = \lim_{r\to 0} \frac{\left[r^{1-\lambda_0} \sigma^1_{\theta|\theta-\frac{\pi}{2}}(r)\right]}{\left[r^{1-\lambda_0} \sigma^2_{\theta|\theta-\frac{\pi}{2}}(r)\right]} = \lim_{r\to 0} \frac{\sigma^1_{\theta|\theta-\frac{\pi}{2}}(r)}{\sigma^2_{\theta|\theta-\frac{\pi}{2}}(r)} \tag{6}$$

Therefore, in this paper, the ratio of intensity of singular stress is mainly considered in the analysis. To obtain the intensity of singular stress from the ratio, a reference problem as shown in Fig. 3 will be used because the intensity of singular stress has been investigated.

3 Interface Stress Distribution and Ratio of the Distributions Obtained by Using FEM

Figure 4 shows the stress distribution on the interface between adhesive and adherent when adhesive thickness $t/W = 1$ and $\alpha = 0.8$, $\beta = 0.1$, $\alpha = 0.9$, $\beta = 0.3$, $\alpha = 0.3$, $\beta = 0$, $\alpha = 0.2$, $\beta = 0.1$, $\alpha = 0.2$, $\beta = 0.2$. It is confirmed that when $\alpha(\alpha - 2\beta) > 0$,

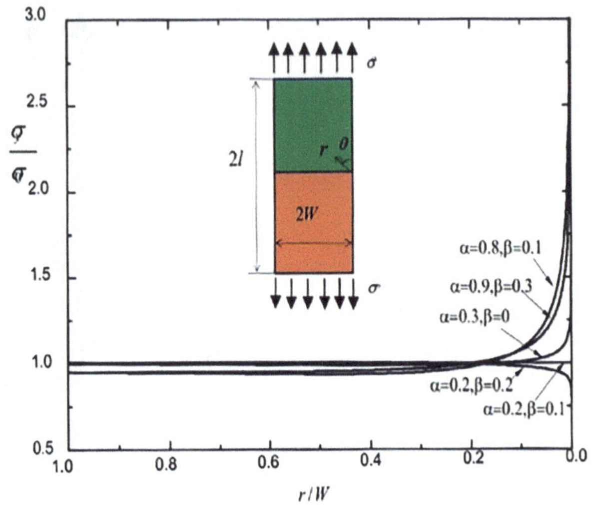

Fig. 4 The stress distribution on the interface with different material combination

the stress at the edge goes to infinite with different intensity depending on α, β; stress goes to constant when $\alpha(\alpha - 2\beta) = 0$; stress goes to 0 when $\alpha(\alpha - 2\beta) < 0$. In this analysis, those stress distributions along the interface are obtained by extrapolation from the results for adherent and adhesive. Usually, those FEM results do not coincide with each other; and therefore, the average values are used to plot the stress distribution.

To understand the effect of adhesive thickness on the intensity of singular stress, the stress distributions with different adhesive thickness are considered. Figure 5a shows the stress distribution on the interface when $\alpha = 0.3$, $\beta = 0$ with adhesive thickness t/W changed from 0.001, to 0.01, 0.1, 0.5, 1, 2, 4 and to see the detail of singular stress distribution at the edge of the interface, the magnified figure is shown in Fig. 5b. It is found that the increase of adhesive thickness causes a significant increase of stress singularity area. When $t/W = 1, 2, 4$, the stress distribution is almost the same. More detail stress distributions at the edge of interface under different adhesive thickness is shown in Fig. 5c. It is seen that the stresses become large suddenly. Real stresses should go to infinity at $r \to 0$, although FEM cannot express the singular stresses. More important discovery is that all the lines are parallel to each other along r/W.

Figure 6 shows the ratio $\sigma_y^1 / \sigma_{y|t/W=1}$ near the edge of adhesive joint when the smallest mesh size is $1/3^8 = 1/6561$ mm with $W = 1000$ mm , and it should be noted that all lines are constant. Although the singularity at the edge is difficult to be described by FEM, the ratio $\sigma_y^1 / \sigma_{y|t/W=1}$ shown in Fig. 6 can be accurate even in the singularity regions.

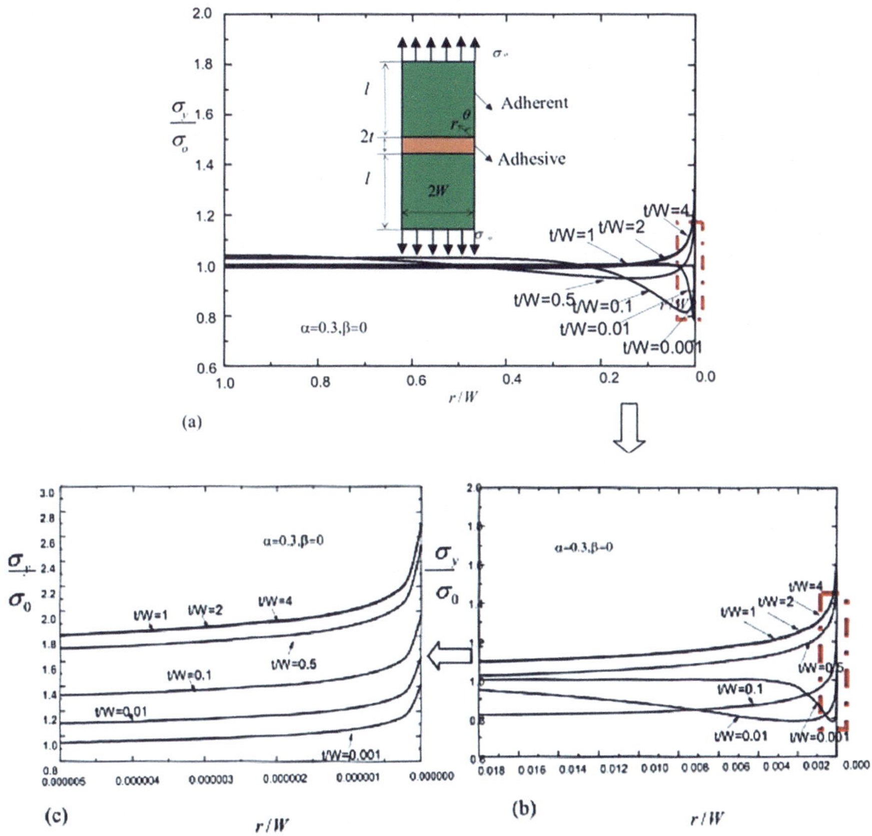

Fig. 5 (**a**) Stress distribution σ_y on the interface with different adhesive thickness when r/w = 0.0~1.0; (**b**) Stress distribution σ_y on the interface with different adhesive thickness when r/w = 0.000~0.018; (**c**) Stress distribution σ_y on the interface with different adhesive thickness when r/w = 0.000000~0.000005

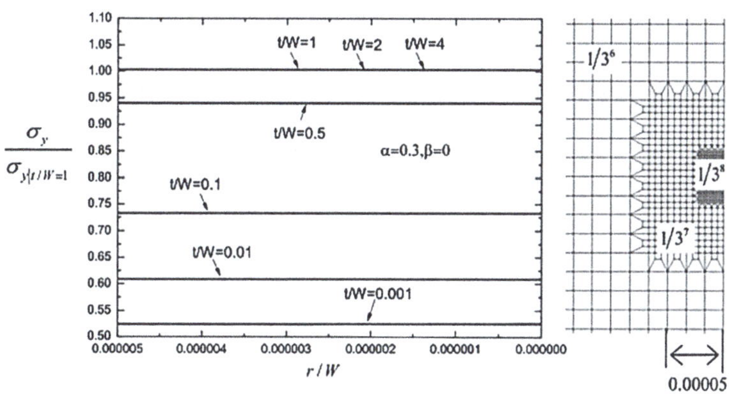

Fig. 6 Ratio of, $\sigma_y^1/\sigma_{y|t/W=1}$ near the edge of adhesive joint

Table 2 Stress distribution σ_y, along the interface when α = 0.3, $\beta = 0$. The ratio of stress distributions $\sigma_y^1/\sigma_{y|t/W=1}$ are indicated in parentheses. σ_y and $\sigma_y^1/\sigma_{y|t/W=1}$ obtained with the smallest mesh size $1/3^8 = 1/6561$ mm and $W = 1000$ mm

t/W r/W	0.001	0.01	0.1	0.5	1	2	4
→ 0	1.414 (0.525)	1.640 (0.609)	1.973 (0.733)	2.530 (0.940)	2.692 (1.000)	2.700 (1.003)	2.670 (1.003)
1/6,561,000	1.177 (0.525)	1.365 (0.609)	1.644 (0.733)	2.108 (0.940)	2.242 (1.000)	2.249 (1.003)	2.249 (1.003)
2/6,561,000	1.138 (0.525)	1.320 (0.609)	1.589 (0.733)	2.038 (0.940)	2.167 (1.000)	2.174 (1.003)	2.174 (1.003)
3/6,561,000	1.109 (0.525)	1.286 (0.609)	1.548 (0.733)	1.985 (0.940)	2.111 (1.000)	2.118 (1.003)	2.117 (1.003)
4/6,561,000	1.088 (0.525)	1.262 (0.609)	1.519 (0.733)	1.948 (0.940)	2.072 (1.000)	2.078 (1.003)	2.078 (1.003)
5/6,561,000	1.071 (0.525)	1.243 (0.609)	1.497 (0.733)	1.919 (0.940)	2.041 (1.000)	2.047 (1.003)	2.047 (1.003)
6/6,561,000	1.058 (0.525)	1.228 (0.609)	1.478 (0.733)	1.896 (0.940)	2.016 (1.000)	2.022 (1.003)	2.022 (1.003)
7/6,561,000	1.047 (0.525)	1.215 (0.609)	1.463 (0.733)	1.876 (0.940)	1.995 (1.000)	2.002 (1.003)	2.001 (1.003)
8/6,561,000	1.038 (0.525)	1.205 (0.609)	1.450 (0.733)	1.859 (0.940)	1.978 (1.000)	1.984 (1.003)	1.984 (1.003)
9/6,561,000	1.030 (0.525)	1.195 (0.609)	1.439 (0.733)	1.845 (0.940)	1.962 (1.000)	1.968 (1.003)	1.968 (1.003)

To see the discovery more deeply, Table 2 compares the stress distributions and the ratio of the results obtained by using FEM with the smallest mesh size $1/3^8 = 1/6561$ mm. It is found that the ratio is almost constant by four digit independent of r. Table 3 shows the results with the smallest mesh size $1/3^4 = 1/81$ mm. In this case, it is found that the values are almost constant by three digit independent of r. It is also found that the ratio in Tables 2 and 3 coincide each other by three digit. Although real interface singular stresses cannot be expressed easily by using the FEM because the values of stress largely depend on the mesh size, it is found that the ratio of stress can be obtained vary accurately as shown in Table 2. In other words, the ratio of interface stress is nearly independent of mesh size.

As explained in the Sect. 2, the ratio K_σ^1/K_σ^2 is equal to the ratio σ_y^1/σ_y^2 along r, and as shown in Tables 2 and 3 since the ratio σ_y^1/σ_y^2 along r is independent of r, only the stress σ_y of the first element should be considered.

In the following of the paper, the stress intensity factors for known reference problem 2 K_σ^2 will be shown and the stress intensity factors for unknown problem 1 K_σ^1 will be discussed from the ratio K_σ^1/K_σ^2.

Table 3 Stress distribution σ_y, along the interface when $\alpha = 0.3$, $\beta = 0$. The ratio of stress distributions $\sigma_y^1/\sigma_{y|t/W=1}$ are indicated in parentheses. σ_y and $\sigma_y^1/\sigma_{y|t/W=1}$ obtained with the smallest mesh size $1/3^4 = 1/81$ mm and $W = 1000$ mm

t/W r/W	0.001	0.01	0.1	0.5	1	2	4
→ 0	1.072 (0.524)	1.246 (0.609)	1.499 (0.733)	1.923 (0.940)	2.045 (1.000)	2.051 (1.003)	2.051 (1.003)
1/81,000	0.889 (0.522)	1.036 (0.609)	1.249 (0.733)	1.601 (0.940)	1.703 (1.000)	1.708 (1.003)	1.708 (1.003)
2/81,000	0.859 (0.522)	1.001 (0.608)	1.207 (0.733)	1.548 (0.940)	1.647 (1.000)	1.652 (1.003)	1.652 (1.003)
3/81,000	0.838 (0.522)	0.975 (0.608)	1.176 (0.733)	1.508 (0.940)	1.604 (1.000)	1.609 (1.003)	1.608 (1.003)
4/81,000	0.824 (0.523)	0.956 (0.608)	1.154 (0.733)	1.480 (0.940)	1.574 (1.000)	1.579 (1.003)	1.579 (1.003)
5/81,000	0.813 (0.525)	0.942 (0.607)	1.137 (0.733)	1.458 (0.940)	1.551 (1.000)	1.555 (1.003)	1.555 (1.003)
6/81,000	0.806 (0.526)	0.930 (0.607)	1.123 (0.733)	1.440 (0.940)	1.532 (1.000)	1.536 (1.003)	1.536 (1.003)
7/81,000	0.800 (0.528)	0.920 (0.607)	1.111 (0.733)	1.425 (0.940)	1.516 (1.000)	1.521 (1.003)	1.520 (1.003)
8/81,000	0.795 (0.529)	0.912 (0.607)	1.092 (0.733)	1.403 (0.940)	1.502 (1.000)	1.507 (1.003)	1.507 (1.003)
9/81,000	0.792 (0.531)	0.904 (0.607)	1.084 (0.733)	1.401 (0.940)	1.491 (1.000)	1.495 (1.003)	1.495 (1.003)

4 Intensity of Singular Stress for Bonded Strip as a Reference Solution

In the previous chapters, it is found that the ratio of interface stress distribution can be given very accurately by using FEM. However, to obtain the intensity of singular stress, a reference solution is necessary. Chen-Nisitani [10] and Noda et al. [11] have analyzed the intensity of singular stress in a bonded strip in Fig. 3 accurately by using the body force method. Table 4 and Fig. 7 indicate the results for bonded strip, which are equivalent to the case $t/W \geq 1$. In the previous studies [10, 11], only the results for singular stress $\lambda > 1$ are indicated, and the results [11] are used in Table 4 and Fig. 7. However, in this study, all material combinations are newly considered; and, therefore Fig. 7 includes new results for $F_\sigma > 1$ where no singular stress because singular index $\lambda > 1$. Those new results are obtained easily by FEM because of no singularity stress. However, the dashed lines are extended from solid lines because some special material combinations are difficult to be obtained by using commercial FEM codes. In this paper, intensity of singular stress will be shown as the ratio $F_\sigma/F_{\sigma|t/W=1}$.

Table 4 F_σ at interface edge point in bonded finite plate. [() Extrapolated or interpolated results. Red figures indicate $\lambda < 1$, blue figures indicate $\lambda > 1$, black figures indicate $\lambda = 1$]

α	$\beta = -0.4$	$\beta = -0.3$	$\beta = -0.2$	$\beta = -0.1$	$\beta = 0$	$\beta = 0.1$	$\beta = 0.2$	$\beta = 0.3$	$\beta = 0.4$
1.00	0.540	0.446	0.395	0.357	0.332	-	-	--	-
-0.95	0.643	(0.349)	(0.381)	(0.422)	(0.491)	--	-	--	-
-0.90	0.726	0.534	0.456	0.412	0.381	--	--	--	--
-0.80	1.000	0.636	0.538	0.487	0.45	--	-	--	-
-0.70	(1.855)	0.800	0.626	0.558	0.486	--	-	--	-
-0.60	(3.291)	1.000	0.724	0.638	0.559	(0.505)	-	--	--
-0.50	-	1.264	0.842	0.722	0.635	(0.551)	--	--	-
-0.40	--	1.467	1.000	0.822	0.718	0.615	-	--	-
-0.30	--	(1.609)	1.118	0.913	0.796	0.697	--	-	-
-0.20	--	(1.690)	1.153	1.000	0.889	0.797	(0.404)	--	-
-0.10	--	-	1.103	1.037	0.955	0.890	0.767	--	-
0.00	-	-	1.000	1.000	1.000	1.000	1.000	--	-
0.10	--	-	0.767	0.890	0.955	1.037	1.103	--	-
0.20	--	-	(0.404)	0.797	0.889	1.000	1.153	(1.690)	--
0.30	-	-	--	0.697	0.796	0.913	1.118	(1.609)	-
0.40	--	-	--	0.615	0.718	0.822	1.000	1.467	--
0.50	-	-	-	(0.551)	0.635	0.722	0.842	1.264	--
0.60	--	-	--	(0.505)	0.559	0.638	0.724	1.000	(3.291)
0.70	--	-	--	--	0.486	0.558	0,626	0.800	1.855
0.80	--	-	--	--	0.450	0.487	0.538	0.636	1.000
0.90	--	-	--	--	0.381	0.412	0.456	0.534	0.726
0.95	--	-	-	--	(0.491)	(0.422)	(0.381)	(0.349)	0.643
1.00	-	-	--	--	0.332	0.357	0.395	0.446	0.540

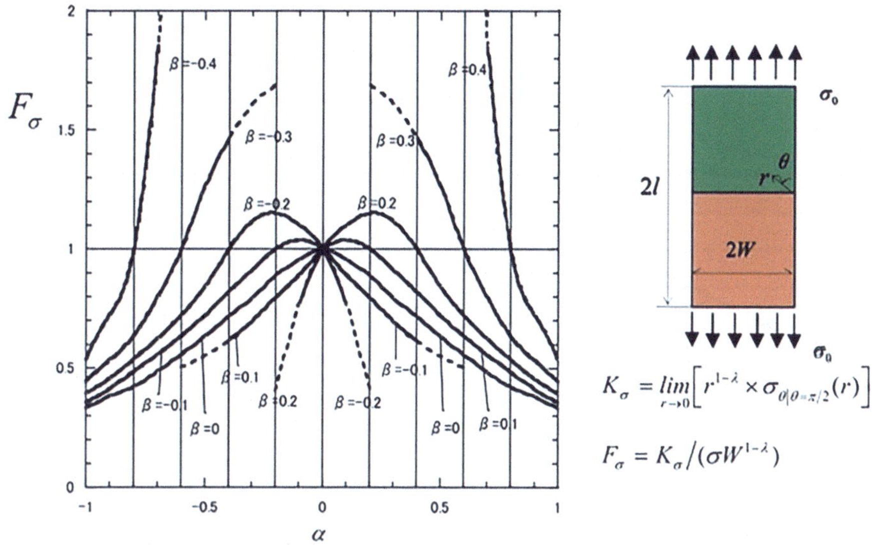

Fig. 7 F_σ for a boned strip in Fig. 3. (Note that $F_\sigma = 1$ when $\alpha(\alpha - 2\beta) = 0$)

5 Results and Discussion

5.1 Effect of Material Combination on Generalized Stress Intensity Factors

As discussed before, it is found that the ratio F_σ^1/F_σ^2 is equivalent to the ratio σ_y^1/σ_y^2 along interface r. By calculating the ratio $\sigma_y/\sigma_{y\,|\,t/W\,=\,1}$ around the edge of interface, the ratio $F_\sigma/F_{\sigma\,|\,t/W\,=\,1}$ has been obtained for all material combinations. Figure 8 shows the map of α and β used for this calculations.

Figure 9 shows $F_\sigma/F_{\sigma\,|\,t/W\,=\,1}$ with varying α and β when (a) $t/W = 0.001$; (b) $t/W = 0.1$ and Table 5 gives the value for Fig. 9. It can be seen that $F_\sigma/F_{\sigma\,|\,t/W\,=\,1}$ increases with increasing of α when α is small. On the other hand, the ratio $F_\sigma/F_{\sigma\,|\,t/W\,=\,1}$ decreases with increasing of α when α is large. Comparing the results of $t/W = 0.001$ and $t/W = 0.1$, it is found that the range of the ratio is different; that is, for $t/W = 0.001$, the ratio $F_\sigma/F_{\sigma\,|\,t/W\,=\,1}$ is widely distributed in the range of 0.025~2.857, while for $t/W = 0.1$, the ratio $F_\sigma/F_{\sigma\,|\,t/W\,=\,1}$ is in the small range of 0.185~1.498.

Using the results $F_\sigma/F_{\sigma\,|\,t/W\,=\,1}$ in Table 4 and $F_\sigma/F_{\sigma\,|\,t/W\,=\,1}$ in Table 3, F_σ are obtained and shown in Fig. 10 for $t/W = 0.001$ and $t/W = 0.1$.

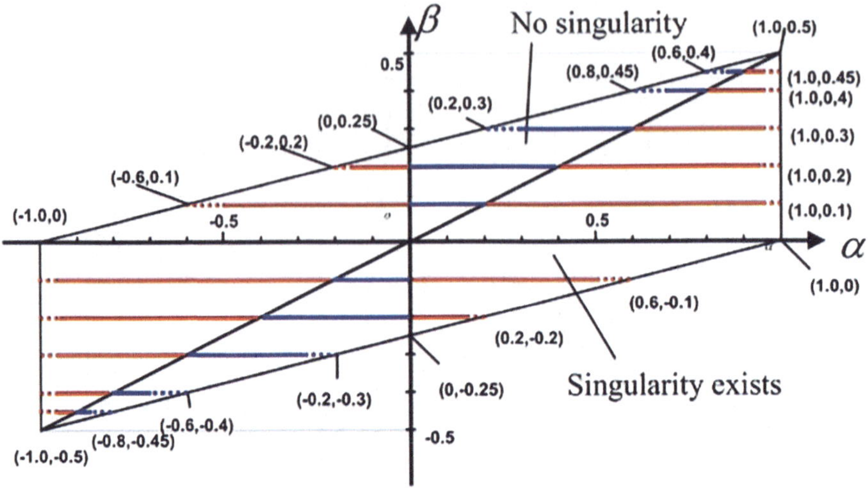

Fig. 8 The map of α and β

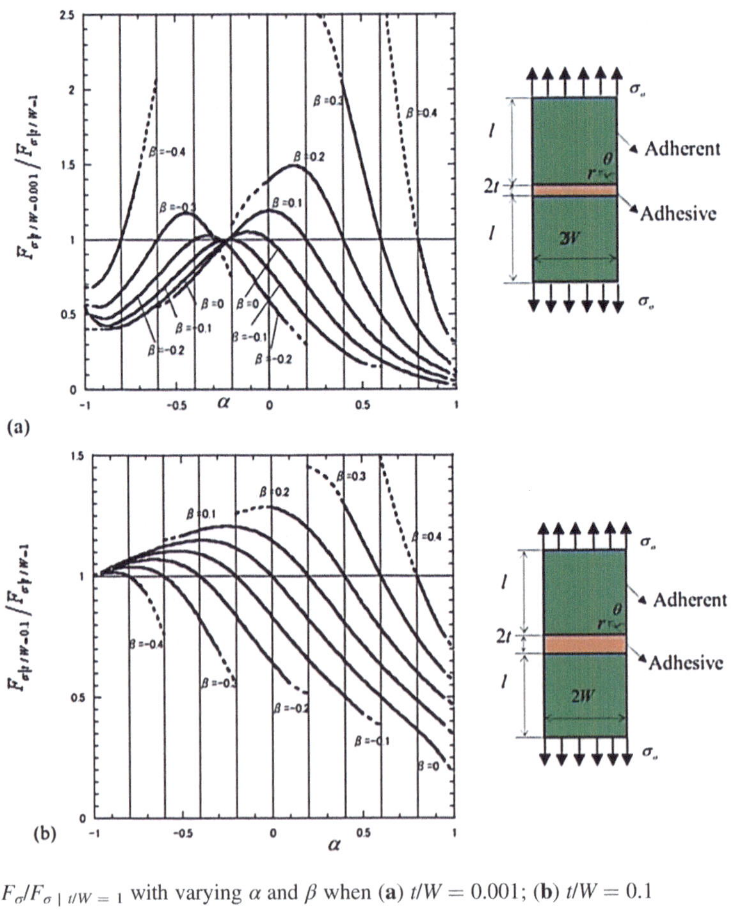

Fig. 9 $F_\sigma/F_{\sigma\mid t/W=1}$ with varying α and β when (**a**) $t/W = 0.001$; (**b**) $t/W = 0.1$

Table 5 $F_{\sigma}/F_{\sigma\ |\ t/W\ =\ 1}$ with varying α and β when (a) $t/W = 0.001$; (b) $t/W = 0.1$. [( ) Extrapolated or interpolated results. Red figures indicate $\lambda < 1$, blue figures indicate $\lambda > 1$, black figures indicate $\lambda = 1$]

α	$\beta = -0.4$	$\beta = -0.3$	$\beta = -0.2$	$\beta = -0.1$	$\beta = 0$	$\beta = 0.1$	$\beta = 0.2$	$\beta = 0.3$	$\beta = 0.4$
-1.0	(0.682)	(0.566)	(0.517)	(0.552)	(0.400)	--	--	--	--
-0.95	0.6864	0.5554	0.4957	0.4629	(0.400)	--	--	--	--
-0.9	0.7420	0.5533	0.4722	0.4252	0.4004	--	--	--	--
-0.8	1.0000	0.6535	0.5254	0.4587	0.4190	--	--	--	--
-0.7	1.4465	0.8130	0.6289	0.5356	0.4812	--	--	--	--
-0.6	(2.073)	1.0000	0.7579	0.6390	0.5690	(0.550)	--	--	--
-0.5	--	1.1509	0.8952	0.7587	0.6769	0.6297	--	--	--
-0.4	--	1.1613	1.0000	0.8794	0.7988	0.7530	--	--	--
-0.3	--	1.0165	1.0232	0.9725	0.9205	0.8924	--	--	--
-0.2	--	(0.750)	0.9346	1.0000	1.0169	1.0203	(1.100)	--	--
-0.1	--	--	0.7716	0.9372	1.0526	1.1374	(1.280)	--	--
0	--	--	0.5912	0.7994	1.0000	1.1925	1.3925	--	--
0.1	--	--	0.4363	0.6331	0.8665	1,1473	1.4837	--	--
0.2	--	--	(0.300)	0.4768	0.6938	1.0000	1.4608	(2.524)	--
0.3	--	--	--	0.3477	0.5253	0.7974	1.2786	(2.443)	--
0.4	--	--	--	0.2478	0.3834	0.5962	1.0000	2.0311	--
0.5	--	--	--	0.1728	0.2729	0.4281	0.7223	1.5100	--
0.6	--	--	--	(0.150)	0.1904	0.2996	0.4984	1.0000	(2.857)
0.7	--	--	--	--	0.1297	0.2058	0.3355	0.6323	(1.825)
0.8	--	--	--	--	0.0852	0.1388	0.2224	0.3942	1.0000
0.9	--	--	--	--	0.0511	0.0913	0.1456	0.2448	0.5173
0.95	--	--	--	--	0.0348	0.0725	0.1172	0.1930	0.3806
1.0	--	--	--	--	(0.025)	(0.050)	(0.080)	(0.110)	(0.300)

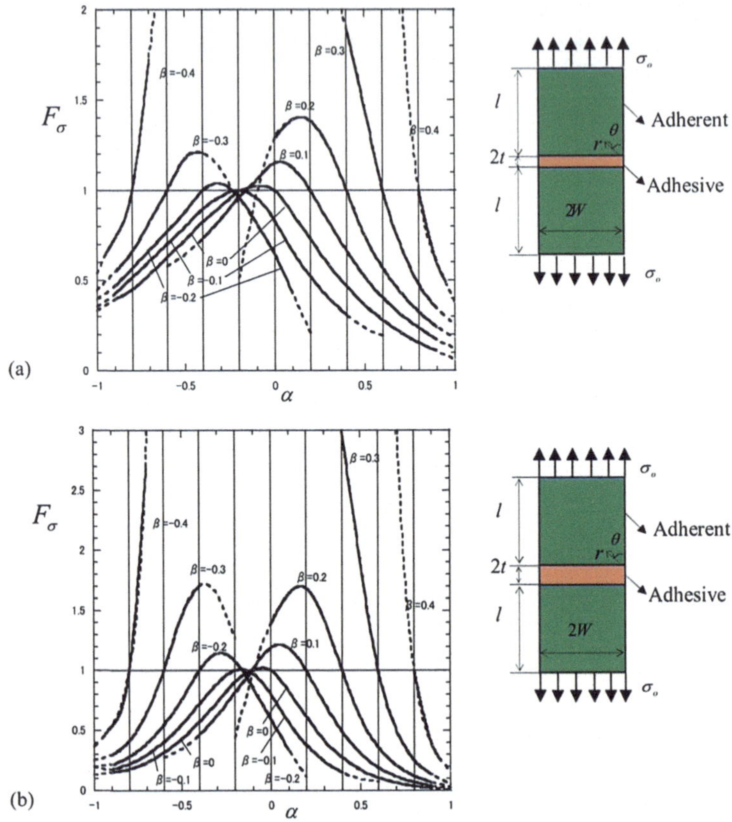

Fig. 10 F_σ with varying material combination β when (**a**) $t/W = 0.001$; (**b**) $t/W = 0.1$

Table 6 Material properties

	Material	Elastic modulus/GPa	Poisson's ratio
Adherent	SUS304 (stainless steel)	206	0.3
	A7075 (aluminum alloys)	71	0.33
	Silicon	166	0.26
	FR-4.5 (IC substrate)	15.34	0.15
Adhesive	Resin	2.74	0.38

5.2 *Effect of Adhesive Thickness on the Intensity of Singular Stress*

To investigate the effect of adhesive thickness on the intensity of singular stresses, stainless steel SUS304, aluminum alloys A7075, silicon and IC substrate FR-4.5 are considered for adherents and resin is considered for adhesive. Table 6 shows the

Table 7 $F_\sigma/F_{\sigma \mid t/W=1}$ with varying adhesive thickness t/W

t/W	SUS304	A7075	Silicon	FR-4.5
0.001	0.100	0.118	0.102	0.229
0.01	0.212	0.236	0.215	0.355
0.1	0.466	0.4884	0.468	0.573
0.5	0.898	0.903	0.898	0.916
1	1.000	1.000	1.000	1.000
2	1.002	1.002	1.002	1.003
4	1.002	1.002	1.002	1.003

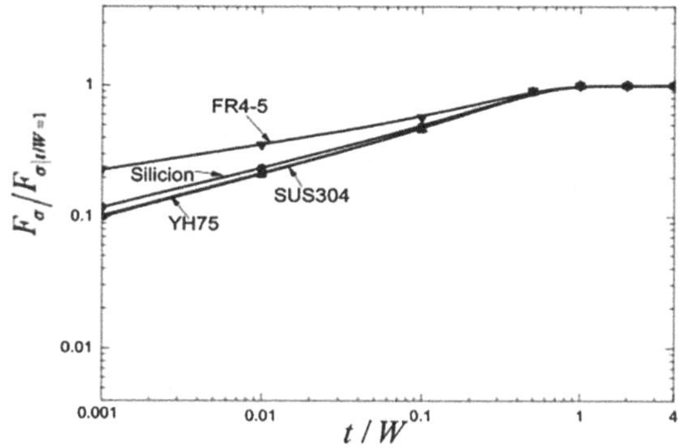

Fig. 11 $F_\sigma/F_{\sigma \mid t/W=1}$ with varying adhesive thickness t/W

material properties of adherents and adhesive. Table 7 shows the ratio $K_\sigma/K_{\sigma \mid t/W=1}$ at the joint of interface when adhesive thickness $t/W = 0.001, 0.01, 0.1, 0.5, 2, 4$. Figure 11 is the figure for Table 7. It is found that $F_\sigma/F_{\sigma \mid t/W=1}$ increases with increasing t/W until $t/W = 1$, and when $t/W > 1$, $F_\sigma/F_{\sigma \mid t/W=1}$ keeps constant 1.0, whatever the material combination is.

Generally, the Young's modulus E_2 of adhesive is smaller than the Young's modulus E_1 of adherent: $E_2 \leq E_1$ and the Poisson's ratio v_2 of adhesive is larger than the Poisson's ratio v_1 of adherent: $v_2 \geq v_1$. In this case, it is found that $\alpha \geq 0$ and $\alpha - 2\beta \geq 0$ and therefore singularity stress exists around the edge of interface. According to the values in Table 5, Fig. 12 shows the variation of logarithmic $F_\sigma/F_{\sigma \mid t/W=1}$ and t/W from $\beta = -0.2$ to $\beta = 0.4$ with different α. It is found that $F_\sigma/F_{\sigma \mid t/W=1}$ increases with increasing t/W until $t/W = 1$ for all the material combinations when $\alpha \geq 0$ and $\alpha - 2\beta \geq 0$. To improve the interface strength, thin adhesive layers are desirable because the intensity of singular stress decreases with decreasing the thickness.

Also, It should be noted that $F_\sigma/F_{\sigma \mid t/W=1}$ when $\alpha - 2\beta = 0$ and $F_\sigma/F_{\sigma \mid t/W=1}$ decreases with decreasing t/W when $\alpha \geq 0$ and $\alpha - 2\beta > 0$.

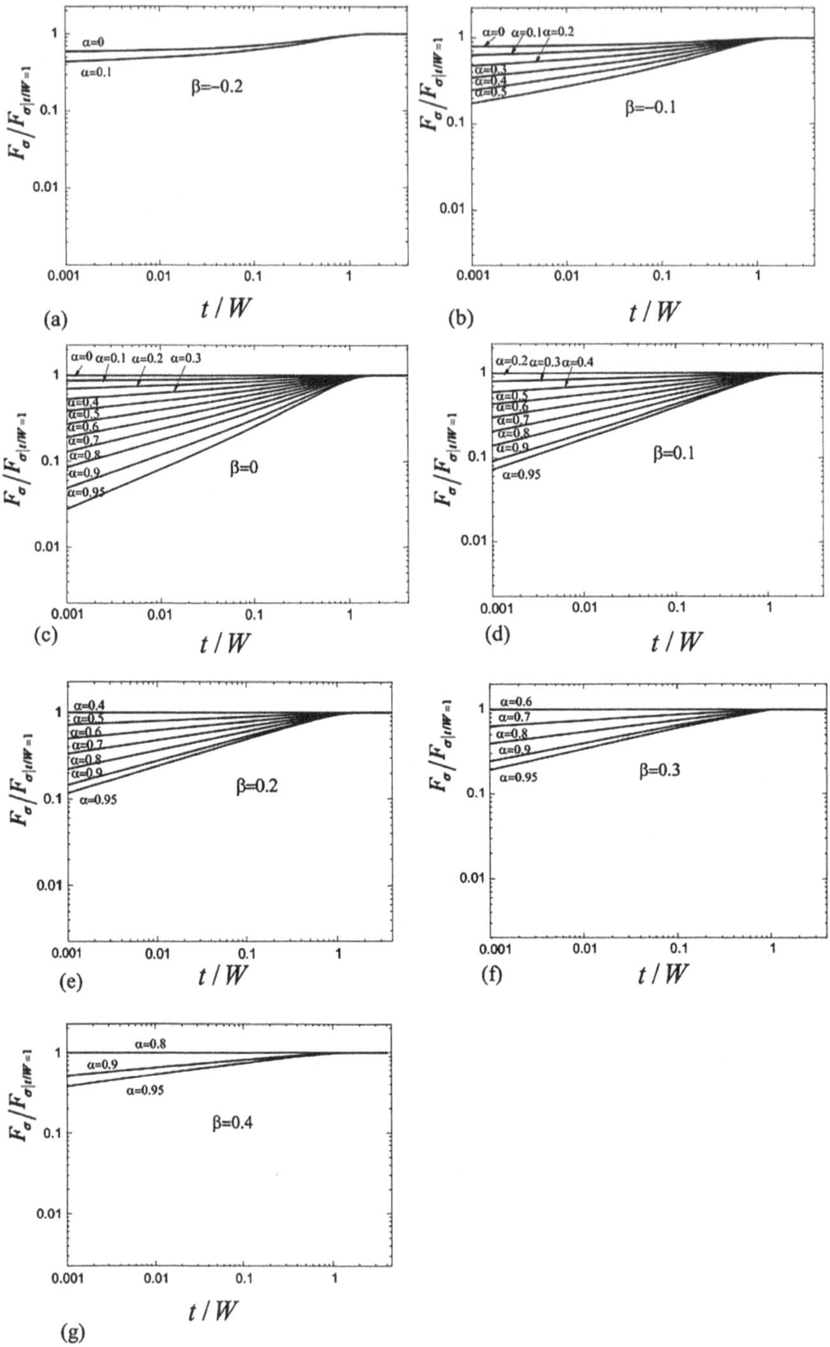

Fig. 12 $F_\sigma/F_{\sigma\,|\,t/W=1}$ t/W. **(a)** $\beta = -0.2$; **(b)** $\beta = -0.1$; **(c)** $\beta = 0$; **(d)** $\beta = 0.1$; **(e)** $\beta = 0.2$; **(f)** $\beta = 0.3$; **(g)** $\beta = 0.4$

6 Conclusions

In this paper the intensity of singular stress at the edge of adhesive dissimilar joint was discusses with varying the adhesive thickness and material combinations. The conclusions can be made in the following way.

1. Accurate method for calculating the intensity of singular stress was proposed by using FEM. It is found that the ratio of intensity of singular stress K_σ^1/K_σ^2 can be obtained from the ratio $\lim_{r\to 0}\left(\sigma^1_{\theta|\theta-\pi/2}/\sigma^2_{\theta|\theta-\pi/2}\right)$. It is also found that the ratio $K_\sigma/K_{\sigma\ |\ t/W\ =\ 1}$ is constant along the interface if suitable FEM mesh is applied. Therefore, only the first node can be considered when the ratio of $K_\sigma/K_{\sigma\ |\ t/W\ =\ 1}$ is calculated.
2. For a fixed value of β, it is fount that K_σ increases with increasing α when α is small. On the other hand, K_σ decreases with increasing α when α is large. The range of intensity of singular stress K_σ is different depending on the adhesive thickness t/W.
3. To improve the interface strength, thin adhesive layers are desirable because the intensity of singular stress Kσ decreases with decreasing the thickness. The increment is different depending on material combination. The ratio $K_\sigma/K_{\sigma\ |\ t/W\ =\ 1}$ when $t/W \geq 1$ whatever the material combination is.

References

1. Kanno T, Ogata M, Foxton RM, et al. Microtensile bond strength of dual-cure resin cement to root canal dentin with different curing strategies. Dent Mater. 2004;23(4):550–6.
2. Kitasako Y, Burrow MF, Nikaido T, et al. Shear and tensile bond testing for resin cemented with SixLluting agents. J Prosthet Density. 1995;80:423–8.
3. Van Meerbeek B, De Munck J, Yoshida Y, et al. Buonocore memorial lecture. Adhesion to enamal and dentin: current status and future challenges. Oper Dent. 2003;28:215–35.
4. Park JH, Choi JH, Kweon JH. Evaluating the strengths of thick aluminum-to-aluminum joint with different adhesive lengths and thicknesses, composite structures. Compos Struct. 2009;92: 2226–35.
5. Arenas J, Narbon JJ, Alia C. Optimum adhesive thickness in structural adhesive joints using statistical techniques based on Weibull distribution. Int J Ahesion Dhesives. 2009;30:160–5.
6. Neves AA, Courinho E, Poitevin A. Influence of joint component mechanical properties and adhesive layer thickness on stress distribution in micro-tensile bond strength specimens. Dent Mater. 2009;25:4–12.
7. Afendi M, Teramoto T. Fracture toughness test of epoxy adhesive dissimilar joint with various adhesive thickness. In: Proceedings of Asian pacific conference for materials and mechanics; 2009. p. a55.
8. Bogy DB. Edge-bonded dissimilar orthogonal elastic wedges under normal and shear loading. J Appl Mech. 1968;35:460–6.

9. Bogy DB. Two edge-bonded elastic wedges of different and wedge angles under surface tractions. J Appl Mech. 1971;38:377–86.
10. Chen DH, Nisitani H. Intensity of singular stress field near the interface edge point of a bonded strip. Trans Jpn Soc Mech Eng. 1993;A-59(567):2682–6.
11. Noda NA, Shirao R, Li J, Sugimoto JS. Intensity of singular stress at the end of a fiber under pull-out force. Int J Solids Struct. 2007;44(13):4472–91.

A Mesh-Independent Technique to Evaluate Stress Singularities in Adhesive Joints

1 Introduction

Due to their low manufacturing cost and ease of maintenance, adhesive joints are most frequently used in numerous industrial sectors, such as automobile, shipbuilding, and aeronautics, to replace or supplement traditional joining technologies, such as welding or riveting. With the extensive use of adhesive joining, research has been conducted to evaluate its strength through experimental data [1–3]. However, as the experiments are often time-consuming and costly, finite element analysis (FEA) of adhesive joints is preferred, which can help to investigate the strength of adhesive joints as a function of varying geometrical features and material properties that can be changed numerically [4]. Most research concentrates on numerical methods, and the validity of these studies has been proved through experimental results [5–8].

The mismatch of different material properties may cause stress singularities at the end of an interface, which easily leads to failures from the bonded interface. Therefore, it is important to analyze the stress singularity to evaluate the strength of adhesive joints. Much valuable research has been conducted to analyze the stress singularity at the end of the interface between adhesive and adherent sections, such as Koguchi et al. [9], Kilic et al. [10], Van Tooren et al. [11], and Goglio and Rossetto [12]. Da Silva et al. made a detailed and comprehensive comparison review about the studies of adhesive joints [13, 14]. Moreover, Mintzas and Nowell [15] presented the fracture criterion and its validation for adhesive joints on the basis of the intensity of singular stress K_σ as the fracture initiation parameter, which was similar to the stress intensity factor K of a crack. The critical values of the intensity of singular stress $K_{\sigma,\, c}$ were obtained under various material combinations, and geometries, as well as mechanical and thermal loadings [16–20].

However, the intensity of singular stress is difficult to obtain directly from stress values of elements in the singular stress field, as these values are sensitive to element mesh sizes. Previously, the intensity of singular stress was calculated by extrapolation, in which needed a fine mesh size to ensure its precision. However, results were

Y. Zhang, *Interfacial Mechanics of Offshore Engineering Equipment*,
https://doi.org/10.1007/978-981-95-3200-1_5

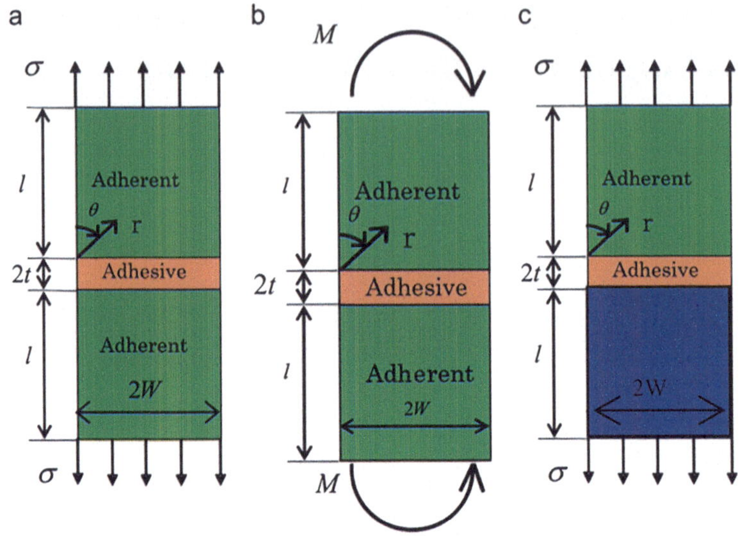

Fig. 1 Adhesive joint under (**a**) tension, (**b**) bending, and (**c**) tension with dissimilar adherents

still not particularly accurate. In this paper, a new method is proposed, in which only the first element stress values at the end of the interface are used to obtain the intensity of singular stress. This method was proved to be valid by comparing the peeling stresses between numerical and experimental results. The intensity of the singular stress values was obtained with different material combinations for the adhesive joint under tension (Fig. 1a) and bending (Fig. 1b). Additionally, the effect of the adhesive thickness on the intensity of the singular stress was discussed. The joint strength under tension and bending is compared based on the intensity of singular stress. Moreover, the interaction between singular stress fields of the adhesive joint with dissimilar adherents (Fig. 1c) was investigated.

2 Analytical Method

For the adhesive joint shown in Fig. 1, it is known that the interface stress $\sigma_{ij}(ij = rr, \theta\theta, r\theta)$ goes to infinity at the end of the joint and has a singularity of $\sigma_{ij} \propto 1/r^{1-\lambda}\sigma$ when $\alpha(\alpha - 2\beta) > 0$. Here, α and β are Dundurs' parameters, which depend on Poisson's ratio ν and the shear modulus G. Moreover, when $\theta = \pi/2$, the stress singularity index λ could be expressed by the following equations [21, 22]:

$$\left[\sin^2\left(\frac{\pi}{2}\lambda\right) - \lambda^2\right]^2\beta^2 + 2\lambda^2\left[\sin^2\left(\frac{\pi}{2}\lambda\right) - \lambda^2\right]\alpha\beta + \lambda^2\left(\lambda^2 - 1\right)\alpha^2 + \frac{\sin^2(\lambda\pi)}{4} = 0 \tag{1}$$

$$\alpha = \frac{G_1(k_2+1) - G_2(k_1+1)}{G_1(k_2+1) + G_2(k_1+1)} \quad \beta = \frac{G_1(k_2-1) - G_2(k_1-1)}{G_1(k_2+1) + G_2(k_1+1)} \tag{2}$$

$$k_j = \begin{cases} \dfrac{3-\nu_j}{1+\nu_j} & \text{(plane stress)} \\ 3-4\nu_j & \text{(plane strain)} \end{cases}, \quad k_j = (j = 1, 2) \tag{3}$$

It is found that $\lambda > 1$ when $\alpha(\alpha - 2\beta) > 0$; $\lambda = 1$ when $\alpha(\alpha - 2\beta) = 0$; and $\lambda < 1$ when $\alpha(\alpha - 2\beta) > 0$. The strength of adhesive joints with interfaces depends on the singular stress fields, which exist at the end of interface, and it is often evaluated by the parameter of intensity of singular stress K_σ. Furthermore, $K_{\sigma,\ T}$ refers to the case of tension and $K_{\sigma,\ M}$ to bending

$$K_{\sigma,T} = \lim_{r \to 0} \left[r^{1-\lambda} \times \sigma_{\theta|\theta=\pi/2,T}(r) \right] \quad \text{(for tension)} \tag{4}$$

$$K_{\sigma,M} = \lim_{r \to 0} \left[r^{1-\lambda} \times \sigma_{\theta|\theta=\pi/2,M}(r) \right] \quad \text{(for bending)} \tag{5}$$

The dimensionless intensities of the singular stress $F_{\sigma,\ T}$ and $F_{\sigma,\ M}$ are defined by

$$F_{\sigma,T} = \frac{K_{\sigma,T}}{\sigma(2W)^{1-\lambda}} = \frac{\lim_{r \to 0} \left[r^{1-\lambda} \sigma_{\theta|\theta=\pi/2,T}(r) \right]}{\sigma(2W)^{1-\lambda}} \quad \text{(for tension)} \tag{6}$$

$$F_{\sigma,M} = \frac{K_{\sigma,M}}{(3M/2W^2)(2W)^{1-\lambda}} = \frac{\lim_{r \to 0} \left[r^{1-\lambda} \sigma_{\theta|\theta=\pi/2,M}(r) \right]}{(3M/2W^2)(2W)^{1-\lambda}} \quad \text{(for bending)} \tag{7}$$

where σ is the tension stress, M is the moment per unit length at the adhesive joint, and $\sigma = 3M/2W^2$ is the maximum bending stress (stress at the furthest point from the neutral axis) caused by M equals the tensile stress σ.

In this paper, the adhesive thickness t/W in the analytical model varies from 0.001 to 4 (0.01, 0.1, 0.5, 1, 2, and 4), and the actual width of the FEM model, which does not affect the results, is $W = 1000$ mm. The length is $l = 2W$, because it is known that the interface stresses do not change for length values of more than twice the width.

It is difficult to obtain the values of the intensity of singular stress ($K_{\sigma,\ T}, K_{\sigma,\ M}$) directly from FEM calculations, as the stresses go to directly from FEM calculations, as the stresses go to infinity at the end of the interface due to the singularity. In this paper, the ratios of the intensity of the singular stress $\left(K^1_{\sigma,T}/K^2_{\sigma,T}, K^1_{\sigma,M}/K^2_{\sigma,M}\right)$ will be considered. Here, the superscripts 1 and 2 refer to different thickness to width ratios (t/W); the adhesive thickness t/W has two cases. As shown in Eqs. (3) and (4), the dimensionless intensity of singular stress is related to the distance r, the singularity index λ, the applied load (σ or M), the width W and the limiting stress $\lim_{r \to 0} \sigma_{\theta|\theta=\pi/2,T}$ or $\lim_{r \to 0} \sigma_{\theta|\theta=\pi/2,M}$.Two models with different adhesive thicknesses

(t_1, t_2) are considered (problem 1 and problem 2), both which have the same applied load at infinity σ or M, the same model width W, and the same material combinations (which means that the singularity index is $\lambda_1 = \lambda_2$). Therefore, the ratios of intensity of singular stress $\left(K^1_{\sigma,T}/K^2_{\sigma,T}, K^1_{\sigma,M}/K^2_{\sigma,M}\right)$ are only related to the ratios of stress $\lim_{r\to 0}\left(\sigma^1_{\theta|\theta=\pi/2,T}/\sigma^2_{\theta|\theta=\pi/2,T}\right)$, $\lim_{r\to 0}\left(\sigma^1_{\theta|\theta=\pi/2,M}/\sigma^2_{\theta|\theta=\pi/2,M}\right)$, which only depend on the adhesive thickness t. The ratios of intensity of singular stress are expressed as

$$\frac{K^1_{\sigma,T}}{K^2_{\sigma,T}} = \frac{\sigma^1(2W)^{1-\lambda_1}F^1_{\sigma,T}}{\sigma^2(2W)^{1-\lambda_1}F^2_{\sigma,T}} = \frac{F^1_{\sigma,T}}{F^2_{\sigma,T}} = \lim_{r\to 0}\frac{\left[r^{1-\lambda_1}\sigma^1_{\theta|\theta=\pi/2,T}(r)\right]}{\left[r^{1-\lambda_1}\sigma^2_{\theta|\theta=\pi/2,T}(r)\right]}$$
$$= \lim_{r\to 0}\frac{\sigma^1_{\theta|\theta=\pi/2,T}(r)}{\sigma^2_{\theta|\theta=\pi/2,T}(r)} \text{ (for tension)} \tag{8}$$

$$\frac{K^1_{\sigma,M}}{K^2_{\sigma,M}} = \frac{(6M/4W^2)^1(2W)^{1-\lambda_1}F^1_{\sigma,M}}{(6M/4W^2)^2(2W)^{1-\lambda_1}F^2_{\sigma,M}} = \frac{F^1_{\sigma,M}}{F^2_{\sigma,M}} = \lim_{r\to 0}\frac{\left[r^{1-\lambda_1}\sigma^1_{\theta|\theta=\pi/2,M}(r)\right]}{\left[r^{1-\lambda_1}\sigma^2_{\theta|\theta=\pi/2,M}(r)\right]}$$
$$= \lim_{r\to 0}\frac{\sigma^1_{\theta|\theta=\pi/2,M}(r)}{\sigma^2_{\theta|\theta=\pi/2,M}(r)} \text{ (for bending)} \tag{9}$$

Therefore, only calculating $\lim_{r\to 0}\sigma_{\theta|\theta=\pi/2,T}$ and $\lim_{r\to 0}\sigma_{\theta|\theta=\pi/2,M}$ is enough to solve the ratio of intensity of singular stress. In this paper, the problem involving an adhesive joint under tension and bending shown in Fig. 1 is referred to as problem 1 and as the unknown problem. The problem involving a bonded strip (just two bonded layers) under tension and bending shown in Fig. 2 is referred to as problem 2 and as the reference problem. To obtain the intensity of singular stress for the unknown problem from the ratios, the results for reference problem shown in Fig. 2 will be introduced and discussed in Sect. 4.

3 Calculation Details by the Finite Element Method

To discuss the stresses on the interface of the adhesive joint, a 2D plane stress FEM model was constructed with the software MSC MARK 2007. The element type chosen was quad 9. Figure 3 shows the detailed stress distributions near the end of the inter-face with different adhesive thicknesses. The smallest element size at the end is $1/3^8 = 1/6561$ mm. It should be noted that although stresses at the end have finite values calculated by FEM, they have the tendency to become infinite. Furthermore, the stress distribution lines at the end for different adhesive thicknesses are parallel to each other (because of the same singularity).

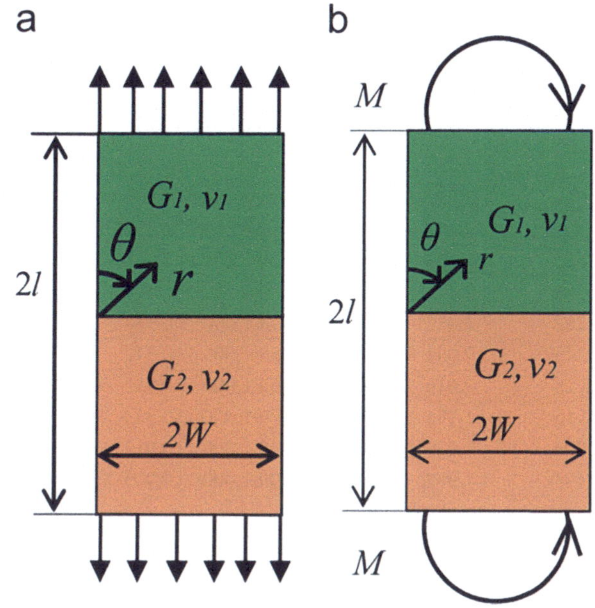

Fig. 2 Bonded strip under (**a**) tension and (**b**) bending

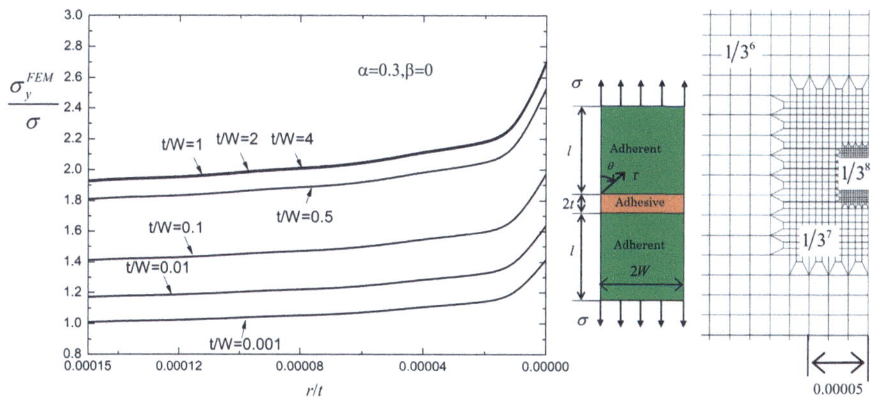

Fig. 3 Stress distributions at the end of interface in adhesive joint with different adhesive thicknesses t/W

Table 1 shows the stress values and their ratios from the end of the interface when $t/W = 0.01$ and $\alpha = 0.3$, $\beta = 0$ for an adhesive joint under tension, using the minimum mesh sizes of 1/38 = 1/6561 mm and 1/34 = 1/81 mm. It is found that although the values of stresses at the end of the interface change with different minimum mesh sizes, the ratios of stresses of problems 1 and 2 stay constant within a four-digit accuracy. This observation indicates that stress ratios are independent of the mesh size, and only the first element's stress values are enough to obtain the

Table 1 Stress distributions at the end of interface for $\sigma^{FEM}_{y|t/W=0.01}$ of the adhesive joint shown in Fig. 1a and σ^{FEM}_T of the bonded strip shown in Fig. 1a and their ratios when $\alpha = 0.3$, $\beta = 0$ with different smallest mesh sizes ($W = 1000$)

Smallest mesh size = 1/6561 mm				Smallest mesh size = 1/81 mm			
r/w	$\sigma^{FEM}_{y\|t/W=0.01}$	σ^{FEM}_T	$\frac{\sigma^{FEM}_{y\|t/W=0.01}}{\sigma^{FEM}_T}$	r/w	$\sigma^{FEM}_{y\|t/W=0.01}$	σ^{FEM}_T	$\frac{\sigma^{FEM}_{y\|t/W=0.01}}{\sigma^{FEM}_T}$
0	1.64	2.692	0.609	0	1.246	2.045	0.609
1/6,561,000	1.365	2.242	0.609	1/81,000	1.036	1.703	0.609
2/6,561,000	1.32	2.167	0.609	2/81,000	1.001	1.647	0.608
3/6,561,000	1.286	2.111	0.609	3/81,000	0.975	1.604	0.608
4/6,561,000	1.262	2.072	0.609	4/81,000	0.956	1.574	0.608
5/6,561,000	1.243	2.041	0.609	5/81,000	0.942	1.551	0.607
6/6,561,000	1.228	2.016	0.609	6/81,000	0.93	1.532	0.607
7/6,561,000	1.215	1.995	0.609	7/81,000	0.92	1.516	0.607
8/6,561,000	1.205	1.978	0.609	8/81,000	0.912	1.502	0.607
9/6,561,000	1.195	1.962	0.609	9/81,000	0.905	1.491	0.607

Table 2 Stress distributions at the end of interface for $\sigma^{FEM}_{y|t/W=0.01}$ of the adhesive joint shown in Fig. 1a and σ^{FEM}_T of the bonded strip shown in Fig. 2a and their ratios when $\alpha = 0.3$, $\beta = 0$ with different smallest mesh sizes ($W = 1000$)

Smallest mesh size = 1/6561 mm				Smallest mesh size = 1/81 mm			
r/w	$\sigma^{FEM}_{y\|t/W=0.01}$	σ^{FEM}_T	$\frac{\sigma^{FEM}_{y\|t/W=0.01}}{\sigma^{FEM}_T}$	r/w	$\sigma^{FEM}_{y\|t/W=0.01}$	σ^{FEM}_T	$\frac{\sigma^{FEM}_{y\|t/W=0.01}}{\sigma^{FEM}_T}$
$\to 0$	2.486	2.692	0.923	$\to 0$	1.889	2.045	0.923
1/6,561,000	2.071	2.242	0.924	1/81,000	1.573	1.703	0.924
2/6,561,000	2.002	2.167	0.924	2/81,000	1.521	1.647	0.923
3/6,561,000	1.95	2.111	0.924	3/81,000	1.482	1.604	0.924
4/6,561,000	1.914	2.072	0.924	4/81,000	1.454	1.574	0.924
5/6,561,000	1.885	2.041	0.924	5/81,000	1.432	1.551	0.923
6/6,561,000	1.862	2.016	0.924	6/81,000	1.415	1.532	0.924
7/6,561,000	1.843	1.995	0.924	7/81,000	1.400	1.516	0.923
8/6,561,000	1.827	1.978	0.924	8/81,000	1.388	1.502	0.924
9/6,561,000	1.812	1.962	0.924	9/81,000	1.377	1.491	0.924

stress ratios. This result could be explained by the fact that errors in problems 1 and 2 produced by the FEM method could be canceled when ratios of stresses are considered. This is possible, because two problem models have the same geometry and material combinations, except for different adhesive thicknesses.

It is known from Eq. (5) that the ratio of intensity of singular stress K^1_σ/K^2_σ is equal to the ratio of stresses $\lim_{r\to 0}\left(\sigma^1\theta|\theta=\pi/2/\sigma^2\theta|\theta=\pi/2\right)$; consequently, the ratio of intensity of singular stress could be obtained from FEM calculations accurately.

To investigate whether the method for the case of bending is valid, stress ratios between bending and tension are considered for the bonded strip shown in Fig. 2. Table 2 shows the stresses and their ratios between bending and tension at the end of the interface with different minimum mesh sizes of $1/3^8 = 1/6561$ mm and $1/3^4 = 1/81$ mm when $\alpha = 0.3$ and $\beta = 0$. It is shown that element stress ratios stay constant near the end, and are independent of element sizes, which means that the analytical method is also valid for the adhesive joint under bending.

4 Reference Problem

However, to obtain the intensity of singular stress of the adhesive joint, a reference solution is necessary. Here, the problem of a bonded strip shown in Fig. 2 is considered to be the reference problem. Chen and Nisitani [23] and Noda et al. [24] calculated accurately the intensity of singular stress in a bonded strip under tension and bending by the body force method. Figure 4 shows the results of the intensity of singular stress for the bonded strip under tension and bending for all of the material combinations, except when $\beta = -0.5,\ 0.5$, because the results are not convergent in these cases. The dotted lines in Fig. 4 mean that those results are obtained by an extrapolation method due to the divergence of the results.

5 Results

5.1 *Results for the Adhesive Joint Under Tension*

Using the constant ratios calculated by the FEM and the results of intensity of the singular stress for the bonded strip shown in Fig. 4, the intensities of singular stress for the adhesive joints shown in Fig. 1 are obtained. Figure 5 shows the dimensionless intensity of singular stress F_σ with varying α and β when (a) $t/W = 0.001$ and (b) $t/W = 0.1$. It can be seen that F_σ increases with increasing α when α is small

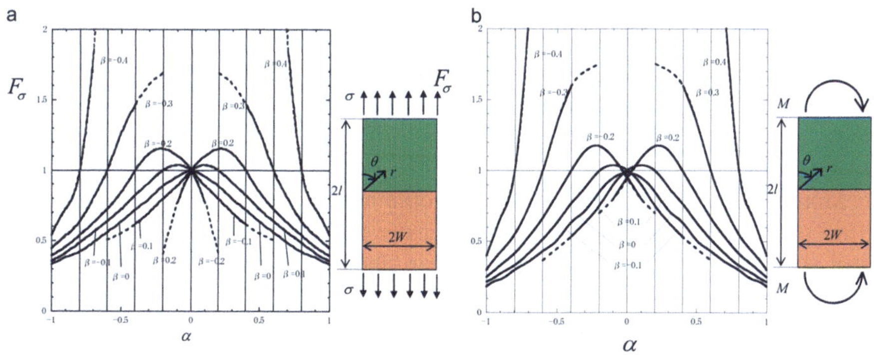

Fig. 4 F_σ for a bonded strip in Fig. 2. (**a**) Tension and (**b**) bending

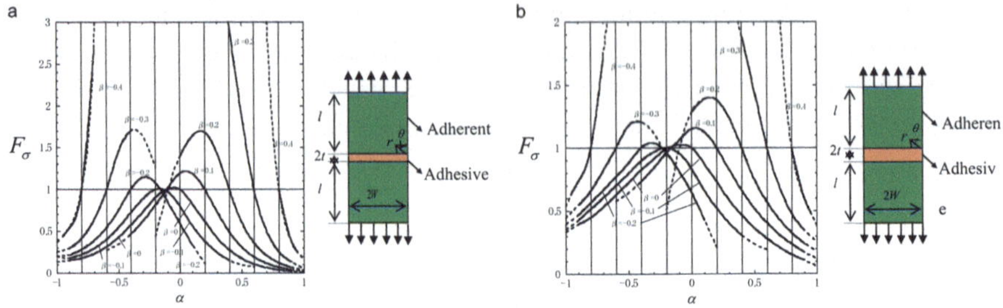

Fig. 5 F_σ of adhesive joint under tension with varying material combinations α; β (**a**) $t/W = 0.001$; (**b**) $t/W = 0.1$

Table 3 Material properties of the adhesive joint under tension

	Material	Elastic modulus/GPa	Poisson's ratio
Adherent	SUS304 (stainless steel)	206	0.3
	A7075 (aluminum alloys)	71	0.33
	Silicon	166	0.26
	FR-4.5 (IC substrate)	15.34	0.15
Adhesive	Resin	2.74	0.38

Table 4 $F_\sigma/F_{\sigma\,|\,t/W=1}$ with varying adhesive thickness t/W

t/W	SUS304	A7075	Silicon	FR-4.5
0.001	0.1	0.118	0.102	0.229
0.01	0.212	0.236	0.215	0.355
0.1	0.466	0.4884	0.468	0.573
0.5	0.898	0.903	0.898	0.916
1	1	1	1	1
2	1.002	1.002	1.002	1.003
4	1.002	1.002	1.002	1.003

(smaller than 0). On the other hand, F_σ decreases with increasing α when α is large (larger than 0). Comparing the results for $t/W = 0.001$ and $t/W = 0.1$, it is noticed that the ranges of F_σ are different. F_σ for $t/W = 0.001$ is widely distributed in the range of 0.025~2.857, whereas for $t/W = 0.1$, F_σ is in the small range of 0.185~1.498.

Moreover, some examples of material combinations were used to investigate the effect of the adhesive thickness on the intensity of singular stresses. Here, stainless steel SUS304, aluminum alloy A7075, silicon, and IC substrate FR-4.5 are chosen as the adherents, and resin is chosen as the adhesive. Table 3 lists the properties of those materials. Table 4 shows the ratio $F_\sigma/F_{\sigma\,|\,t/W=1}$ for those material combinations when the adhesive thickness is $t/W = 0.001$, 0.01, 0.1, 0.5, 2 and 4. The results are also given in Fig. 6. It is found that $F_\sigma/F_{\sigma\,|\,t/W=1}$j increases with increasing t/W until

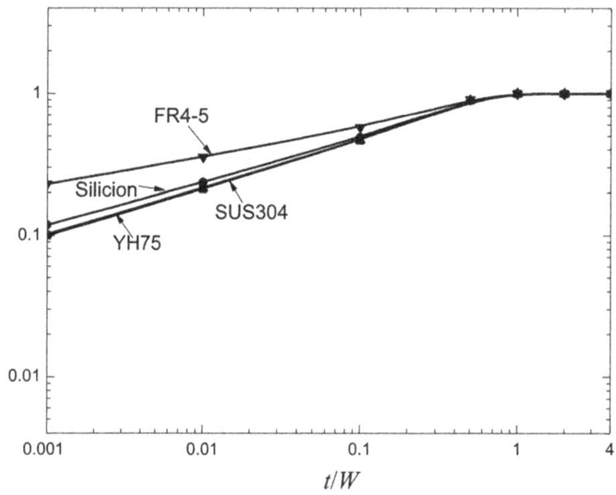

Fig. 6 $F_\sigma/F_{\sigma\ |\ t/W\ =\ 1}$ with varying adhesive thickness $t = W$

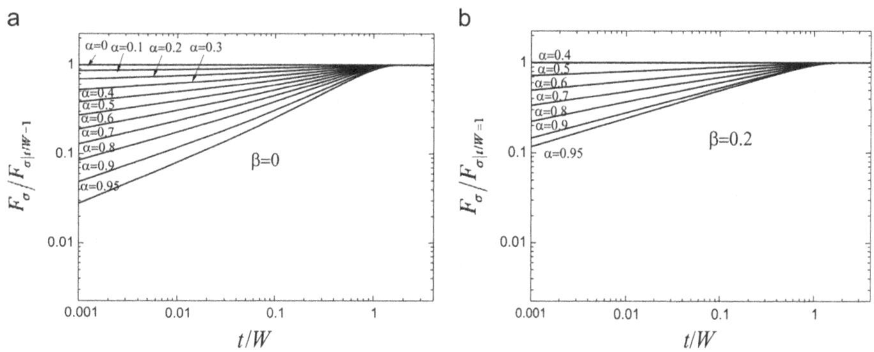

Fig. 7 $F_\sigma/F_{\sigma\ |\ t/W\ =\ 1}$ of adhesive joint under tension with varying adhesive thicknesses t/W (**a**) $\beta = 0$, (**b**) $\beta = 0.2$

$t/W = 1$, and when $t/W > 1$, $F_\sigma/F_{\sigma\ |\ t/W\ =\ 1}$ remain constant at 1.0, for all material combinations.

Generally, Young's modulus of the adhesive is smaller than that of the adherent: $E_2 \leq E_1$. Additionally, Poisson's ratio of the adhesive is larger than that of the adherent: $\nu_2 \geq \nu_1$. Therefore, the parameters α and β satisfy the conditions $\alpha \geq 0$ and $\alpha - 2\beta \geq 0$. Figure 7 shows two examples of the logarithmic variation of $F_\sigma/F_{\sigma\ |\ t/W\ =\ 1}$ and t/W for $\beta = 0$ and $\beta = 0.2$ when $\alpha \geq 0$ and $\alpha - 2\beta > 0$. It is found that $F_\sigma/F_{\sigma\ |\ t/W\ =\ 1}$ increases with increasing t/W until $t/W = 1$ for any material combination. To improve the adhesive joint strength, thin adhesive layers are desirable because the intensity of singular stress decreases with decreasing adhesive thickness.

Previous research suggested that a thicker adhesive layer contained more defects and cavities when a larger amount of glue was used, which meant that the thicker the adhesive was, the more easily failures happened. As an example, Suzuki [3] evaluated the debonding strength experimentally using the adhesive joint shown in Fig. 8, in which S35C JIS medium carbon steel plates were bonded by epoxy resin Epikote 871. The adhesive strength was determined for varying adhesive thickness. Figure 8a shows the relationship between the peeling stress and the adhesive thickness. It is found that the peeling stress at failure increases with decreasing adhesive thickness, as shown in Fig. 8a.

For the failure criteria of adhesive joints, the maximum principal stress and Mises equivalent stress are widely used. However, sometimes it is found that the existing failure criteria cannot accurately estimate the experimental results [5]. Recently, much attention has been paid to a fracture initiation criterion that is based on the critical intensity of singular stress. This criterion states that fracture occurs at an interface corner when the magnitude of the intensity of singular stress K_σ attains a critical value $K_{\sigma c}$. Groth reported a correlation between the stress intensity factor and the strength of lap joints with a certain degree of approximation [25]. Akisanya assessed the validity of the criterion by a combination of theoretical and non-linear finite element analyses, and it was found that this criterion is valid if the plastic zone around the singular point is smaller than the zone dominated by the elastic singular solution [20]. Lefebvre used the critical stress intensity factor criterion to characterize the prediction of fatigue crack theoretically and experimentally [26, 27]. Liechti checked the suitability of the using stress intensity factor as the fracture criterion [28].

The conclusion of the experiment mentioned above could also be explained by the analysis method of this paper using the $K_{\sigma c}$-based fracture initiation criterion. As explained above, $K_\sigma = F_\sigma \sigma (2W)^{1-\lambda}$ and here, $K_{\sigma c} = F_\sigma \sigma_c (2W)^{1-\lambda}$ is defined as the critical intensity of singular stress when the adhesive joint begins to have some failures. The model of Suzuki [3] shown in Fig. 8 includes $2W = 12.7$ mm. The

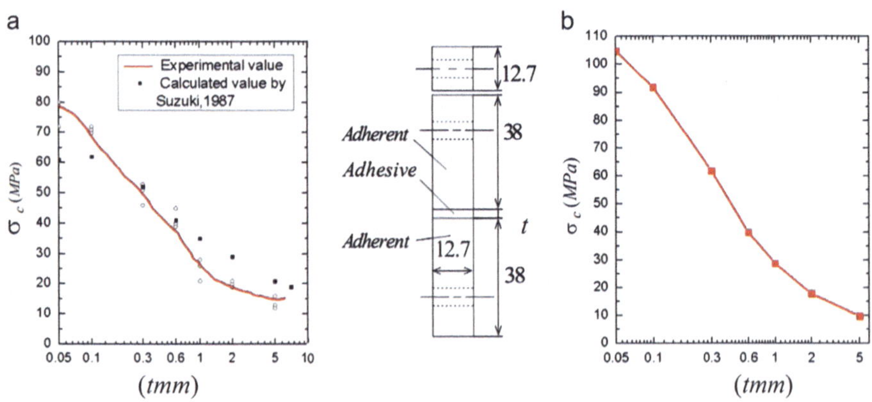

Fig. 8 The relationship between the peeling stress σ_c and the adhesive thickness t. (**a**) Experimental data by (Suzuki [3]) and (**b**) calculated data by this paper

value $F_\sigma = 0.48$ could be obtained from the results shown in Figs. 5 and 7 by extrapolation. Additionally, λ, which is equal to 0.7389, could be derived from Eq. (1). Considering a varying adhesive thickness, it is noticed that critical values of the intensity of singular stress $K_{\sigma c}$ are almost constant and independent of the adhesive thickness. $K_{\sigma c}$ is $12.74 \pm K_{\sigma c}$. Substituting the constant value $K_{\sigma c}$ into $K_{\sigma c} = F_{\sigma c}\sigma_c(2W)^{1-\lambda}$, the peeling stress σ_c is obtained and shown in Fig. 8b. Comparing with the results calculated by Suzuki [3], it should be found that the tendency of the peeling stress obtained by this paper is almost in agreement with the experiment results, although a little deviation happens.

5.2 Results for the Adhesive Joint Under Bending

When the adhesive joint is under bending, the dimensionless intensity of singular stress F_σ as a function of adhesive thickness and material combinations has been obtained. Figure 9 shows F_σ with varying α and β when (a) $t/W = 0.001$ and (b) $t/W = 0.1$. It can be seen that F_σ increases with increasing α when α is small and F_σ decreases with increasing α when α is large. Comparing the results of $t/W = 0.001$ and $t/W = 0.1$, it is found that F_σ changes somewhat quickly for $t/W = 0.001$, whereas for $t/W = 0.1$, it changes somewhat slowly (the slopes are smaller than the case of $t/W = 0.001$). Moreover, the variation range for $t/W = 0.001$ is large, and that for $t/W = 0$ is small. These conclusions are almost the same, as in the case of the adhesive joint under tension. Figure 10 shows two examples of the logarithmic variation of $F_\sigma/F_{\sigma|t/W=1}$ and t/W for $\beta = 0$ and $\beta = 0.2$ with $\alpha \geq 0$ and $\alpha - 2\beta > 0$. It is found that $F_\sigma/F_{\sigma|t/W=1}$ increases with increasing t/W until $t/W = 1$ for any material combination, which means that thinner adhesive layers should be perferred when considering singular stress fields.

Although the maximum bending stress (stress at the furthest point from the neutral axis) caused by M is equal to the subjected tensile stress σ, which means that stresses applied locally to the singular point are the same, the intensity of singular stress under tension and bending is different. This result could be explained by the effect of stress gradient.

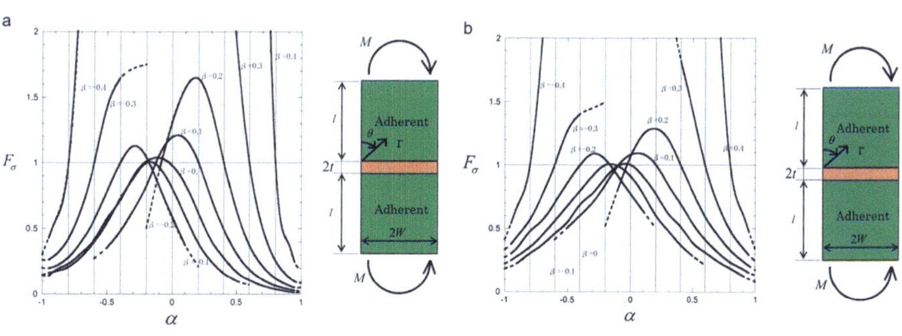

Fig. 9 F_σ of adhesive joint under bending with varying material combinations α, β (**a**) $t/W = 0.001$; (**b**) $t/W = 0.1$

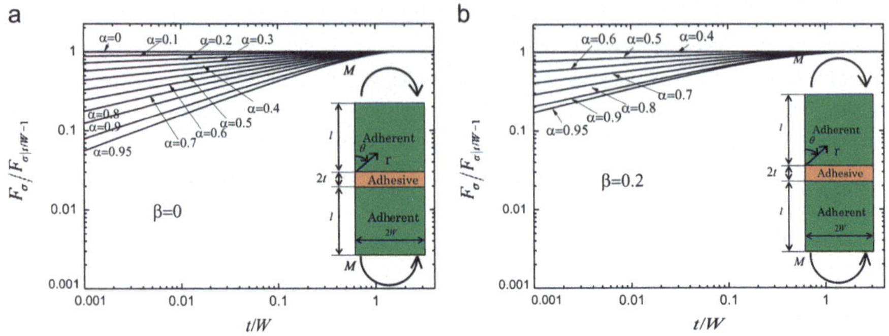

Fig. 10 $F_\sigma/F_{\sigma\,|\,t/W\,=\,1}$ of adhesive joint under bending with varying adhesive thicknesses $F_\sigma/F_{\sigma\,|\,t/W\,=\,1}$, (**a**) $\beta = 0$ and (**b**) $\beta = 0.2$

5.3 Comparison of the Intensity of Singular Stress Between Tension and Bending

Table 5 shows the dimensionless intensity of singular stress F_σ for the bonded strip (Fig. 2) under tension and bending. It should be noticed that $F_{\sigma,\,T} > F_{\sigma,\,M}$ is always true when $\lambda < 1$; $F_{\sigma,\,M} = F_{\sigma,\,T} = 1.000\lambda = 1$; and $F_{\sigma,\,T} < F_{\sigma,\,M}$ when $\lambda > 1$. As discussed previously, stress singularity fields only exist when $\lambda < 1$ (caused by $\alpha(\alpha - 2\beta) > 0$), a condition under which failures are prone to happen from the end of the interface. Consequently, the bonded strip is more endangered under tension σ than under bending M when $\sigma = 3M/2W^2$ (the maximum stresses are the same at the two ends).

Table 6 shows the dimensionless intensity of singular stress F_σ for the adhesive joint (Fig. 1) under tension and bending when $t/W = 0.1$. It is seen that the relation F_σ between values for tension and bending is not only related to $\alpha(\alpha - 2\beta)$ or λ, it is also dependent on $\alpha - 2\beta$. $F_{\sigma,\,T} > F_{\sigma,\,M}$ when $\alpha - 2\beta < 0$ ($\alpha(\alpha - 2\beta) < 0$ or $\lambda > 1$ when $\alpha > 0$ and $\alpha(\alpha - 2\beta) > 0$ or $\lambda < 1$ when $\alpha < 0$); $F_{\sigma,\,M} = F_{\sigma,\,T} = 1000$ when $\alpha - 2\beta = 0$; and $F_{\sigma,\,T} < F_{\sigma,\,M}$ when $\alpha - 2\beta > 0$ ($\alpha(\alpha - 2\beta) > 0$ or $\lambda < 1$ when $\alpha > 0$ and $\alpha(\alpha - 2\beta) < 0$ or $\lambda > 1$ when $\alpha < 0$). As discussed previously, stress singularity fields only exist when $\lambda < 1$ (when by $\alpha(\alpha - 2\beta) > 0$), a condition in which failures are prone to happen from the extremes of the interface because of singular stress fields. From the comparison, it is noted that $F_{\sigma,\,T} > F_{\sigma,\,M}$ for $\alpha < 0$ and $F_{\sigma,\,T} < F_{\sigma,\,M}$ for $\alpha > 0$ when singular stress fields exist at the extremes of the interfaces. Which situation endangers the adhesive joint under tension and bending is not determined, because this would depend on the sign of α.

Comparison of criticality (the tendency to have failures when the maximum stresses are the same at both sides) between tension and bending for the homogeneous plate, bonded strip and adhesive joint is shown in Fig. 11.

Table 5 F_σ for the bonded strip with varying α and β under tension and bending (Fig. 2) [( ): Extrapolated or interpolated results. Red figures indicate $\lambda < 1$, blue figures indicate $\lambda > 1$, black figures indicate $\lambda = 1$]. $K_\sigma = F_\sigma \sigma (2W)^{1-\lambda}$, *T*: Tension: $\sigma = P/2W$, *M*: Bending: $\sigma = 6M/(2W)^2$

α		$\beta = -0.4$	$\beta = -0.3$	$\beta = -0.2$	$\beta = -0.1$	$\beta = 0$	$\beta = 0.1$	$\beta = 0.2$	$\beta = 0.3$	$\beta = 0.4$
-1	T	0.540	0.446	0.395	0.357	0.332	--	--	--	--
	M	0.400	0.300	0.250	0.217	0.184	--	--	--	--
-0.9	T	0.726	0.534	0.456	0.412	0.381	--	--	--	--
	M	0.667	0.425	0.333	0.275	0.233	--	--	--	--
-0.8	T	1.000	0.636	0.538	0.487	0.450	--	--	--	--
	M	1.000	0.55	0.417	0.333	0.300	--	--	--	--
-0.7	T	(1.855)	0.800	0.626	0.558	0.486	--	--	--	--
	M	2.097	0.742	0.533	0.417	0.358	--	--	--	--
-0.6	T	(3.29)	1.000	0.724	0.638	0.559	(0.505)	--	--	--
	M	(3.96)	1.000	0.667	0.517	0.433	(0.373)	--	--	--
-0.5	T		1.264	0.842	0.722	0.635	(0.551)	--	--	--
	M	--	1.334	0.833	0.633	0.508	(0.449)	--	--	--
-0.4	T		1.467	1.000	0.822	0.718	0.615	--	--	--
	M	--	1.595	1.000	0.767	0.633	0.533	--	--	--
-0.3	T M	--	(1.61)	1.118	0.913	0.796	0.697	--	--	--
		--	(1.70)	1.141	0.883	0.717	0.633	--	--	--
-0.2	T	--	(1.69)	1.153	1.000	0.889	0.797	(0.404)	--	--
	M	--	(1.75)	1.173	1.000	0.850	0.750	(0.702)	--	--
-0.1	T	--	--	1.103	1.037	0.955	0.89	0.767	--	--
	M	--	--	1.089	1.038	0.95	0.900	0.800	--	--
0	T	--	--	1.000	1.000	1.000	1.000	1.000	--	--
	M	--	--	1.000	1.000	1.000	1.000	1.000	--	--
0.1	T	--	--	0.767	0.890	0.955	1.037	1.103	--	--
	M	--	--	0.800	0.900	0.950	1.038	1.089	--	--
0.2	T	--	--	(0.404)	0.797	0.889	1.000	1.153	(1.69)	--
	M	--	--	(0.702)	0.750	0.850	1.000	1.173	(1.75)	--
0.3	T	--	--	--	0.697	0.796	0.913	1.118	(1.61)	--
	M	--	--	--	0.633	0.717	0.883	1.141	(1.70)	--
0.4	T	-	--	--	0.615	0.718	0.822	1.000	1.467	--
	M	--	--	--	0.533	0.633	0.767	1.000	1.595	--
0.5	T	--	--	--	(0.551)	0.635	0.722	0.842	1.264	--
	M	--	--	--	(0.449)	0.508	0.633	0.833	1.334	--

(continued)

Table 5 (continued)

0.6	T	--	--	--	(0.505)	0.559	0.638	0.724	1.000	(3.29)
	M	--	--	--	(0.373)	0.433	0.517	0.667	1.000	(3.96)
0.7	T	--	--	--	--	0.486	0.558	0.626	0.800	1.855
	M	--	--	--	--	0.358	0.417	0.533	0.742	2.097
0.8	T	--	--	--	--	0.450	0.487	0.538	0.636	1.000
	M	--	--	--	--	0.300	0.333	0.417	0.550	1.000
0.9	T	--	--	--	--	0.381	0.412	0.456	0.534	0.726
	M	--	--	--	--	0.233	0.275	0.333	0.425	0.667
1	T	--	--	--	--	0.332	0.357	0.395	0.446	0.54
	M	--	--	--	--	0.184	0.217	0.250	0.300	0.400

5.4 *Results for the Adhesive Joint with Dissimilar Adherents*

In fact, the adhesive joint is often used to bond different adherent materials. In this study, singular stress fields in the adhesive joints with dissimilar adherent materials shown in Fig. 1c are also discussed.

Here, some examples are presented. For the first example, SUS 304 and YH75 are considered as adherent materials, and epoxy is considered as the adhesive material. For the second example, Silicon and FR-4.5 are considered as adherent materials and a resin underfill is considered as the adhesive material. For the two examples of material combinations, the elastic modulus of the top adherent is larger than the elastic modulus of the bottom adherent. Table 7 shows all of these material properties [29].

Using the analytical method of this paper, the intensities of singular stress for the model shown in Fig. 1c have been obtained and plotted in Figs. 12 and 13. Because the singular stress fields of the upper and bottom ends of the interfaces are different because of dissimilar material combinations, the results are shown in separate figures (a and b).

The intensity of singular stress also decreases with decreasing adhesive thickness until the adhesive thickness is equal to the width of the model, staying constant beyond that point. This is the same result as that for the adhesive joint with the same adherent material.

To compare the intensities of singular stress for the adhesive joints with one and two adherents, the results are plotted together in Fig. 12a, b. The problem of dissimilar adherents with SUS304 and YH-75 is considered. Figure 12a indicates that the intensity of singular stress at the upper end is larger than that for the adhesive

Table 6 F_σ for the bonded strip with varying α and β under tension and bending (Fig. 1) [( ): Extrapolated or interpolated results. Red figures indicate $\lambda < 1$, blue figures indicate $\lambda > 1$, black figures indicate $\lambda = 1$]. $K_\sigma = F_\sigma \sigma (2W)^{1-\lambda}$, T: Tension: $\sigma = P/2W$, M: Bending: $\sigma = 6M/(2W)^2$

-0.8	T	1.000	0.661	0.567	0.518	0.481	--	--	--	--
	M	1.000	0.567	0.476	0.351	0.301	--	--	--	--
-0.7	T	1.721	0.827	0.669	0.607	0.534	--	--	--	--
	M	1.998	0.762	0.561	0.445	0.369	--	--	--	--
-0.6	T	(2.51)	1.000	0.773	0.703	0.628	(0.581)	--	--	--
	M	(3.33)	1.000	0.700	0.556	0.458	(0.415)	--	--	--
-0.5	T	--	1.175	0.881	0.798	0.725	0.644	--	--	--
	M	--	1.270	0.858	0.677	0.549	0.505	--	--	--
-0.4	T	--	1.207	1.000	0.897	0.825	0.732	--	--	--
	M	--	1.404	1.000	0.871	0.687	0.601	--	--	--
-0.3	T	--	(1.12)	1.036	0.965	0.910	0.84	--	--	--
	M	--	(1.46)	1.087	0.914	0.777	0.704	--	--	--
-0.2	T	--	(0.933)	0.962	1.000	0.993	0.960	(0.467)	--	--
	M	--	(1.50)	1.057	1.000	0.900	0.843	(0.520)	--	--
-0.1	T	--	--	0.812	0.956	1.022	1.058	0.982	--	--
	M	--	--	0.909	0.989	0.988	0.994	0.809	--	--
0	T	--	--	0.643	0.832	1.000	1.150	1.286	--	--
	M	--	--	0.721	0.889	1.000	1.065	1.124	--	--
0.1	T	--	--	0.428	0.66	0.873	1.126	1.388	--	--
	M	--	--	0.633	0.769	0.906	1.083	1.244	--	--
0.2	T	--	--	(0.207)	0.522	0.732	1.000	1.383	(2.46)	--
	M	--	--	(0.501)	0.599	0.770	1.000	1.286	(2.40)	--
0.3	T M	--	--	--	0.401	0.584	0.825	1.24	2.267	--
		--	--	--	0.480	0.615	0.840	1.203	(2.25)	--
0.4	T	--	--	--	0.308	0.466	0.663	1.000	1.902	--
	M	--	--	--	0.368	0.514	0.692	1.000	1.819	--
0.5	T	--	--	--	0.237	0.363	0.517	0.748	1.456	--
	M	--	--	--	0.283	0.388	0.539	0.789	1.424	--
0.6	T	--	--	--	(0.193)	0.279	0.403	0.567	1.000	(4.93)
	M	--	--	--	(0.212)	0.311	0.412	0.594	1.000	(4.21)
0.7	T	--	--	--	--	0.209	0.310	0.431	0.691	2.271
	M	--	--	--	--	0.240	0.312	0.447	0.694	2.302
	T	--	--	--	--	0.163	0.236	0.325	0.475	1.000

(continued)

Table 6 (continued)

0.8	M	--	--	--	--	0.182	0.231	0.328	0.483	1.000
	T	--	--	--	--	0.109	0.172	0.241	0.346	0.980
0.9	M	--	--	--	--	0.124	0.187	0.241	0.347	0.614
	T	--	--	--	--	(0.061)	(0.121)	(0.183)	(0.25)	(0.376)
1	M	--	--	--	--	(0.081)	(0.128)	(0.17)	(0.245)	(0.335)

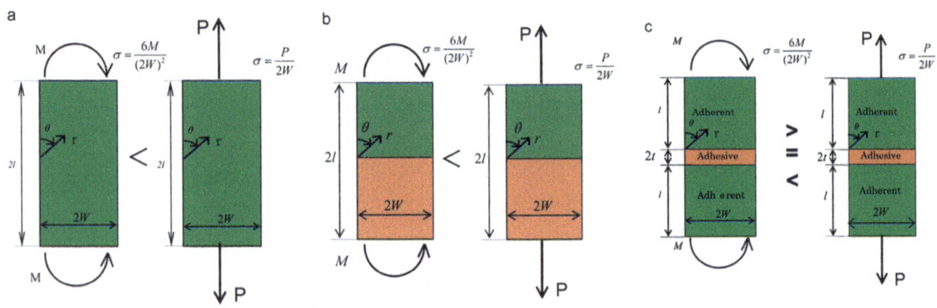

Fig. 11 Comparison of criticality between tension and bending (**a**) homogeneous plate, (**b**) bonded strip, and (**c**) adhesive joint

Table 7 Material properties of the adhesive joint with dissimilar adherents

		Material	Elastic modulus/GPa	Poisson's
1	Adherent(up)	SUS304 (stainless steel)	206	Ratio
	Adhesive	Epoxy1	3.4	0.396
	Adherent(down)	YH75 (aluminum)	71	0.33
2	Adherent (up)	Silicon	166	0.26
	Adhesive	Resin	2.74	0.38
	Adherent(down)	FR-4.5 (substrate)	15.34	0.15

joint whose adherent is only SUS304 (see Fig. 12a). On the other hand, the intensity of singular stress at the bottom end of the interface is smaller than that of the adhesive joint whose adherent is only YH-75 (see Fig. 12b). In Fig. 13a, b, the problem of dissimilar adherents with Silicon and FR-4.5 is considered. The intensity of singular stress at the upper end is larger than that of the adhesive joint with only Silicon as adherent (see Fig. 13a). On the other hand, the intensity of singular stress at the bottom end has no clear trend when compared to that of the adhesive joint with only FR-4.5 as adherent; the intensity of singular stress is dependent on the adhesive thickness, as shown in Fig. 13b.

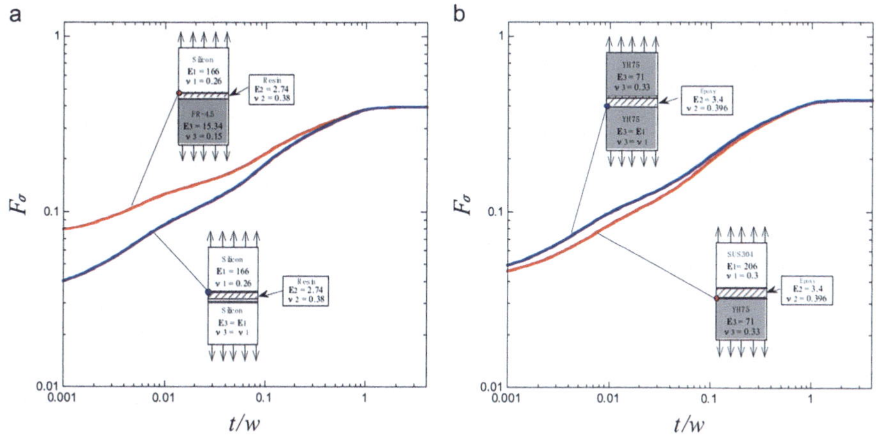

Fig. 12 Interaction of singular stress fields for the adhesive joint with dissimilar adherents (adherents: SUS 304, YH75; adhesive: epoxy) (**a**) F_σ at the above end of interface and (**b**) F_σ at the below end of interface

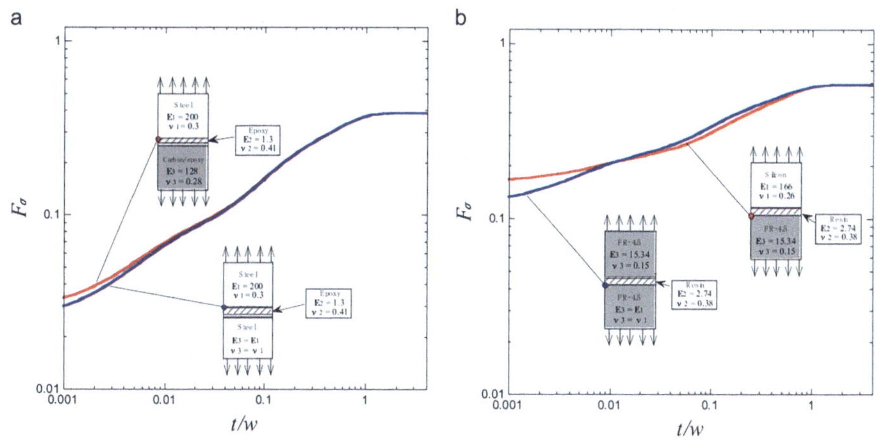

Fig. 13 Interaction of singular stress fields for the adhesive joint with dissimilar adherents (adherents: Silicon, FR-4.5; adhesive: resin) (**a**) F_σ at the above end of interface and (**b**) F_σ at the below end of interface

From those comparisons, it could be concluded that owing to the interaction between the top and bottom adherents, the intensity of singular stress becomes smaller or larger when compared with the result for the adhesive joint made of one adherent. However, because only two examples of material combinations are studied here, the interaction rules between the top and bottom stress singular fields are not clear yet. More studies about these interactions should be undertaken.

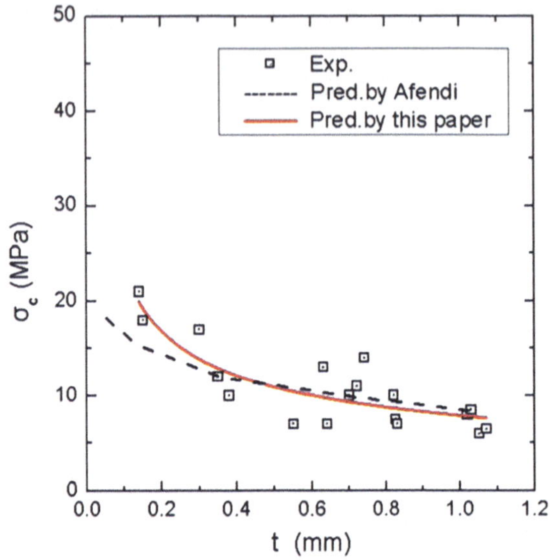

Fig. 14 The relationship between the failure stress and the adhesive thickness for the dissimilar adherents

The results of the intensity of singular stress for the case of dissimilar adherents could also be examined from the comparison with experimental results. The failure tensile tests of adhesive joints were conducted using a universal testing machine [5]. The material combination of the adhesive joint was the same as in example 1, which is shown in Table 7. Failure stresses were obtained experimentally (see Fig. 14), and it was found that the average value of the intensity of singular stress for failure was 4.155 with a stand deviation of 0.876. This finding suggests that $K_{\sigma c} = F_{\sigma c}\sigma_c(2W)^{1-\lambda}$ is a constant that is independent of adhesive thickness. The dimensionless intensity of singular stress F_σ could be derived from Fig. 12, and the calculated failure stress σ_c (the calculation method is the same as in Fig. 8) is plotted in Fig. 14 together with the experimental stress. It was be found that the failure stress agree with the experiment results better than the numerical method results in Afendi's paper, in which some components of K_σ were assumed to be constants.

From the comparison with experimental results in Figs. 8 and 14, failure stresses calculated based on the critical intensity of singular stress are in good agreement with experimental results. Therefore, it could be said that the proposed method is valid for adhesive joints (Fig. 1) made of one and two adherents.

6 Discussion

In this paper, the method to obtain the intensity of singular stress involves only calculating the first element stress at the end of interface for the unknown and reference problems. Adopting the same material combinations for the two problems and the same mesh sizes around the singularity fields, the stress singularities are the same. It is proved that errors caused by the FEM when calculating stresses could be canceled (the same errors produced by the FEM remain) under these conditions when stress ratios are considered. These ratios are equivalent to the ratios of intensity of singular stress. It is also shown that this analytical method is valid for both tension and bending.

Nisitani et al. [30] and Oda et al. [31] proposed a similar method to calculate the stress intensity factors of a crack. Nisitani et al. [30] proposed a method to calculate the stress intensity factor of a crack in a homogenous material, in which only stresses of the first element at the crack tip were used. Then, this method was extended and applied by Oda et al. [31] to calculate the stress intensity factors of a crack at the interface in a bonded plate. Both methods are only suitable to the crack problem, but the proposed method in this paper could be applied to bonded joints without a crack.

It should be noted that the methods proposed by this paper, and the ones by Nisitani et al. [30] and Oda et al. [31] have some similarity, which is that the unknown problem is solved by considering the solutions of the reference problem. Furthermore, the two problems are the same material combinations and the same mesh size around the singularity fields. It gives us such inspiration that the known results of the singular stress fields could help solving singular stress fields of unknown problems by constructing a mathematical relationship if both material combinations and mesh sizes are the same for the two problems.

7 Conclusions

In this paper, the intensity of singular stress at the end of adhesive joint under tension and bending is discussed with varying adhesive thicknesses and material combinations, and the problem of dissimilar adherents is also considered. The conclusions drawn from this work are stated below.

1. A new method is proposed to calculate the intensity of singular stress for the adhesive joint, and only stresses of the first elements at the end of interface between the adhesive and adherent materials are used. This method is proved to be valid by comparing these paper's results and experimental results. Additionally, the two results are found to be almost consistent with each other.
2. Using this method, results of the intensity of singular stress are obtained for the adhesive joint under tension and bending. The influences of material combinations and adhesive thickness on the intensity of singular stress are discussed. For both tension and bending, the intensities of singular stress of the adhesive joint $K_{\sigma,T}$ and $K_{\sigma,M}$ are found to increase with increasing α when α is small. On the other hand, $K_{\sigma,T}$ and $K_{\sigma,M}$ decrease with increasing α when α is large. The range of the intensity of singular stress K_{σ} varies depending on the adhesive thickness t/W. To improve the interface strength, thin adhesive layers are desirable because $K_{\sigma,T}$ and $K_{\sigma,M}$ decrease with decreasing thickness. The increment is different

depending on the material combination. The adhesive thickness has no effect on the intensity of singular stress when $t/W \geq 1$ ($K_{\sigma,\,T}/K_{\sigma\,|\,t/W=1,\,T} = 1$ and $K_{\sigma,\,M}/K_{\sigma\,|\,t/W=1,\,M} = 1$ when $t/W \geq 1$) for any material combination.
3. The dimensionless intensity of singular stress is compared for the bonded strip and adhesive joint under tension and bending. For the bonded strip with stress singularity fields at the end of the interface (under the condition $\alpha(\alpha - 2\beta) > 0$), it is found that $F_{\sigma,\,T} > F_{\sigma,\,M}$, which means that the bonded strip is more endangered under tension σ than under bending M when $\sigma = 3M/2W^2$. For the adhesive joint with stress singularity fields at the end of the interface, it is found that $F_{\sigma,\,T} > F_{\sigma,\,M}$ when $\alpha - 2\beta < 0$, and $F_{\sigma,\,T} < F_{\sigma,\,M}$ when $\alpha - 2\beta > 0$.
4. Comparisons between the intensities of singular stress for tension and bending with a bonded strip and an adhesive joint are presented. For the bonded strip, $F_{\sigma,\,T} > F_{\sigma,\,M}$ when singular stress fields exist, which means that the bonded strip is more endangered under tension σ than under bending M (the maximum stresses are the same at the end for tension and bending). For the adhesive joint, $F_{\sigma,\,T} > F_{\sigma,\,M}$ for $\alpha < 0$, and $F_{\sigma,\,T} < F_{\sigma,\,M}$ for $\alpha > 0$ when singular stress fields exist at the end of the interfaces. Which cases endanger the adhesive joint under tension and bending is not determined, because they will depend on the sign of α.
5. Singular stress fields on the top interface interact with singular stress fields on the bottom interface. The dimensionless intensity of singular stress Fσ for the adhesive joint with dissimilar adherents is found to become larger or smaller than that for the adhesive joint with the same adherent. More studies should be performed to investigate the interaction of stress singular fields.

References

1. Arenas J, Narbon JJ, Alia C. Optimum adhesive thickness in structural adhesive joints using statistical techniques based on Weibull distribution. Int J Adhes Adhes. 2010;30:160–5.
2. Azari S, Papini M, Spelt JK. Effect of adhesive thickness on fatigue and fracture of toughened epoxy joints—part I: experiments. Eng Fract Mech. 2011;78:153–62.
3. Suzuki Y. Adhesive tensile strength of scarf and butt joints of steel joints (3rd report), relation between adhesive layer thickness and adhesive strength of joints. Trans Jpn Soc Mech Eng. 1987;86:514–22.
4. He XC. A review of finite element analysis of adhesively bonded joints. Int J Adhes Adhes. 2011;31:248–64.
5. Afendi M, Teramoto T, Bakri HB. Strength prediction of epoxy adhesively bonded scarf joints of dissimilar adherends. Int J Adhes Adhes. 2011;31:402–11.
6. Hunston D, Miyagi Z, Schultheisz C, Zaghi S. The sandwich bending specimen for characterizing adhesive properties. Mech Time-Depend Mater. 2003;7:71–88.
7. Park JH, Choi JH, Kweon JH. Evaluating the strength of thick aluminum-to-aluminum joint with different adhesive lengths and thicknesses. Compos Struct. 2010;92:2226–35.
8. de A. Neves A, Courinho E, Poitevin A, van der Sloten J, van der Sloten J, van Meerbeek B, et al. Influence of joint component mechanical properties and adhesive layer thickness on stress distribution in micro-tensile bond strength specimens. Dent Mater. 2009;25:4–12.
9. Koguchi H, Hoshi K. Evaluation of joining strength of silicon-resin interface at a vertex in a three-dimensional joint structure. J Electron Packag. 2012;132:134.

10. Kilic B, Madenci E, Ambur DR. Influence of adhesive spew in bonded single-lap joints. Eng Fract Mech. 2006;73:1472–90.
11. Van Tooren MJL, Gleich DM, Beukers A. Experimental verification of a stress singularity model to predict the effect of bond line thickness on joint strength. J Adhes Sci Technol. 2004;18(4):395–412.
12. Goglio L, Rossetto M. Stress intensity factor in bonded joints: influence of the geometry. Int J Adhes Adhes. 2010;30:313–21.
13. Da Silva LFM, Das Neves PJC, Adams RD, Spelt JK. Analytical models of adhesively bonded joints—part I: literature survey. Int J Adhes Adhes. 2009;29:319–30.
14. Da Silva LFM, Das Neves PJC, Adams RD, Wang A, Spelt JK. Analytical models of adhesively bonded joints—part II: comparative study. Int J Adhes Adhes. 2009;29:331–41.
15. Mintzas A, Nowell D. Validation of an Hcr-based fracture initiation criterion for adhesively bonded joints. Eng Fract Mech. 2012;80:13–27.
16. Reedy ED, Guess TR. Comparison of butt tensile strength data with interface corner stress intensity factor prediction. Int J Solids Struct. 1993;30:2929–36.
17. Reedy ED, Guess TR. Interface corner failure analysis of joint strength: effect of adherend stiffness. Int J Fract. 1997;88:305–14.
18. Qian Z, Akisanya AR. An experimental investigation of failure initiation in bonded joints. Acta Mater. 1998;46:4895–904.
19. Qian Z, Akisanya A. Analysis of free-edge and displacement fields in scarf joints subjected to a uniform change in temperature. Fatigue Fract Eng Mater Struct. 1998;21:687–703.
20. Akisanya AR, Meng CS. Initiation of fracture at the interface corner of bimaterial joints. J Mech Phys Solids. 2003;51:27–46.
21. Bogy DB. Edge-bonded dissimilar orthogonal elastic wedges under normal and shear loading. J Appl Mech. 1968;35:460–6.
22. Bogy DB. Two edge-bonded elastic wedges of different materials and wedge angles under surface tractions. J Appl Mech. 1971;38:377–86.
23. Chen DH, Nisitani H. Intensity of singular stress field near the interface edge point of a bonded strip. Trans Jpn Soc Mech Eng A. 1993;59:2682–6.
24. Noda N-A, Shirao R, Li J, Sugimoto J-S. Intensity of singular stress at the end of a fiber under pull-out force. Int J Solids Struct. 2007;44:4472–91.
25. Groth HL. Stress singularities and fracture at interface corners in bonded joints. Int J Adhes Adhes. 1988;8:107–13.
26. Lefebvre DR, Dillard DA. A stress singularity approach for the prediction of fatigue crack initiation in adhesive bonds. Part I: theory. J Adhes. 1999;70:119–38.
27. Lefebvre DR, Dillard DA, Dillard JG. A stress singularity approach for the prediction of fatigue crack initiation in adhesive bonds. Part 2: experimental. J Adhes. 1999;70:139–54.
28. Mohammed I, Liechti KM. The effect of corner angles in bimaterial structures. Int J Solids Struct. 2001;38:4375–94.
29. Nakajima M, Koguchi H. Influence of interlayer thickness on the intensity of singular stress field in 3D three-layered joints under an external load. In: Proceedings of Asian pacific conference for materials and mechanics; 2009. p. a-112.
30. Nisitani H, Teranisi T, Kukuyama F. Stress intensity factor analysis of a bimaterial plate based on the crack tip stress method. Trans Jpn Soc Mech Eng. 2003;69:1203–8.
31. Oda K, Kamisugi K, Noda NA. Analysis of stress intensity factor for interface cracks based on proportional method. Trans Jpn Soc Mech Eng A. 2009;75:476–82.

Analytical Study of the Strength of Adhesive Joints of Riser Pipes

1 Introduction

Quite recently, composite pipes have become popular in the offshore oil and gas industry as drilling depths have increased and the weight of the riser pipe has become an important factor in deep sea drilling. As is well-known, because of the limitation of manufacturing, transportation and installation, pipes are produced in sections and are connected by a joint. Pipe joining methods can be categorised into three main types: (1) mechanical fastening; (2) welding and (3) adhesive bonding. Compared with the mechanical fastening and welding methods, adhesively bonded joints offer several advantages including, lower stress concentration, more uniform stress distribution, lighter weight, water tightness, corrosion-resistance and better fatigue properties [1, 2]. Recently, with the advancement of adhesive material science and bonding techniques, adhesively bonded composite piping systems have been increasingly adopted in the marine industries, including drilling platforms and pipelines [3].

From previous work on joint systems, it was found that the stress concentration in the end regions of the adhesive layer was the lethiferous reason for premature failure of adhesively bonded joints. Therefore, pipe joints are the focus of failures in piping systems. To prevent the joint systems from premature failure, reduction or elimination of stress concentration is the critical challenge for engineers and researchers.

Modern adhesive pipe joint constructions are also used in the repair of offshore pipes such as risers (see Fig. 1a), which are subjected to complex combined loads including tension, bending, as well as internal and external pressure. A schematic diagram of a typical repaired system is shown in Fig. 1b, where the corroded region in the pipelines is wound with an adhesive and composite coupler [4, 5].

Considering the wide use of adhesive pipe joints, the successful application of adhesive pipe joint repairing systems in the offshore pipelines and the importance of stress concentration in the adhesive pipe joint ends, a model of typical adhesive pipe joint was constructed and analyzed in this study. The adhesive joint consists of two

Y. Zhang, *Interfacial Mechanics of Offshore Engineering Equipment*,
https://doi.org/10.1007/978-981-95-3200-1_6

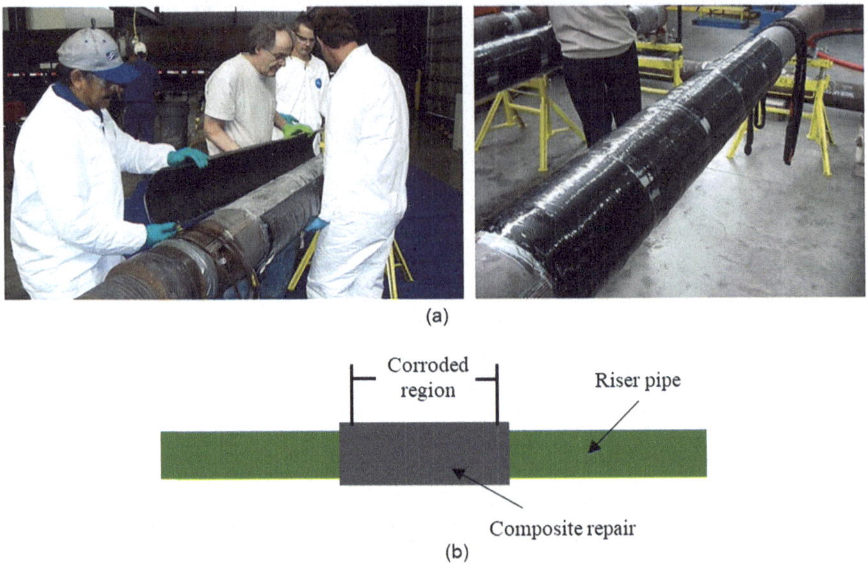

Fig. 1 Installation of the carbon half shells. (**a**) Final CRA composite repair. (**b**) Schematic diagram. (This figure is available in colour online)

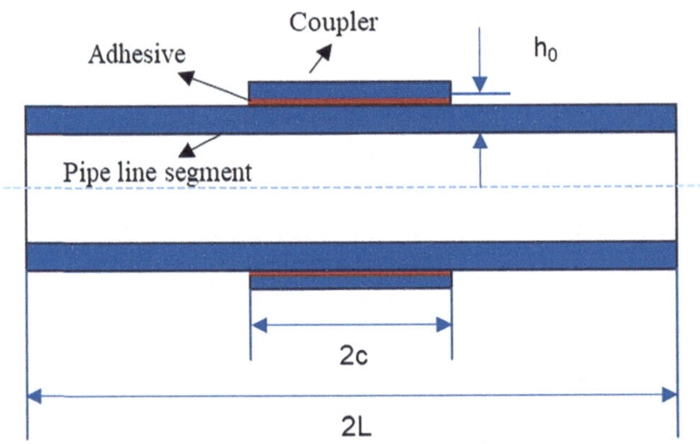

Fig. 2 The proposed construction of adhesive joint used in pipe lines in deep sea. (This figure is available in colour online)

pieces of joined pipe, a coupler to carry the load at the connection, and a medium (adhesive) that transfers the load from the pipes to the coupling (see Fig. 2).

In reported of studies of pipe joints, the problem of torsional stress in tubular lap joints was first investigated by Volkersen [6]. Following Volkersen's work, Adams and Peppiatt [7] improved Volkersen's analysis by considering the effect of the thickness of the adhesive layer.

Based on the variational principle of the complementary energy method, Chen and Cheng [8] presented a stress distribution formulation for the adhesively bonded tubular lap joint under torsion. Zhou and Taheri [9] analysed the stress distribution of adhesively bonded sandwich pipe joints subject to torsional loading. Quyang and Li [10] provided an analytical solution for adhesively bonded pipe joints under torsional loading. Most of the analyses were conducted to obtain basic numerical stress distributions by the Finite Element Method [11, 12]. There have been only a few theoretical analyses conducted by the analytical solution methods in terms of the classical/first-order shear laminated plate/shell theory for torsion loading case. For example, Zou and Taheri [9] obtained a simple analytical solution based on the classical lamination theory. Oh [13] established a first-order shear deformation theory-based model and used the stress-based solution method to study a nonlinear effect.

There has been some work conducted for the case of adhesive pipe joints under tension and bending and other loading cases. Knox et al. [14] analysed stress distributions in the adhesive layer of the bonded connections in glass reinforced epoxy pipes under tension and bending loads using finite element analysis and an experimental method. Yang [15] and Yang et al. [16] analysed the peel and shear stress of an adhesive in pipe joints under tensile loading and bending using the first-order laminated anisotropic plate theory, but the effects of the adhesive thickness on the peel and shear stress were not considered and average stresses were derived only on the pipe-adhesive and coupler-adhesive interfaces.

In this study, the pipe-adhesive and coupler-adhesive interface stresses were considered respectively for the adhesive joint under tension, and comparison of the stress distributions along the interface were conducted between numerical method and finite element method (FEM). Moreover, the strength of adhesively bonded joints of pipelines was studied under external pressure, internal pressure, tension, bending in the deep sea and the combination of these loadings and the effects of adhesive thickness and coupler length on the strength of adhesive joint were also investigated.

2 Numerical Analysis

2.1 Kinematics Relations

Because both the coupling and the pipe have the same tubler geometry, only the kinematics relations of the pipe are provided in this section. The same kinematics relations were used for coupling. The coordinate systems used in the derivations are shown in Fig. 3. The displacements fields u, v and w, which are in the longitudinal, tangential, and radial directions, respectively, are represented as

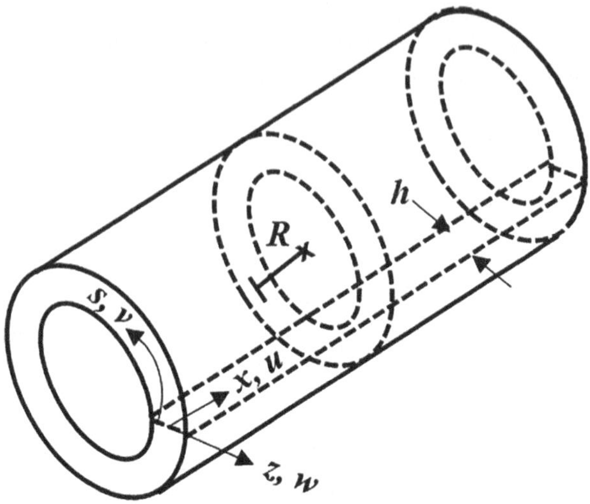

Fig. 3 Dimensions and coordinate system

$$u = u^o(x) + z\psi(x), \tag{1}$$

$$v = 0, \tag{2}$$

$$w = w(x), \tag{3}$$

where u^o represents the displacements of the mid-plane and ψ in the bending slope.

The strain fields can be written in terms of u^o, ψ and w as

$$\varepsilon_x = \frac{\partial u}{\partial x} = \frac{du^o}{dx} + z\frac{d\psi}{dx}, \tag{4}$$

$$\varepsilon_s = \frac{w}{R+z}, \tag{5}$$

$$\varepsilon_z = \frac{\partial w}{\partial z} = 0, \tag{6}$$

$$\varepsilon_{xs} = \varepsilon_{sz} = 0, \tag{7}$$

$$\varepsilon_{xz} = \frac{\partial u}{\partial z} + \frac{\partial w}{\partial x} = \psi + \frac{dw}{dx}, \tag{8}$$

where R is the radius of the mid-plane of the pipe wall as shown in Fig. 3.

2.2 *Constitutive Relations*

The resultant stress in the x-direction (or axial force per unit circumference), N_x, the resultant stress in the s-direction (or hoop force per unit axial length), N_s, and the resultant moment in the s-direction (or bending moment per unit circumference), M_s, are related to only the mid-plane strain and the plate curvature and not to the in-plane shear strain:

$$N_x = A\frac{du^o}{dx} + B\frac{d\psi}{dx} + Ew, \tag{9}$$

$$N_s = N_x, \tag{10}$$

$$M_s = B\frac{du^o}{dx} + D\frac{d\psi}{dx} + Fw, \tag{11}$$

where A, B, D, E, F are the matrices of the equivalent modulus:

$$(A, B, D) = \int_{-h/2}^{h/2} \frac{R+z}{R} Q(1, z, z^2)dz, \tag{12}$$

$$(E, F) = \int_{-h/2}^{h/2} \frac{Q}{R+z}(1, z)dz, \tag{13}$$

$$Q_x = kA_G\varepsilon_{xz} = kA_G\psi + kA_G\frac{dw}{dx}, \tag{14}$$

$$A_G = \int_{-h/2}^{h/2} Gdz. \tag{15}$$

2.3 Force Equilibrium

Adhesive stresses exist on the pipe-adhesive interface and coupler-pipe interface as shown in Fig. 4, and the equilibrium equations result in the following:

$$\frac{dN_{x,c}}{dx} = -\frac{R_{i,c}}{R_c}\tau_c, \tag{16}$$

$$\frac{dN_{x,p}}{dx} = -\frac{R_{o,p}}{R_p}\tau_p, \tag{17}$$

$$\frac{dM_{s,c}}{dx} = Q_{x,c} + \frac{h_c R_{i,c}}{2R_c}\tau_c, \tag{18}$$

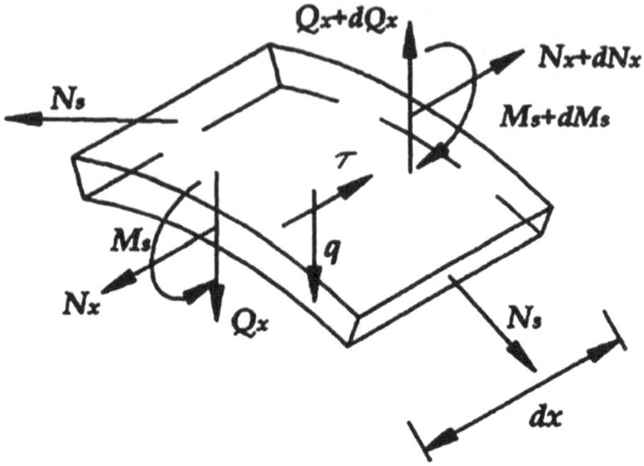

Fig. 4 Free-body diagram and sign conventions (a segment of coupling in joint region)

$$\frac{dM_{s,p}}{dx} = Q_{x,p} + \frac{h_p R_{o,p}}{2R_p}\tau_p, \tag{19}$$

$$\frac{dQ_{x,c}}{dx} = \frac{R_{i,c}}{R_c}q_c + \frac{N_{s,c}}{R_c}, \tag{20}$$

$$\frac{dQ_{x,p}}{dx} = -\frac{R_{o,p}}{R_p}q_p + \frac{N_{s,p}}{R_p}. \tag{21}$$

2.4 Adhesive Stresses

The shear stress at the pipe-adhesive interface, τ_p, and the coupling-adhesive interface τ_c arise from the displacement of the inner surface of the coupling and the outer surface of the pipe and from the first-order axial derivative of the radial deflection:

$$\tau_p = \frac{G}{\eta}\left(u_p^o + h_p\psi_p\right) - G\frac{dw_p}{dx}, \tag{22}$$

$$\tau_c = \frac{G}{\eta}\left(-u_c^o + h_c\psi_c\right) - G\frac{dw_c}{dx}, \tag{23}$$

where G, η are the shear modulus and the thickness of the adhesive, respectively. Similarly, the adhesive peel stress is related to the radial displacements of the pipe and coupler:

$$q_c = \frac{E}{\eta} w_c, \tag{24}$$

$$q_p = -\frac{E}{\eta} w_p. \tag{25}$$

The relations between τ_p and τ_c, q_P and q_c can be obtained from the following adhesive equilibrium equations:

$$\tau_p = -\frac{R_{i,c}}{R_{o,p}} \tau_c, \tag{26}$$

$$q_P = -\frac{R_{i,c}}{R_{o,p}} q_c. \tag{27}$$

Substitution of the constitutive relations, Eqs. (9)–(11) and (14) of the pipe and the coupler, and the adhesive shear and peel stress, Eqs. (22)–(25), the relations of shear and peel stress in the pipe-adhesive interface and coupler-adhesive interface (26), (27) into the equations of force equilibrium, Eqs. (16)–(21), yields the governing equations.

At the centre of the joint, the cross-section of the pipe is a free surface, so the axial stress resultant, bending moment, and transverse shear stress results must be zero. Moreover, the end of the cross-section of the coupling and the end of the pipe are both free surfaces, and the symmetry conditions of the joint geometry and loading conditions for the coupling and the continuity of the pipe from the joint region to the bare pipe region could also be used. Using all the above boundary conditions, the governing equations can be solved by the software Maple14, and the exact solutions of shear and peel stresses on the pipe-adhesive interface τ_p and q_P, and coupling-adhesive interface τ_c and q_c can be obtained.

3 Stress Analysis by FEM

The proposed pipe line material was steel with elastic properties $E = 206$ GPa, $v = 0.3$ and the proposed adhesive material was epoxy resin $E = 3.33$ GMPa, $v = 0.33$. Generally, the coupler material was a composite material, but to simplify the analytical model, the proposed coupler was the same material as the pipe line with elastic properties $E = 206$ GPa, $v = 0.3$,which means that the existence of a singular stress fields condition was satisfied at the end of adhesive-pipe interface and adhesive-coupler interface [17, 18]. The proposed adhesive pipe joint model is shown in Table 1.

For the calculation of offshore loadings, Hirdaris et al. [19] discussed the loading for use in the design of ships and offshore structures, and in this paper the internal pressure, external pressure, tension and bending loadings were considered. The supposed underwater depth was 1200 m, and the loadings subjected to pipelines were calculated as following [20]:

Table 1 The dimension of the adhesive pipe joint

	Length (mm)	Thickness (mm)	External diameter (mm)
Pipe line	290	15	144
Adhesive	250	1	146
Coupler	250	15	176

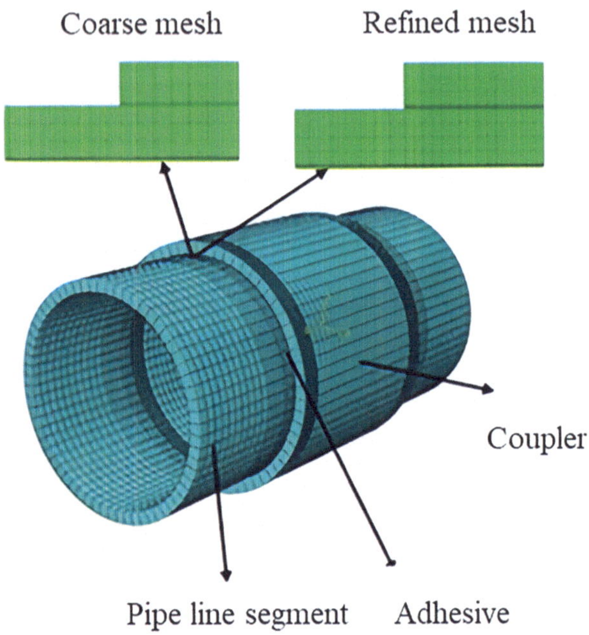

Fig. 5 The 3D meshing model of adhesive joint. (This figure is available in colour online)

$$\textbf{internal pressure } P_i = P_i'(1 + a_p) + \rho_w g H = 27.02 \text{ MPa},$$

$$\textbf{external pressure } P_o = \rho_w g H = 12.0 \text{ MPa},$$

$$\textbf{tension loading } T = W_{\text{wet}} f_{\text{dyn}} \tan(\theta) = 5.04 \times 10^5 \text{N},$$

$$\textbf{bending moment } M = \frac{EI}{[\text{L}\tan(\theta)] f_{\text{dyn}}} = 2.41 \times 10^5 \text{Nm},$$

where $H = 1200$ m is the depth of seawater, $P_i' = 15.31$ MPa is the delivery pressure, $\rho_w = 1025$ kg/m^3 is the density of seawater, $f_{\text{dyn}} = 1.2$ is the safety factor of dynamic load, $a_p = 10\%$ is pressure fluctuation coefficient, and $\theta = 12$ is top hanging angle, the hanging length $L = H \cos \theta/(1 - \sin \theta) = 1482$ m, and the pipeline underwater weight $W_{wet} = 1.97 \times 106$ N.

The three-dimensional (3D) meshing analytical model shown in Fig. 5 was constructed using the software ABAQUS. Castello and Estefen [21] analysed sandwich pipes by a finite element model, where three different interface types were employed: slip model, friction model, bonded model, and it was found that

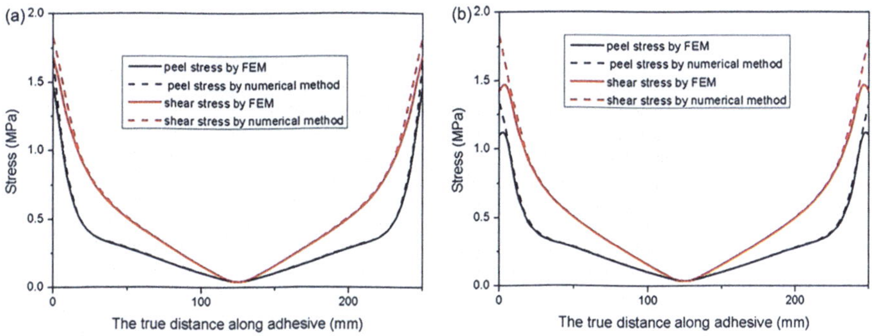

Fig. 6 Adhesive peel shear stress distributions on the (**a**) pipe-adhesive interface and (**b**) coupler-adhesive interface. (This figure is available in colour online)

the stress results were nearly the same. In this study, the bonded model was used at the interface between adhesive and pipe and the adhesive and coupler. The stresses are dependent on the mesh size using the finite element analysis, especially at the end of the interface where stress singular field exists. In this paper, the element type is C3D8R, and two types of mesh sizes were used: the coarse mesh and the refined mesh at the end of interface, and the coarse mesh size in the radius direction was twice that of the refined mesh size. The element number of the coarse mesh is 4096, and it is 16,384 for the case of refined mesh.

Figure 6 shows the peel and shear stress distributions at the pipe-adhesive interface and coupler-adhesive interface of the adhesive joint under tension, which were obtained using the numerical method proposed above this paper and FEM. Stress distributions are nearly the same along the interfaces, except at the end of interface, where singular stress fields exists.

4 Results and Discussion

4.1 Interfacial Stress Distributions in the Adhesive Pipe Joint Under Tension, Bending, Internal and External Pressure

Gleich et al. [22] and da Silva et al. [23] found that the adhesive joint strength can be explained by the interfacial adhesive-adherend stresses. For example, in the case of concrete beams strengthened with an FRP plate, premature failure modes are caused by interfacial stress concentrations in the adhesive layer. Tounsi and Benyoucef [24] gave an analytical solution for the interface stresses. However, for the adhesive pipe joint, there are few reports detailing inter-facial stresses. Stress nephograms of adhesive pipe joints show that maximum stresses occur along the interfaces, especially at the end of the interface, which is coincident with the singular stress field

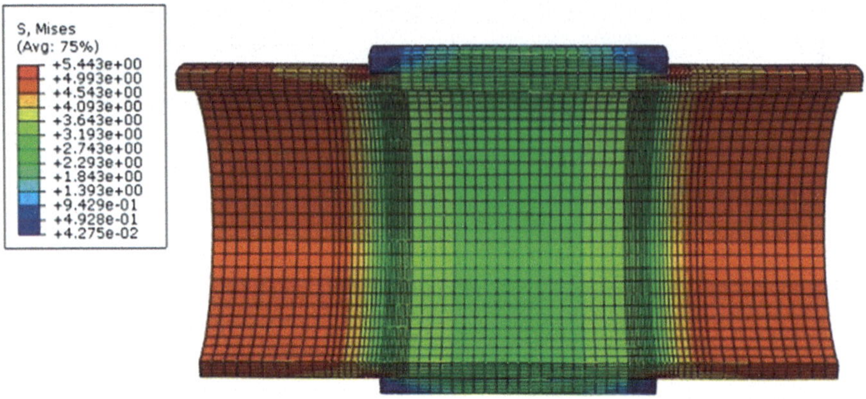

Fig. 7 Stress nephogram for the adhesive pipe joint under tension. (This figure is available in colour online)

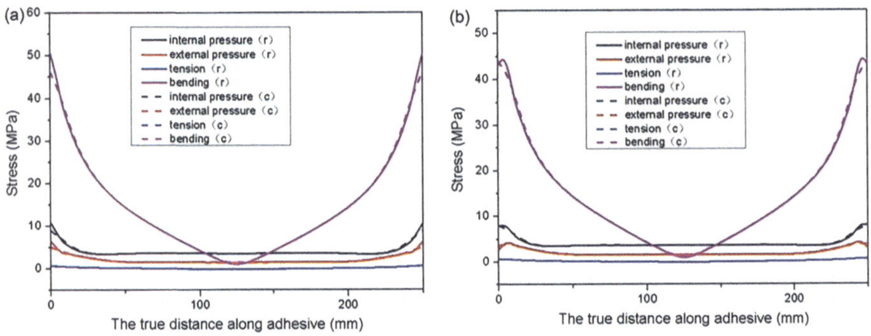

Fig. 8 (**a**) The maximum shear stress distributions on the adhesive-pipe interface for the adhesive joint under tension, bending, internal and external pressure; (**b**) The maximum shear stress distributions on the adhesive-cover interface for the adhesive joint under tension, bending, internal and external pressure. (This figure is available in colour online)

theory [17, 18]. For example, Fig. 7 shows the stress nephogram for the case of tension and it can be seen that the maximum stresses are at the end of the interface.

The failure criterion for an adhesively bonded joint could be based on the maximum principle stress or Misses stress [25]. Figure 8 shows the maximum shear stress on the adhesive-pipe interface and adhesive-coupler interface for the adhesive pipe joint under deep sea loading, i.e. tension, bending, internal and external pressure. In all the analytical models, two kinds of mesh size coarse mesh size and refined mesh size were used, while the solid lines show the results of refined mesh size and the dotted lines represent the results of rough mesh size. It was found that stress results are almost the same for both kinds of mesh size. The largest shear stress on the interface is caused by the bending moment and stresses on the adhesive-pipe interface are larger than the stresses on adhesive-coupler interface. Stresses at the end of adhesive-pipe interface have tend toward infinity according to the theory of Body [17, 18], while the maximum stress occurs at a distance from the end of adhesive-coupler interface, due to zero shear stress at the pipe end.

Figure 9 shows the maximum principle stress distributions on the adhesive-pipe interface and adhesive-cover interface for the adhesive joint under tension, bending, internal and external pressure, respectively. Almost the same conclusions can be obtained, except that the maximum values of shear stresses and principle stresses are different. The bending moment produces the largest shear stress and principle stress, while stresses caused by tension, internal and external pressure are very small. The bending moment under deep sea is often caused by wave and current action, so it can be predicted that wave and currents cause larger stresses than other loadings for the adhesive joint of pipe lines.

Figure 10 shows Misses stress distributions on the adhesive-pipe interface and the adhesive-coupler interface for the joint under the combination of tension, bending, internal and external pressure. It was found that stresses on the adhesive-pipe interface are larger than stresses on adhesive-coupler interface and the maximum

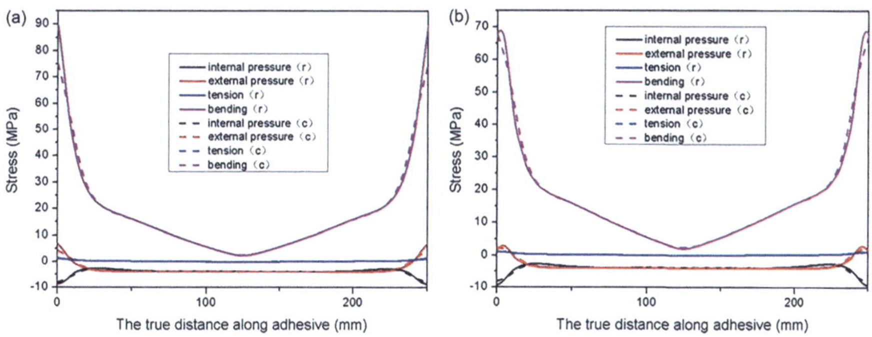

Fig. 9 (**a**) The maximum principle stress distributions on the adhesive-pipe interface for the adhesive joint under tension, bending, internal and external pressure; (**b**) The maximum principle stress distributions on the adhesive-cover interface for the adhesive joint under tension, bending, internal and external pressure. (This figure is available in colour online)

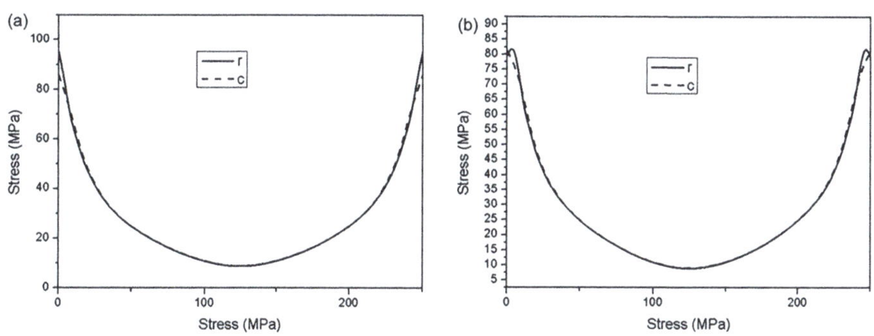

Fig. 10 (**a**) The Misses stress distributions on the adhesive-pipe interface for the adhesive joint under mixed loadings; (**b**) The Misses stress distributions on the adhesive-cover interface for the adhesive joint under mixed loadings

stress occurs at a distance from the end of interface, due to zero stress at the end of adhesive-coupler interface.

From Figs. 8, 9, and 10, it can be concluded that the bending moment causes the largest stresses on the interface compared with tension, internal and external pressure and failure is more likely to occur on the adhesive-pipe interface in comparison to the adhesive-coupler interface. In addition, a short distance point from the end of adhesive-coupler interface is more prone to have failures in comparison to the end of the interface due to zero stress boundary condition at the end.

4.2 The Effect of Adhesive Thickness and Adhesive Length on the Strength

To investigate the effect of adhesive thickness on stress distributions for the adhesive pipe joint under a combination of loads, the adhesive thickness in the model was changed from 2 to 4 mm and the effect of adhesive thickness on stress distributions is shown in Fig. 11. Because it has been proven that stresses obtained with coarse mesh size are almost the same as the case of refined mesh size, only the coarse mesh size was used here. From Fig. 11 it can be seen that stresses decrease with increasing of adhesive thickness and the increase rate is smaller for larger adhesive thickness, which means that the strength of adhesive joint of the pipe lines increases with increasing of adhesive thickness. Additionally, the maximum Misses stress does not happen at the end of interface, and it happens a distance from the end.

Figure 12 shows the stress distributions as a function of coupler length $2c = 200$, 250, 300 mm, and it is found that stress distributions are nearly the same, which means that the coupler length has no effect on stress distributions. Perhaps further investigation is needed to determine the effect of adhesive length using shorter coupler lengths.

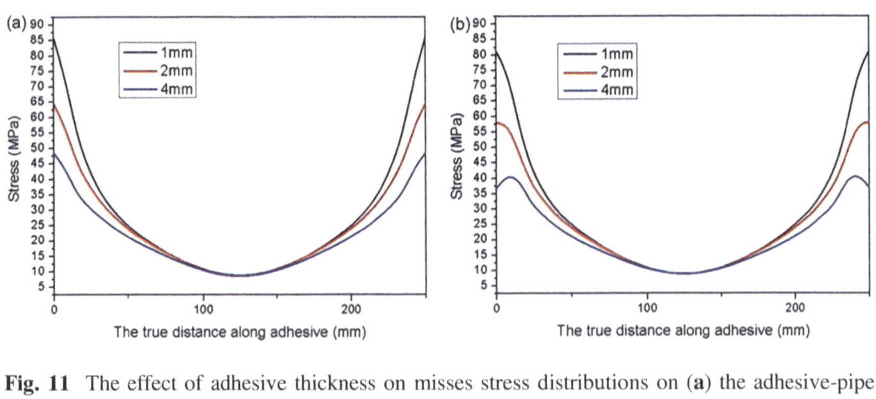

Fig. 11 The effect of adhesive thickness on misses stress distributions on (**a**) the adhesive-pipe interface and (**b**) the adhesive-coupler interface. (This figure is available in colour online)

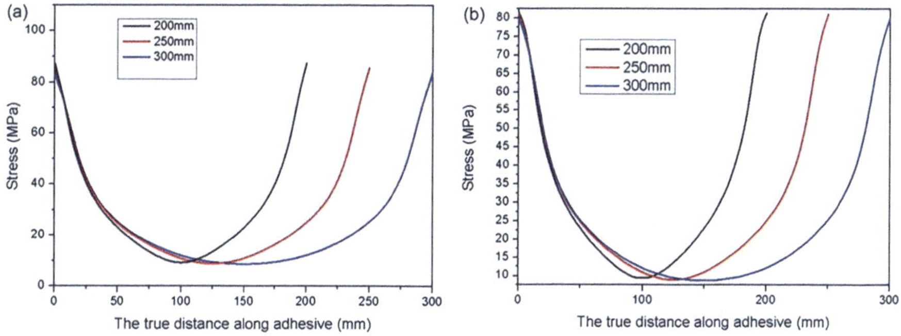

Fig. 12 The effect of coupler length on misses stress distributions on (**a**) the adhesive-pipe interface and (**b**) the adhesive-coupler interface. (This figure is available in colour online)

5 Conclusions

Stress distributions were investigated for an adhesive joint of a riser pipe in deep sea under tension, bending, internal and external pressure respectively, as well as combination of these loadings. The conclusions of this study are as follows:

1. The first-order plate theory was used to simulate the stresses of adhesive joint under tension taking the thickness of the adhesive layer into account. Adhesive peel and shear stresses on the pipe-adhesive and coupler-adhesive interfaces were derived. It was found that the numerical results are nearly the same as the FEM results, with the exception of the end of interface where singular stress fields exist;
2. The bending moment, spawned by the sea wave and current action, generates the largest stresses on interfaces in comparison to tension, internal and external pressure, so it can be predicted that waves and current actions are more dangerous loadings than other loadings for the adhesive joint of pipe lines;
3. Failures are easily happen on the adhesive-pipe interface in comparison to the adhesive-coupler interface. Additionally, failures occur a short distance from the end of adhesive-coupler interface;
4. The strength of the adhesive joint of a pipe lines in deep sea increases with increasing of adhesive thickness, while the coupler length has no effect on stress distributions under the conditions of this study, further discussion is concerning the case of adhesive joints with shorter coupler lengths.

References

1. Cheng JQ, Li GQ. Stress analyses of a smart composite pipe joint integrated with piezoelectric composite layers under torsion loading. Int J Solids Struct. 2008;45:1153–78.
2. Hashim S, Berggreen C, Tsouvalis N, et al. Fabrication, testing and analysis of steel/composite DLS adhesive joints. Ships Offshore Struct. 2011;6:115–26.

3. Mejri M, Cognard JY. Time-variant assessment for adhesively bonded assemblies with non-linear behaviour in naval applications. Ships Offshore Struct. 2009;4:275–86.
4. Alexander C, Ochoa OO. Extending onshore pipeline repair to offshore steel risers with carbon-fiber reinforced composites. Compos Struct. 2010;92:499–507.
5. Shamsuddoha M, Islam MM, Aravinthan T, Manalo A, Lau K. Effectiveness of using fibre-reinforced polymer composites for underwater steel pipeline repairs. Compos Struct. 2013;100: 40–54.
6. Volkersen O. Recherches Sur La Theorie Des Assemblages Colles. Constr Métal. 1965;4:3–13.
7. Adams RD, Peppiatt NA. Stress analysis of adhesive bonded tubular lap joint. J Adhes. 1977;9: 1–18.
8. Chen D, Cheng S. Torsional stress in tubular lap joints. Int J Solids Struct. 1992;29:845–53.
9. Zhou GP, Taheri F. Stress analysis of adhesively bonded sandwich pipe joints subjected to torsional loading. Int J Solids Struct. 2006;43:5953–68.
10. Ouyang ZY, Li GQ. Cohesive zone model based analytical solutions for adhesively bonded pipe joints under torsional loading. Int J Solids Struct. 2009;46:1205–17.
11. Hashim SA, Cowling MJ, Lafferty S. The integrity of bonded joints in large composite pipes. Int J Adhes Adhes. 1998;18(6):421–9.
12. Harte AM, McNamara JF, Roddy ID. Application of optimisation methods to the design of high performance composite pipelines. J Mater Process Technol. 2003;142(1):58–64.
13. Oh JH. Nonlinear analysis of adhesively bonded tubular single-lap joints for composites in torsion. Compos Sci Technol. 2007;67:1320–9.
14. Knox EM, Lafferty S, Cowling MJ, Hashim SA. Design guidance and structural integrity of bonded connections in GRE pipes. Compos A: Appl Sci Manuf. 2001;32:231–41.
15. Yang C. Design and analysis of composite pipe joints under tensile loading. J Compos Mater. 2000;34:332–48.
16. Yang C, Huang H, Guan ZD. Stress model of composite pipe joints under bending. J Compos Mater. 2002;36:1331–48.
17. Bogy DB. Edge bonded dissimilar orthogonal wedges under normal and shear loading. J Appl Mech. 1968;35:460–6.
18. Bogy DB. Two edge-bonded elastic wedges of different materials and wedge angles under surface tractions. J Appl Mech. 1971;38:377–86.
19. Hirdaris SE, Bai W, Dessi D, et al. Loads for use in the design of ships and offshore structures. Ocean Eng. 2014;78:131–74.
20. Sha Y, Cao J, Zheng XR, Wu YX. Minimum wall thickness calculation and analysis of deepwater steel catenary riser based on the principles of working stress. China Offshore Platform. 2009;24:46–56.
21. Castello X, Estefen SF. Limit strength and reeling effects of Sandwich pipes with bonded layers. Int J Mech Sci. 2007;49:577–88.
22. Gleich DM, Van Tooren MJL, Beukers A. Analysis and evaluation of bondline thickness effects on failure load in adhesively bonded structures. J Adhes Sci Technol. 2001;15:1091–101.
23. da Silva LFM, Rodrigues TNSS, Figueiredo MAV, et al. Effect of adhesive type and thickness on the lap shear strength. J Adhes. 2006;82:1091–115.
24. Tounsi A, Benyoucef S. Interfacial stresses in externally FRP-plated concrete beams. Int J Adhes Adhes. 2007;27:207–15.
25. da Silva LFM, das Neves PJC, Adams RD, et al. Analytical models of adhesively bonded joints—part II: comparative study. Int J Adhes Adhes. 2009;29:331–41.

Research on Interface Viscoelastic Mechanical Model Based on Modified Cohesive Zone and Generalized Maxwell Models

1 Introduction

Composite structures are widely used in aerospace, ocean, automotive, construction, and infrastructure industries due to their advantages such as high strength-to-weight and stiffness-to-weight ratios, excellent corrosion resistance and durability, and design flexibility [1]. However, the interface within composite structures is often the weakest point, and its strength is a critical factor affecting the overall performance of the composite structure [2, 3]. Marine composite pipes, as a typical example in the marine engineering field, serve as lifelines for offshore resource exploitation [4–7]. Depending on the manufacturing process, marine composite pipes can be classified into bonded and unbonded, with their typical joint and pipe body structures illustrated in Fig. 1 [8, 9]. In bonded offloading hoses, the flange and the inner reinforcement layer are joined using acrylate adhesive and a vulcanization process, incorporating a steel-rubber interface. In unbonded pipe terminations, the end fitting body and the tensile armor are bonded with epoxy resin adhesive, forming a steel-epoxy resin adhesive interface.

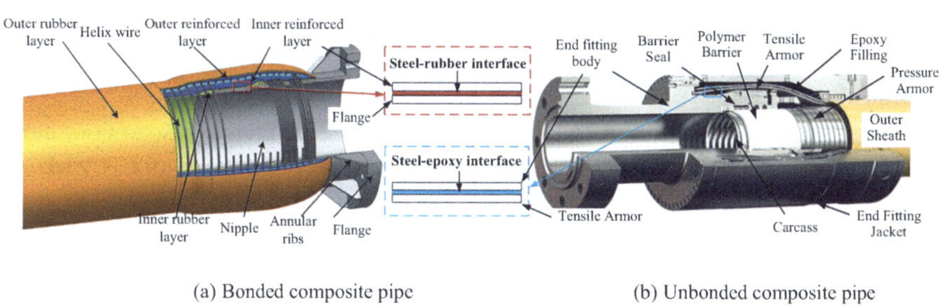

Fig. 1 Typical structure of marine composite flexible pipes [8, 9]. (**a**) Bonded composite pipe; (**b**) Unbonded composite pipe

Y. Zhang, *Interfacial Mechanics of Offshore Engineering Equipment*,
https://doi.org/10.1007/978-981-95-3200-1_7

The interface layer is often identified as the weakest region in marine composite pipes [3] making it susceptible to debonding failures under complex environmental loads [10]. Existing studies have demonstrated that many interfaces, such as steel-rubber and steel-epoxy adhesive interfaces, exhibit viscoelastic mechanical behavior [11–14]. These interfaces are prone to creep and stress relaxation, and their mechanical response is significantly influenced by the loading rate [15]. The harsh marine environment often leads to composite flexible pipes failure, which is frequently attributed to sudden and significantly large load with a rapid loading rate. However, the viscoelastic characteristics of interfaces have not been adequately considered in studies on the ultimate load-bearing capacity of composite flexible pipes [16–19]. Consequently, investigating viscoelastic constitutive and damage models for interfaces is of great importance, as it provides essential insights for analyzing the ultimate load-bearing characteristics of marine composite pipes.

The interface failure in complex bonding structures is challenging to characterize experimentally and to describe comprehensively through theoretical approaches. The finite element numerical method offers a precise simulation of interface mechanical behavior. Representative finite element methods include the virtual crack closure technique (VCCT), grounded in fracture mechanics [20, 21]; the cohesive zone model (CZM), based on damage mechanics; and the extended finite element method (XFEM) [22]. Among these, the CZM is particularly suitable and widely applied for analyzing interface failure, as it effectively accounts for the constrained propagation paths of bonding interface cracks and ensures comprehensive crack description.

The CZM has been extensively utilized to characterize the constitutive relationships and damage evolution at interfaces. Originally proposed by Dugdale [23] and later implemented via the finite element method by Hillerborg [24], the CZM defines the traction-separation relationship at the interface through specific traction-separation criteria. Commonly employed traction-separation models include exponential [25], polynomial [26], trilinear [27], linear [28], and bilinear forms [29]. Among these, the bilinear model stands out for its simplicity and high accuracy, making it a popular choice for simulating interface mechanical behavior in composite laminates [30, 31], ceramic matrix composites [32, 33], asphalt concrete [34], and rubber composites [35]. Key parameters in the bilinear cohesive model include damage initiation separation, critical failure separation, and critical fracture energy, all of which are typically obtained through experimental tests such as the double cantilever beam (DCB) test, end-notched flexure (ENF) test, and mixed-mode bending (MMB) test.

In the numerical simulation of viscoelastic interface behavior, researchers have modified the traction-separation criterion in the cohesive zone model (CZM) to account for viscoelastic properties. This approach involves defining parameters such as interface fracture energy, interface strength, and final crack opening length as functions of the crack opening rate, thereby incorporating the relationship between mechanical response and loading rate. Marzi et al. [36] developed a rate-dependent CZM, verified through experimental tests, where traction is proportional to the square of the logarithm of the strain rate. Valoroso et al. [37] proposed an enhanced CZM by introducing a rate-dependent critical energy release rate into the damage function. Similarly, May et al. [38] proposed an exponential traction-separation criterion that accounts for loading rate effects. To better represent the nonlinear viscoelastic mechanical behavior of interfaces, some researchers have combined viscoelastic rheological models with CZM. Musto et al. [11] developed a CZM that integrates viscoelasticity and interface damage, where the rheological model comprises a Maxwell model in parallel with a spring. The resulting numerical model is validated using double cantilever beam (DCB) tests. While the model effectively simulates viscoelastic behavior within a loading rate range of 0.1–100 mm/min using a simplified discrete spectrum and single relaxation time, it demonstrates reduced accuracy at very low or high loading rates. Zhao et al. [39] employed the Kohlrausch-Williams-Watts function to describe viscosity in the Maxwell model, assuming that damage initiation and critical failure separation at the interface vary with loading speed. Yang et al. [40] proposed a novel CZM incorporating a rheological model consisting of a Maxwell model in parallel with a spring and an additional spring connected in series, which successfully simulates the unique mechanical properties of pressure-sensitive adhesives.

According to the literature review, it is found that currently scholars have proposed mechanical models to describe the viscoelastic nonlinear mechanical behavior of the interface based on the CZM, but most of the models are not accurate enough to describe the interface viscoelastic mechanical behavior, especially at high and low loading speeds. To address this limitation, this study proposes an interface viscoelastic model that is applicable across a wide range of loading rates. The model is formulated by combining the generalized Maxwell model with a modified rate-dependent CZM. The proposed mechanical model is discretized, and a corresponding VUMAT subroutine is developed. Using ABAQUS, a double cantilever beam (DCB) numerical model is established to simulate the mechanical response of specimens under varying loading rates. The simulation results are then compared with experimental data to validate the accuracy and effectiveness of the proposed model.

2 Interface Viscoelastic Mechanical Model Considering Interface Damage

Due to the harsh marine environment, the failure of composite flexible pipes is frequently associated to their sudden exposure to large loads with high loading rates. The interface, being the weakest component of marine composite pipes, exhibits viscoelastic mechanical properties that are significantly influenced by the loading rate. This study focuses on the viscoelastic interfaces in marine composite pipes, such as rubber-steel and rubber-epoxy interfaces (as shown in Fig. 1). A viscoelastic constitutive model for the interface is proposed based on the generalized Maxwell model. Furthermore, a rate-dependent cohesive zone model (CZM) is employed to establish the interface damage model. The proposed interfacial viscoelastic mechanical model accounts for the rate dependence of the intrinsic spring and the viscous dissipation of multiple Maxwell elements, enabling the description of interface mechanical behavior over a wide range of loading rates. To facilitate the numerical simulation of interfacial mechanical behavior, the mechanical model is discretized presented in this section.

2.1 Interface Viscoelastic Constitutive Model

The generalized Maxwell model, which is able to capture more complex rheological behaviors, is employed to describe the viscoelastic mechanical behavior of the interface. The rheological representation of the generalized Maxwell model is illustrated in Fig. 2. Taking the case of the interface under a normal load as an example, the relationship between interface stress and displacement may be derived.

The free energy potential of the interface includes the intrinsic fracture energy and the dissipated energy [41], and its calculation formula is shown in Eq. (1), where E_0^{n} is the stiffness of the intrinsic spring, E_i^{n} is the stiffness of the Maxwell ith spring, δ_{n} is the total displacement, α_i^{n} is the displacement of the ith dashpot, $\delta_{ei}{}^{\mathrm{n}}$ is the displacement of the ith spring, K_c is the compression factor, symbols $\langle\cdot\rangle_+$ and $\langle\cdot\rangle_-$ represent the positive and negative parts of the displacement, respectively. The last term of Eq. (1) describes the mechanical behavior of the interface under compression. Destrade et al. [42] demonstrated through experimental research that damage does not affect the mechanical response of the interface due to crack closure during compression, the rate dependence of the mechanical response of the interface could be ignored during compression, and the ratio of the interface stiffness during compression to that during tension is K_c.

$$\Psi_{\mathrm{n}} = \frac{1}{2}E_0^{\mathrm{n}}\langle\delta_{\mathrm{n}}\rangle_+^2 + \frac{1}{2}\sum_{i=1}^{n} E_i^{\mathrm{n}}\langle\delta_{\mathrm{n}} - \alpha_i^{\mathrm{n}}\rangle_+^2 + K_c\cdot\left(\frac{1}{2}E_0^{\mathrm{n}}\langle\delta_{\mathrm{n}}\rangle_-^2 + \frac{1}{2}\sum_{i=1}^{n} E_i^{\mathrm{n}}\langle\delta_{\mathrm{n}}\rangle_-^2\right) \tag{1}$$

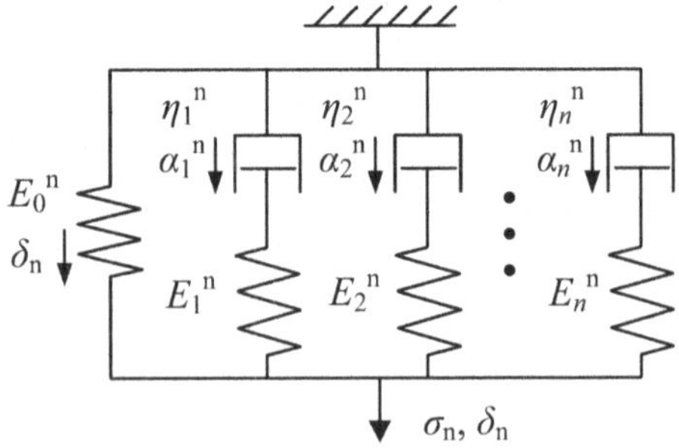

Fig. 2 Rheological model of the generalized Maxwell model

The system is assumed to be isothermal and the principle of increase of entropy applies [43]:

$$\sigma_{\mathrm{n}}\dot{\delta}_{\mathrm{n}} - \dot{\Psi}_{\mathrm{n}} \geq 0 \tag{2}$$

Substituting Eq. (1) into Eq. (2) gives:

$$\left\{\sigma_{\mathrm{n}} - \left[\left(E_0^{\mathrm{n}}\langle\delta_{\mathrm{n}}\rangle_+ + \sum_{i=1}^{n} E_i^{\mathrm{n}}\left\langle\delta_{\mathrm{n}} - \alpha_i^{\mathrm{n}}\right\rangle_+\right) + K_c \cdot \left(E_0{}^{\mathrm{n}}\langle\delta_{\mathrm{n}}\rangle_- + \sum_{i=1}^{n} E_i^{\mathrm{n}}\langle\delta_{\mathrm{n}}\rangle_-\right)\right]\right\}\dot{\delta}_{\mathrm{n}} + \sum_{i=1}^{n} E_i^{\mathrm{n}}\left\langle\delta_{\mathrm{n}} - \alpha_i^{\mathrm{n}}\right\rangle_+ \cdot \dot{\alpha}_i^{\mathrm{n}} \geqslant 0 \tag{3}$$

The sufficient conditions for Eq. (3) to hold are as follows:

$$\begin{cases} \sigma_{\mathrm{n}} = E_0^{\mathrm{n}}\langle\delta_{\mathrm{n}}\rangle_+ + \sum\limits_{i=1}^{n} E_i^{\mathrm{n}}\left\langle\delta_{\mathrm{n}} - \alpha_i^{\mathrm{n}}\right\rangle_+ + K_c \cdot \left(E_0^{\mathrm{n}}\langle\delta_{\mathrm{n}}\rangle_- + \sum\limits_{i=1}^{n} E_i^{\mathrm{n}}\langle\delta_{\mathrm{n}}\rangle_-\right) \\ \sum\limits_{i=1}^{n} E_i^{\mathrm{n}}\left\langle\delta_{\mathrm{n}} - \alpha_i^{\mathrm{n}}\right\rangle_+ \cdot \dot{\alpha}_i^{\mathrm{n}} \geqslant 0 \end{cases} \tag{4}$$

The viscoelastic dissipation of viscoelastic parts can be expressed by the Boltzmann superposition principle, which indicates that stress relaxation is the linear sum of various relaxation processes throughout the entire load history, and the stress response can be written as:

$$\sum_{i=1}^{n} E_i^{\mathrm{n}}\left\langle\delta_{\mathrm{n}} - \alpha_i^{\mathrm{n}}\right\rangle_+ = \left\langle\int_{-}^{t} \mathcal{I}(t-\tau)\dot{\delta}_{\mathrm{n}}(\tau)d\tau\right\rangle_+ \tag{5}$$

The generalized Maxwell model is composed of multiple Maxwell units in parallel, and the following equation can be obtained:

$$\sigma_{\mathrm{n}} = \sigma_{\mathrm{n}}^1 + \sigma_{\mathrm{n}}^2 + \ldots + \sigma_{\mathrm{n}}^n \tag{6}$$

in which, σ_{n}^1, σ_{n}^2 and σ_{n}^n represent the stresses of the first, second, and nth units, respectively.

A Laplace variation of Eq. (6) yields Eq. (7), in which $\overline{\sigma}$ and $\overline{\delta}$ respectively represent the Laplace transforms of σ_{n} and δ_{n}.

$$\overline{\sigma} = s\overline{\delta}\left(\frac{1}{\frac{s}{E_1^{\mathrm{n}}} + \frac{1}{\eta_1^{\mathrm{n}}}} + \frac{1}{\frac{s}{E_2^{\mathrm{n}}} + \frac{1}{\eta_2^{\mathrm{n}}}} + \ldots + \frac{1}{\frac{s}{E_n^{\mathrm{n}}} + \frac{1}{\eta_n^{\mathrm{n}}}}\right) = s\overline{\delta}\sum_{i=1}^{n}\frac{1}{\frac{s}{E_i^{\mathrm{n}}} + \frac{1}{\eta_i^{\mathrm{n}}}} \tag{7}$$

Multiplying $\prod_{i=1}^{n} \frac{s}{E_i^{\mathrm{n}}} + \frac{1}{\eta_i^{\mathrm{n}}}$ on both sides of Eq. (7) simultaneously yields:

$$\overline{\sigma}\sum_{i=0}^{n} P_i s^i = s\overline{\delta}\sum_{i=1}^{n}\left(\sum_{j=1}^{n} K_j s^{j-1}\right) \tag{8}$$

where, $\sum_{i=0}^{n} P_i s^i = \prod_{i=1}^{n} \frac{s}{E_i^{\mathrm{n}}} + \frac{1}{\eta_i^{\mathrm{n}}}, \sum_{j=1}^{n} K_j s^{j-1} = \prod_{j=1, j\neq i}^{n} \frac{s}{E_i^{\mathrm{n}}} + \frac{1}{\eta_i^{\mathrm{n}}}$.

The simplification of Eq. (8) is as follows:

$$\overline{\jmath} = \frac{\sum_{i=1}^{n} Q_i s^i}{\sum_{i=0}^{n} P_i s^i} = \frac{1}{P_n}\left(\frac{a_1}{s-\lambda_1} + \frac{a_2}{s-\lambda_2} + \ldots + \frac{a_n}{s-\lambda_n}\right) \tag{9}$$

The inverse Laplace transformation of Eq. (9) is:

$$\jmath(t) = \sum_{i=1}^{n} E_i^{\mathrm{n}} e^{-\frac{t}{\tau_i^{\mathrm{n}}}} \tag{10}$$

where, $E_i^{\mathrm{n}} = a_i/P_n$, $\tau_i^{\mathrm{n}} = -1/\lambda_i$.

The relationship between stress and displacement when the interface is subjected to normal load is obtained from Eqs. (4), (5) and (10):

$$\begin{aligned}\sigma_{\mathrm{n}} = E_0^{\mathrm{n}}\langle\delta_{\mathrm{n}}\rangle_+ + \left\langle \int_-^t \sum_{i=1}^{n} E_i^{\mathrm{n}} e^{-\frac{t-\tau}{\tau_i^{\mathrm{n}}}} \dot{\delta}_{\mathrm{n}}(\tau)d\tau \right\rangle_+ \\ + K_c\left(E_0^{\mathrm{n}}\langle\delta_{\mathrm{n}}\rangle_- + \sum_{i=1}^{n} E_i^{\mathrm{n}}\langle\delta_{\mathrm{n}}\rangle_-\right)\end{aligned} \tag{11}$$

Similarly, the relationship between stress and displacement when the interface is subjected to tangential load, as shown in Eq. (12), can be obtained using the same method described above. The interface generally exhibits similar mechanical properties in the tangential direction, as observed in steel-rubber and steel-epoxy adhesive interfaces. Therefore, the mechanical model assumes that the interface behaves identically in the tangential direction.

$$\sigma_{\mathrm{t}} = E_0^{\mathrm{t}}\delta_{\mathrm{t}} + \int_-^t \sum_{i=1}^{n} E_i^{\mathrm{t}} e^{-\frac{t-\tau}{\tau_i^{\mathrm{t}}}} \dot{\delta}_{\mathrm{t}}(\tau)d\tau \tag{12}$$

Fig. 3 The traction-separation law of the mixed-mode model

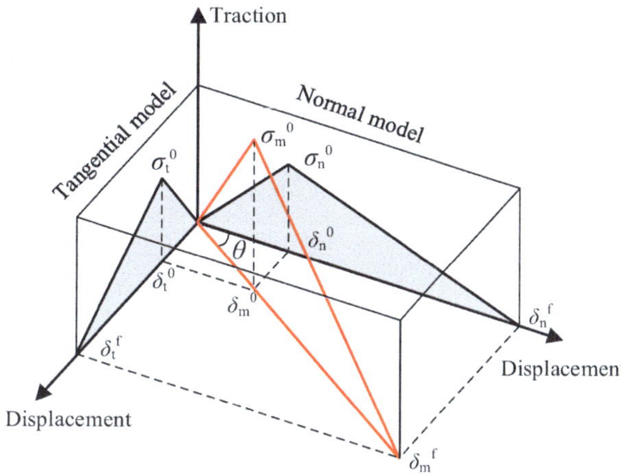

2.2 *Interface Damage Model*

The CZM is widely used to describe the damage evolution of the interface, and the traction-separation law of the mixed-mode model is shown in Fig. 3, where δ_n^0 and δ_t^0 are the normal and tangential damage initiation separations, respectively; δ_n^f and δ_t^f are the normal and tangential critical failure separations, respectively; σ_n^0 and σ_t^0 are the normal and tangential critical damage initiation tractions, respectively; δ_m^0, σ_m^0 and δ_m^f are the equivalent damage initiation separation, equivalent damage initiation traction and equivalent critical failure separation, respectively. In this study, an improved CZM model is used to describe the interfacial mechanical behavior considering damage, with the model parameters varying with the loading rate.

The tangential direction is assumed to have the same interface stiffness and damage parameters in the CZM. The equivalent separation δ_m is calculated using Eq. (13):

$$\delta_m = \sqrt{\delta_n{}^2 + \delta_t{}^2} \tag{13}$$

where, δ_n is the normal separation, δ_t is the equivalent tangential separation.

The commonly used damage initiation criteria for cohesive zone models include the maximum nominal stress or strain criteria, and the quadratic nominal stress or strain criteria. The former criteria do not account for the interaction of stresses or strains in different directions. Additionally, the quadratic nominal stress criterion is generally more conservative than the quadratic nominal strain criterion. Therefore, the former criterion is selected as the damage initiation criterion for the interface layer, and it is expressed in Eq. (14). In Eq. (14), σ_n and σ_t are the normal the tangential stress, respectively; σ_n^c and σ_t^c are the normal and tangential critical damage initiation tractions in the single mode, respectively. The term $\langle\sigma_n\rangle$ is defined in Eq. (15), ensuring that no damage occurs when the interface layer is under compression. By simplifying Eq. (14), Eq. (16) is derived as the calculation formula for the equivalent damage initiation separation, where $\beta = \tan\theta = \delta_t/\delta_n$ represents the separation composite ratio.

$$\left\{\frac{\langle\sigma_n\rangle}{\sigma_n^c}\right\}^2 + \left\{\frac{\sigma_t}{\sigma_t^c}\right\}^2 = 1 \tag{14}$$

$$\langle\sigma_n\rangle = \begin{cases} 0 & \sigma_n \le 0 \\ 1 & \sigma_n > 0 \end{cases} \tag{15}$$

$$\delta_m^0 = (1+\beta^2)\frac{\sigma_n^c\sigma_t^c}{\sqrt{\beta^2 K_t{}^2(\sigma_n^c)^2 + K_n{}^2(\sigma_t^c)^2}} \tag{16}$$

The $K = \sigma_t/\sigma_n$ is defined as the stress composite ratio. The Eq. (17) is used to calculate the normal damage initiation stress σ_n^0 and the tangential damage initiation stress σ_s^0 of the interface at a certain stress composite ratio, which is obtained from Eq. (14).

$$\begin{cases} \sigma_n^0 = \dfrac{\sigma_n^c\sigma_t^c}{\sqrt{(\sigma_t^c)^2 + \kappa^2(\sigma_n^c)^2}} \\ \sigma_t^0 = \dfrac{\kappa\sigma_n^c\sigma_t^c}{\sqrt{(\sigma_t^c)^2 + \kappa^2(\sigma_n^c)^2}} \end{cases} \tag{17}$$

When the equivalent separation exceeds the equivalent initial damage separation, damage begins to occur at the interface, and the damage degree is calculated according to Eq. (18).

$$D(\delta_m(t)) = \begin{cases} 0 & \delta_m(t) < \delta_m^0 \\ \dfrac{\delta_m^f(\delta_m(t) - \delta_m^0)}{\delta_m(t)(\delta_m^f - \delta_m^0)} & \delta_m^0 \le \delta_m(t) < \delta_m^0 \\ 1 & \delta_m(t) > \delta_m^f \end{cases} \tag{18}$$

The main energy-based damage failure criteria are the B-K and the power index criteria. Since the B-K criterion is generally used for interfaces with the same tangential characteristics, this model is here selected. The B-K criterion is shown in Eq. (19), in which G_T, G_n and G_t are respectively the total, the normal and the tangential energy release rates; η is the coefficient of the B-K criterion, G_T^c is the mixed-mode fracture toughness; G_n^c and G_t^c are the normal and tangential fracture toughness, respectively. The damage failure separation can be obtained from Eqs. (17) and (19) yielding Eq. (20), in which $\lambda = G_t/G_n$ is the fracture energy composite ratio.

$$G_{\mathrm{n}}^{\mathrm{c}} + \left(G_{\mathrm{t}}^{\mathrm{c}} - G_{\mathrm{n}}^{\mathrm{c}}\right)\left(\frac{G_{\mathrm{t}}}{G_{\mathrm{T}}}\right)^{\eta} = G_{\mathrm{T}}^{\mathrm{c}} \tag{19}$$

$$\delta_{\mathrm{m}}^{\mathrm{f}} = 2\sqrt{1+\beta^2}\frac{G_{\mathrm{n}}^{\mathrm{c}} + \left(G_{\mathrm{t}}^{\mathrm{c}} - G_{\mathrm{n}}^{\mathrm{c}}\right)\left(\frac{\lambda}{1+\lambda}\right)^{\eta}}{(1+\lambda)\frac{\sigma_{\mathrm{n}}^{\mathrm{c}}\sigma_{\mathrm{t}}^{\mathrm{c}}}{\sqrt{\left(\sigma_{\mathrm{t}}^{\mathrm{c}}\right)^2 + \kappa^2\left(\sigma_{\mathrm{n}}^{\mathrm{c}}\right)^2}}} \tag{20}$$

According to previous research [39, 44], the square root of fracture toughness ($G_{\mathrm{n}}^{\mathrm{c}}$, $G_{\mathrm{t}}^{\mathrm{c}}$), stiffness ($E_0^{\mathrm{n}}$, E_0^{t}), and critical damage initiation traction ($\sigma_{\mathrm{n}}^{\mathrm{c}}$, $\sigma_{\mathrm{t}}^{\mathrm{c}}$) of the intrinsic spring are logarithmically linearly correlated with separation velocity, and the evolution equation of above parameters is written as:

$$\begin{cases} \sqrt{G_{\mathrm{n}}^{\mathrm{c}}} = \gamma_G^{\mathrm{n}}\lg\dot{\delta}_{\mathrm{n}} + b_G^{\mathrm{n}}, & \sqrt{G_{\mathrm{t}}^{\mathrm{c}}} = \gamma_G^{\mathrm{t}}\lg\dot{\delta}_{\mathrm{t}} + b_G^{\mathrm{t}} \\ E_0^{\mathrm{n}} = \gamma_E^{\mathrm{n}}\lg\dot{\delta}_{\mathrm{n}} + b_E^{\mathrm{n}}, & E_0^{\mathrm{t}} = \gamma_E^{\mathrm{t}}\lg\dot{\delta}_{\mathrm{t}} + b_E^{\mathrm{t}} \\ \sigma_{\mathrm{n}}^{\mathrm{c}} = \gamma_T^{\mathrm{n}}\lg\dot{\delta}_{\mathrm{n}} + b_T^{\mathrm{n}}, & \sigma_{\mathrm{t}}^{\mathrm{c}} = \gamma_T^{\mathrm{t}}\lg\dot{\delta}_{\mathrm{t}} + b_T^{\mathrm{t}} \end{cases} \tag{21}$$

where, γ_G^{n}, γ_G^{t}, γ_E^{n}, γ_E^{t}, γ_T^{n}, γ_T^{t}, b_G^{n}, b_G^{t}, b_E^{n}, b_E^{t}, b_T^{n} and b_T^{t} are the fitting coefficients.

Kachanov proposed that the free energy potential is linearly related to the damage variable [45]. Therefore, the expression for the free energy potential considering interface damage can be written as shown in Eq. (22), in which $\delta_{\max}$ is defined as $\delta_{\max} = \max_{\tau \le t} \delta(\tau)$. Note that the mechanical behavior of the interface is not affected by damage due to crack closure when the interface is compressed.

$$\begin{cases} \Psi^{\mathrm{n}} = (1 - D(\delta_{\max}))\left[\frac{1}{2}E_0^{\mathrm{n}}\langle\delta_{\mathrm{n}}\rangle_+^2 + \frac{1}{2}\sum_{i=1}^{n}E_i^{\mathrm{n}}\langle\delta_{\mathrm{n}} - \alpha_i^{\mathrm{n}}\rangle_+^2\right] \\ \qquad + K_c\cdot\left(\frac{1}{2}E_0^{\mathrm{n}}\langle\delta_{\mathrm{n}}\rangle_-^2 + \frac{1}{2}\sum_{i=1}^{n}E_i^{\mathrm{n}}\langle\delta_{\mathrm{n}}\rangle_-^2\right) \\ \Psi^{\mathrm{t}} = (1 - D(\delta_{\max}))\left[\frac{1}{2}E_0^{\mathrm{t}}\delta_{\mathrm{t}}^2 + \frac{1}{2}\sum_{i=1}^{n}E_i^{\mathrm{t}}\left(\delta_{\mathrm{t}} - \alpha_i^{\mathrm{t}}\right)^2\right] \end{cases} \tag{22}$$

According to Eq. (22) and the derivation process in Sect. 2.1, the relationship between interface stress and displacement considering interface damage can be obtained:

$$\begin{cases} \sigma_{\mathrm{n}} = (1 - D(\delta_{\max}))\left[E_0^{\mathrm{n}}\langle\delta_{\mathrm{n}}\rangle_+ + \left\langle\int_-^t \sum_{i=1}^n E_i^{\mathrm{n}} e^{-\frac{t-\tau}{\tau_i^{\mathrm{n}}}} \dot{\delta}_{\mathrm{n}}(\tau)d\tau\right\rangle_+\right] \\ \quad + K_c\left(E_0^{\mathrm{n}}\langle\delta_{\mathrm{n}}\rangle_- + \sum_{i=i}^n E_i^{\mathrm{n}}\langle\delta_{\mathrm{n}}\rangle_-\right) \\ \sigma_{\mathrm{t}} = (1 - D(\delta_{\max}))\left[E_0^{\mathrm{t}}\delta_{\mathrm{t}} + \int_-^t \sum_{i=1}^n E_i^{\mathrm{t}} e^{-\frac{t-\tau}{\tau_i^{\mathrm{t}}}} \dot{\delta}_{\mathrm{t}}(\tau)d\tau\right] \end{cases} \tag{23}$$

2.3 Numerical Implementation

In Sect. 2.2, the expression for the relationship between interfacial stress and displacement, considering interfacial damage, is derived. However, Eq. (23) involves a continuous integral form, which requires discretization to numerically simulate the viscoelastic mechanical behavior of the interface. Assuming that the load is applied at $t = 0$ and ceases at $t = T$, the interval [0, T] is discretized into M time increments:

$$[0, T] = \bigcup_{m=1}^{M} [t_m, t_{m+1}] \quad t_{m+1} = t_m + \Delta t_m \tag{24}$$

Therefore, the dissipative part of σ^n can be written as:

$$\int_-^{t_{m+1}} \sum_{i=1}^n E_i^{\mathrm{n}} e^{-\frac{t-\tau}{\tau_i^{\mathrm{n}}}} \dot{\delta}_{\mathrm{n}}(\tau)d\tau = \int_-^{t_m} \sum_{i=1}^n E_i^{\mathrm{n}} e^{-\frac{t_{m+1}-\tau}{\tau_i^{\mathrm{n}}}} \dot{\delta}_{\mathrm{n}}(\tau)d\tau + \int_{t_m}^{t_{m+1}} \sum_{i=1}^n E_i^{\mathrm{n}} e^{-\frac{t_{m+1}-\tau}{\tau_i^{\mathrm{n}}}} \dot{\delta}_{\mathrm{n}}(\tau)d\tau \tag{25}$$

Let $\chi_i^{\mathrm{n}}(t_m) = \int_-^{t_m} E_i^{\mathrm{n}} e^{-\frac{t_m-\tau}{\tau_i^{\mathrm{n}}}} \dot{\delta}^{\mathrm{n}}(\tau)d\tau$, simplifying the first term of Eq. (25) yields:

$$\int_-^{t_m} \sum_{i=1}^n E_i^{\mathrm{n}} e^{-\frac{t_{m+1}-\tau}{\tau_i^{\mathrm{n}}}} \dot{\delta}_{\mathrm{n}}(\tau)d\tau = \sum_{i=1}^n e^{-\frac{\Delta t_m}{\tau_i^{\mathrm{n}}}} \chi_i^{\mathrm{n}}(t_m) \tag{26}$$

The discrete form of displacement derivative can be written as:

$$\dot{\delta}_{\mathrm{n}} = \frac{\delta_{\mathrm{n}}(t_{m+1}) - \delta_{\mathrm{n}}(t_m)}{\Delta t_m} \tag{27}$$

Bringing Eq. (27) into the second term of Eq. (25) gives:

$$\int_{t_m}^{t_{m+1}} \sum_{i=1}^{n} E_i^{\mathrm{n}} e^{-\frac{t_{m+1}-\tau}{\tau_i^{\mathrm{n}}}} \dot{\delta}_{\mathrm{n}}(\tau) d\tau = \int_{t_m}^{t_{m+1}} \sum_{i=1}^{n} E_i^{\mathrm{n}} e^{-\frac{t_{m+1}-\tau}{\tau_i^{\mathrm{n}}}} \frac{\delta_{\mathrm{n}}(t_{m+1}) - \delta_{\mathrm{n}}(t_m)}{\Delta t_m} d\tau$$
$$= \sum_{i=1}^{n} \tau_i^{\mathrm{n}} E_i^{\mathrm{n}} \left(1 - e^{-\frac{\Delta t_m}{\tau_i^{\mathrm{n}}}}\right) \cdot \frac{(\delta_{\mathrm{n}}(t_{m+1}) - \delta_{\mathrm{n}}(t_m))}{\Delta t_m} \tag{28}$$

The recursive form of σ_{n} at t_{m+1} is obtained from Eqs. (23), (25), (26) and (28):

$$\sigma_{\mathrm{n}}(t_{m+1}) = (1 - D_{m+1}) \Bigg[E_0^{\mathrm{n}} \langle \delta_{\mathrm{n}}(t_{m+1}) \rangle_+$$
$$+ \left\langle \sum_{i=1}^{n} e^{-\frac{\Delta t_m}{\tau_i^{\mathrm{n}}}} \chi_i^{\mathrm{n}}(t_m) + \sum_{i=1}^{n} \tau_i^{\mathrm{n}} E_i^{\mathrm{n}} \left(1 - e^{-\frac{\Delta t_m}{\tau_i^{\mathrm{n}}}}\right) \cdot \frac{(\delta_{\mathrm{n}}(t_{m+1}) - \delta_{\mathrm{n}}(t_m))}{\Delta t_m} \right\rangle_+ \Bigg] \tag{29}$$
$$+ K_c \left(E_0^{\mathrm{n}} \langle \delta_{\mathrm{n}}(t_{m+1}) \rangle_- + \sum_{i=1}^{n} E_i^{\mathrm{n}} \langle \delta_{\mathrm{n}}(t_{m+1}) \rangle_- \right)$$

where, $D_{m+1} = \max(D_m, \min(1, D(\delta(t_{m+1}))))$.

Through the same derivation method, the recursive form of σ_{t} at t_{m+1} can be obtained:

$$\sigma_{\mathrm{t}}(t_{m+1}) = (1 - D_{m+1})$$
$$\times \left[E_0^{\mathrm{t}} \delta_{\mathrm{t}}(t_{m+1}) + \sum_{i=1}^{n} e^{-\frac{\Delta t}{\tau_i^{\mathrm{t}}}} \chi_i^{\mathrm{t}}(t_m) + \sum_{i=1}^{n} \tau_i^{\mathrm{t}} E_i^{\mathrm{t}} \left(1 - e^{-\frac{\Delta t}{\tau_i^{\mathrm{t}}}}\right) \cdot \frac{(\delta_{\mathrm{t}}(t_{m+1}) - \delta_{\mathrm{t}}(t_m))}{\Delta t_m} \right] \tag{30}$$

The VUMAT subroutine in ABAQUS is implemented to simulate the viscoelastic mechanical behavior of the interface considering interface damage. The numerical implementation flowchart is presented in Fig. 4. To compute the stress at the $(i + 1)$th increment step, the subroutine requires the following inputs: material parameter, strain, stress, and damage at the ith increment step, the strain increment and time increment at the $(i + 1)$th incremental step, and the current step time. The strain and the damage are passed in the form of state variables. Within the VUMAT subroutine, the CZM parameters are updated using the input values, and the damage parameters for the $(i + 1)$th increment step are calculated. If the element's damage reaches or exceeds 1, it indicates failure, and the element is deleted. If the damage is less than 1, the stress for the $(i + 1)$th increment step is computed. After updating the state variables, the current increment step is completed. The same process is repeated for subsequent increment steps until the specified time period is reached.

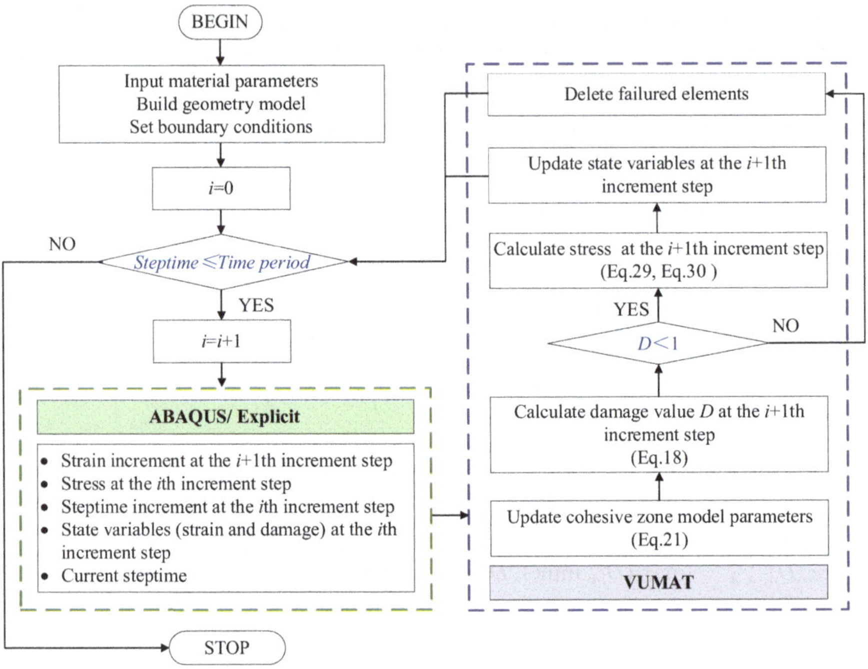

Fig. 4 Flowchart of the numerical implementation

3 Validation of Proposed Interface Mechanical Model

To validate the proposed interface viscoelastic mechanical model, a two-dimensional numerical model of a Double Cantilever Beam (DCB) specimen with a rubber interface is developed using ABAQUS software. This model is used to study the interface's mechanical behavior under different loading rates, and the numerical results are compared with the DCB experiments conducted by Musto and Alfano [11].

3.1 DCB Finite Element Model

The finite element model of the DCB specimen is shown in Fig. 5. The geometric dimensions of the numerical model match those of the experimental setup. The two steel beams, each 200 mm in length and 8 mm in thickness, are bonded with a rubber interface. The interface has a length of 120 mm and a thickness of 1 mm, resulting in an initial crack length of 80 mm in the DCB specimen. In the element setting, the beams are modeled using CPE4R elements (4-node bilinear plane strain quadrilateral elements with reduced integration), and the interface is modeled using COH2D4

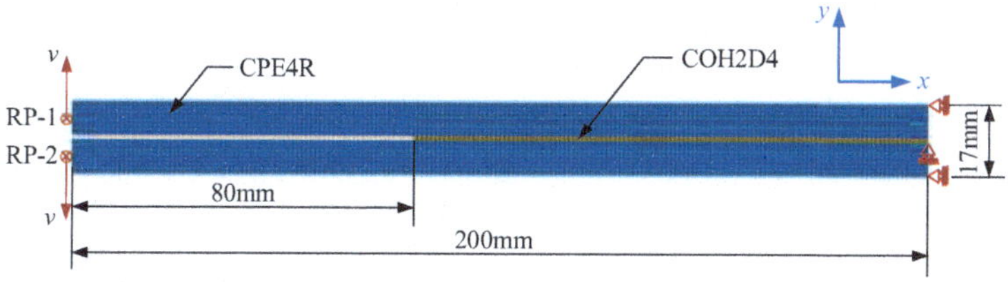

Fig. 5 DCB finite element model

Table 1 The material parameters of the interface

Parameter	Value	Parameter	Value	Parameter	Value
γ_G^n	0.12 mJ s/mm^3	γ_E^n	0.15 MPa s/mm	γ_T^n	0.20 MPa s/mm
b_G^n	2.90 mJ/mm^2	b_E^n	4.9 MPa	b_T^n	3.8 MPa
K_c	10	E_1^n	2.2 MPa	E_2^n	0.5 MPa
τ_1^n	110 s	τ_2^n	390 s		

elements (4-node two-dimensional cohesive elements). Boundary conditions include constraining the right side of the beam in the x-direction and the right side of the interface in the y-direction. Reference Points 1 and 2 are created and coupled to the left ends of the two beams, where displacements with different velocities are applied in the y-direction.

In the model, the Young's modulus and Poisson's ratio of the steel beam are set to 210 GPa and 0.3, respectively. The viscoelastic mechanical behavior of the interface is implemented by invoking the VUMAT subroutine in ABAQUS. Since the DCB analysis involves only normal loading on the interface, only the normal material parameters of the interface are specified. The material parameters of the interface include intrinsic spring parameters (γ_G^n, γ_E^n, γ_T^n, b_G^n, b_E^n, b_T^n, K_c) and Maxwell unit parameters (E_1^n, τ_1^n, E_2^n, τ_2^n). These values are summarized in Table 1. They are derived based on prior studies [39, 45, 46] and determined through a sensitivity analysis of each parameter combined with iterative trial-and-error adjustments to fit the DCB experimental data reported by Musto and Alfano [11].

Taking a loading rate of 10 mm/min as an example, the sensitivity of the model input parameters is illustrated using b_E^n and b_T^n as representative variables. Notably, when analyzing the sensitivity of a given parameter, only that parameter is varied while keeping all other parameters constant. Figure 6a presents the load-displacement curves under varying b_E^n values, demonstrating that as b_E^n increases, the stiffness of the curve gradually increases, and the entire curve shifts forward. This is because the larger the b_E^n, the higher the interface stiffness at the same loading rate. Similarly, Fig. 6b displays the load-displacement curves for different b_T^n values, from which it can be seen that with the increase of b_T^n, the maximum load and damage initiation displacement both increase correspondingly. This is because the larger the b_T^n, the higher the interface strength at the same loading rate.

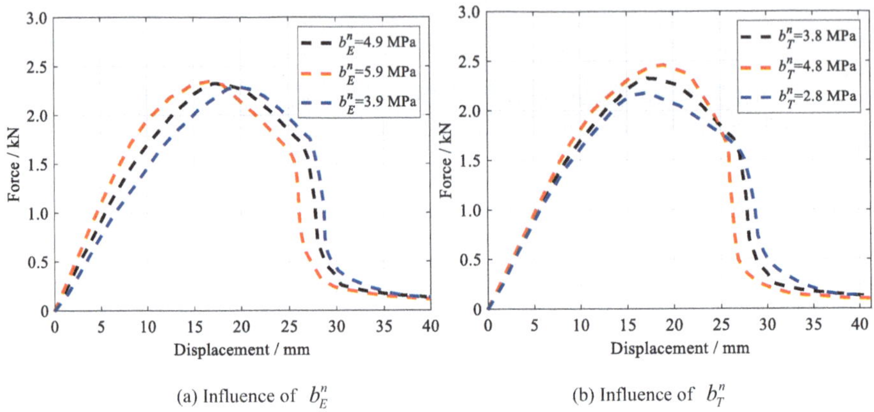

Fig. 6 Load displacement curves with different parameters. (**a**) Influence of b_E^n; (**b**) Influence of b_T^n

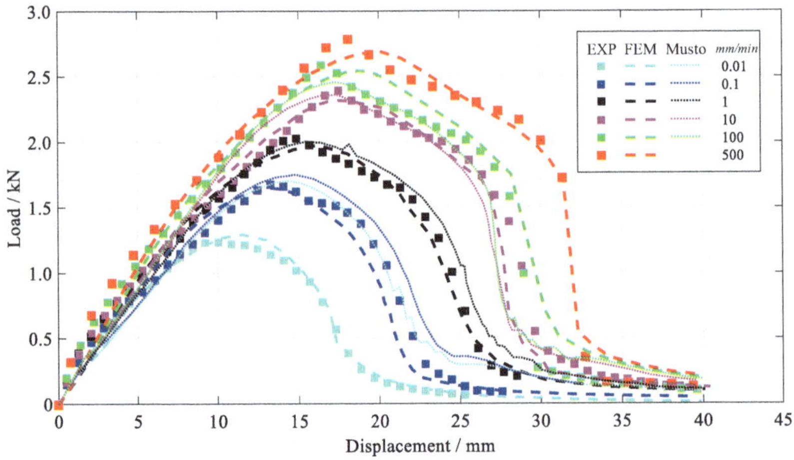

Fig. 7 Comparison between experimental results and finite element analysis results

3.2 *Results and Discussions*

The comparison between the results obtained from the numerical model developed in this study and the experimental and numerical data reported by Musto and Alfano [11] is illustrated in Figs. 7 and 8. As demonstrated in Fig. 7, the load-displacement curve is notably affected by the loading velocity. Figure 8 reveals that the relationship between the maximum load and the loading rate is approximately linear on a semi-logarithmic scale. Table 2 presents the errors in the maximum load between the experiment and numerical model at different loading rates. As shown in this table, the maximum error between the results of the numerical model established in this study and the experimental results is 4.84%, while the maximum error between the

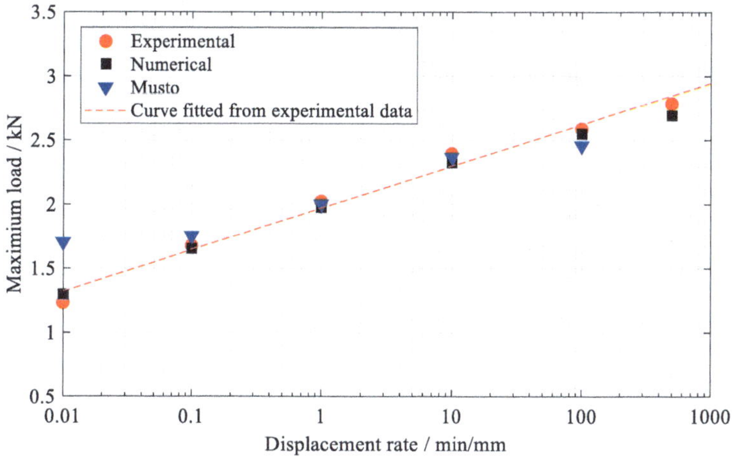

Fig. 8 Experimental and numerical analysis of the maximum load at different loading rates

Table 2 Error of maximum load at different loading rates

Loading rate (mm/min)	Maximum load (Experiment) (kN)	Maximum load (FEM) (kN)	Error (FEM) (%)	Maximum load (Musto) (kN)	Error (Musto) (%)
0.01	1.24	1.30	4.84	1.71	37.83
0.1	1.68	1.65	1.60	1.76	4.52
1	2.03	1.98	2.49	2.00	1.14
10	2.40	2.33	3.07	2.37	1.29
100	2.59	2.55	1.59	2.46	5.02
500	2.79	2.69	3.48	\	\

results of Musto's numerical simulation and the experimental results is 37.83%. The results of the numerical model proposed in this study exhibit good agreement with the experimental results in terms of stiffness, strength, and toughness, demonstrating the validity of the interface viscoelastic mechanical model that accounts for damage based on the generalized Maxwell model and the rate-dependent CZM.

In contrast, the model proposed by Musto and Alfano [11] fails to accurately describe the mechanical behavior of the interface under faster loading speeds (100 mm/min) and slower loading speeds (0.01 mm/min). This limitation arises because their model accounts for the rate dependence only through the dissipative component, without considering the rate-dependent nature of intrinsic strain energy. Specifically, their model assumes that the interface damage initiation separation and critical failure separation are invariant with the loading rate. Moreover, the standard linear model (SLM) used in their study does not sufficiently capture the complex mechanical behavior of the interface. The interfacial mechanical model proposed in this study addresses these limitations by incorporating both the rate dependence of the intrinsic spring and the viscous dissipation effects of the multiple Maxwell units.

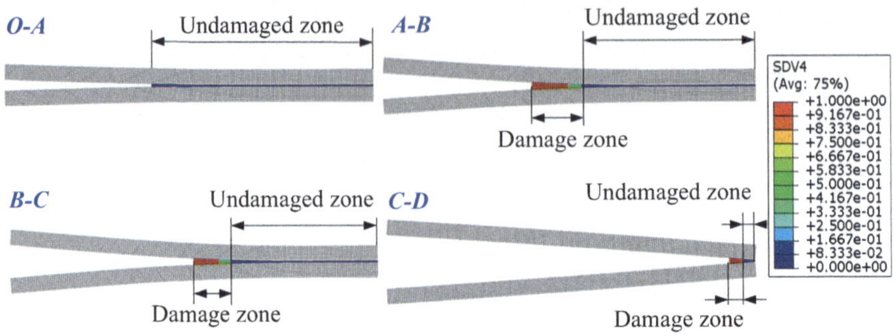

(a) Interface damage nephogram for a loading speed of 10 mm/min

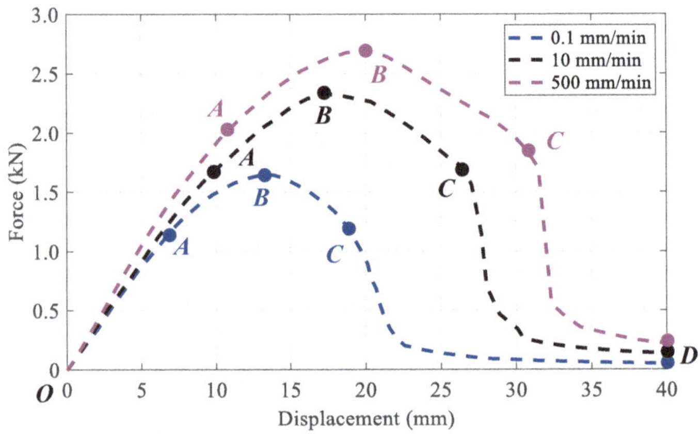

(b) Force-displacement curve

Fig. 9 The debonding process of the DCB specimen. (**a**) Interface damage nephogram for a loading speed of 10 mm/min. (**b**) Force-displacement curve

This approach enables a more accurate description of the interfacial mechanical behavior across a wide range of loading rates, including extreme conditions.

The debonding process of DCB specimens is categorized into four stages: the undamaged stage (O-A section), the damage initiation stage (A-B section), the stable crack extension stage (B-C section), and the rapid crack extension stage (C-D section), as shown in Fig. 9. The interfacial damage contour of the DCB specimen under a loading speed of 10 mm/min are used to illustrate the debonding process. The variable SDV4 in Fig. 9a represents the fourth state variable in the subroutine, which indicates the damage degree of the interface, where a value of 1 signifies complete interface failure, and 0 represents no damage. The O-A section is the undamaged stage, in which the interface begins to deform with the application of load, but the deformation of the interface does not reach the damage initiation separation, and no damage occurs to the interface. The A-B section is the damage initiation stage, in which the interface deformation exceeds the damage initiation separation, resulting in the onset of damage. The damaged zone starts to grow gradually. At point B, interface failure begins, and both the reaction force, and the length of the damage zone reach their maximum value. The B-C section is the stable crack extension stage, in which the length of the damage zone remains constant, and the crack expands at a steady rate. The C-D section is the rapid crack extension stage, in which the distance between the region of the interface subjected to tensile load and the region subjected to compressive load gradually decreases. This leads to a rapid decrease in the combined moment, which subsequently causes a rapid reduction in the reaction force. As a result, the slope of the load-displacement curve is very steep. In this stage, the overall strength of the DCB specimen sharply declines, and finally the structure undergoes debonding failure.

From Fig. 9b, it can be observed that as the loading rate increases, point A rises, indicating that a greater displacement is required to initiate interface damage. Higher loading rates result in an increased ultimate load capacity of the DCB specimen (point B rises), signifying that the interface strength gradually increases with the loading rate. Furthermore, as the loading rate increases, a larger displacement is required for the interface to enter the rapid crack extension stage, indicating that the fracture toughness of the interface also increases with the loading rate, which aligns with the findings of Chaudhury's work [44].

4 Conclusions

Marine composite flexible pipes often fail due to sudden application of large loads with high loading rates. The interface represents the weakest link in composite flexible pipes. However, existing studies on the ultimate load-bearing capacity of such pipes generally neglect the rate dependency of the interface. To more accurately investigate the ultimate load-bearing characteristics of marine composite flexible pipes using numerical methods, this study introduces a viscoelastic interface mechanical model incorporating damage.

The proposed model is built upon a modified rate-dependent cohesive zone model (CZM) and the generalized Maxwell model, accounting for both the rate dependence of the intrinsic spring and the viscous dissipation of multiple Maxwell units. A VUMAT subroutine is developed based on this interface mechanical model. A numerical model of a double cantilever beam (DCB) specimen is established in ABAQUS, utilizing the VUMAT subroutine to explore the interface's mechanical behavior under various loading rates. By taking b_E and b_T as examples, the influence of different parameters on the interface's mechanical behavior is analyzed. The results of the numerical model exhibit good agreement with the experimental results in terms of stiffness, strength, and toughness over a wide range of loading rates (0.01–500 mm/min), and the maximum error between the results of the numerical model and the experimental results is 4.84%, which proves the accuracy of the viscoelastic interface mechanical model proposed in this paper. The debonding process of DCB specimens under different loading rates is analyzed, and the debonding process is divided into four stages: the undamaged stage, the damage initiation stage, the stable crack extension stage, and the rapid crack extension stage.

The model proposed in this study can effectively simulate the interfacial mechanical behavior over a wide range of loading rates. In the subsequent numerical analysis of the ultimate bearing capacity of composite pipes, the subroutine developed based on this model can be employed to simulate the interfacial behavior, enabling the calculation of the ultimate bearing capacity under different loading rates. This allows for a more accurate assessment of the load-bearing performance of composite pipes. However, the current model requires numerous input parameters, making the parameter calibration process relatively cumbersome. Therefore, future research should focus on developing interfacial mechanical models that demand fewer parameters while offering improved simulation accuracy.

References

1. Gong JL, Ma K, Zhang J, Zhang Z, Bian C, Zhu G, Yan N. Experiment study of failure mode of HTPB/AP interfacial debonding. In: 2021 12th international conference on mechanical and aerospace engineering (ICMAE). IEEE; 2021. p. 1–6.
2. Jones SC, Civjan SA. Application of fiber reinforced polymer overlays to extend steel fatigue life. J Compos Constr. 2003;7(4):331–8.
3. Zhang Y, Zhang P, Yang P, Yao S, Wang F. The analysis of debonding damage between steel tension wire and epoxy resin in flexible end fitting under cyclic loading. Appl Ocean Res. 2023;137:103606.
4. Liu Q, Qu Z, Chen F, Liu X, Wang G. Review of the development of an unbonded flexible riser: new material, types of layers, and cross-sectional mechanical properties. Materials. 2024;17 (11):2560.
5. Wei D, An C, Zhang J, Huang Y, Gu C. Effect of winding steel wire on the collapse pressure of submarine hose. J Mar Sci Eng. 2022;10(10):1365.
6. Yan S, Jia K, Xu W, Ma Y. An experimental study on vortex-induced vibration suppression for submarine multispan pipelines. Ocean Eng. 2023;271:113678.
7. Chen NZ, Zhang J, Feng A, Ma Y. Experimental study on vibration responses of flexible riser transporting spiral flow in deep sea mining: part I-liquid single-phase transportation. Ocean Eng. 2024;298:117068.
8. Fernando US, Karabelas G. Analysis of end fitting barrier seal performance in high pressure unbonded flexible pipes. In: International conference on offshore mechanics and arctic engineering, vol. 45462. American Society of Mechanical Engineers; 2014. p. V06AT04A038.
9. Wei D, An C, Zhang Z, Zhang J, Zhang A. Collapse response of a single carcass offshore rubber hose under bending moment. In: ISOPE international ocean and polar engineering conference. ISOPE; 2022. p. ISOPE-I.
10. Chen B, Zhang Y, Wang Q, Zhang H, Wang Y. The signal analysis of lamb wave in steel plates strengthened by CFRP with interface debonding. Ocean Eng. 2023;283:115061.
11. Musto M, Alfano G. A novel rate-dependent cohesive-zone model combining damage and visco-elasticity. Comput Struct. 2013;118:126–33.
12. Majda P, Skrodzewicz J. A modified creep model of epoxy adhesive at ambient temperature. Int J Adhes Adhes. 2009;29(4):396–404.
13. Liu D, Ai S, Sun L, Soares CG. Vortex-induced vibrations of catenary risers in varied flow angles. Int J Mech Sci. 2024;269:109086.

14. Rocha IBCM, Van Der Meer FP, Raijmaekers S, Lahuerta F, Nijssen RPL, Sluys LJ. Numerical/experimental study of the monotonic and cyclic viscoelastic/viscoplastic/fracture behavior of an epoxy resin. Int J Solids Struct. 2019;168:153–65.
15. Musto M, Alfano G. A fractional rate-dependent cohesive-zone model. Int J Numer Methods Eng. 2015;103(5):313–41.
16. Lassen T, Eide AL, Meling TS. Ultimate strength and fatigue durability of steel reinforced rubber loading hoses. In: International conference on offshore mechanics and Arctic engineering, vol. 49132; 2010. p. 277–86.
17. Lassen T, Lem AI, Imingen G. Load response and finite element modelling of bonded loading hoses. In: International conference on offshore mechanics and arctic engineering, vol. 45462. American Society of Mechanical Engineers; 2014. p. V06AT04A034.
18. Wei D, An C, Wu C, Duan M, Estefen SF. Torsional structural behavior of composite rubber hose for offshore applications. Appl Ocean Res. 2022;128:103333.
19. Wei D, An C, Zhang Z, Xu Z, Gao Q, Li Y. Study on damage mechanism and debonding of adhesive surface of floating hose joint. China Offshore Oil Gas. 2023;35(4):173–80.
20. Krueger R. Virtual crack closure technique: history, approach, and applications. Appl Mech Rev. 2004;57(2):109–43.
21. Tay T. Characterization and analysis of delamination fracture in composites: an overview of developments from 1990 to 2001. Appl Mech Rev. 2003;56(1):1–32.
22. Belytschko T, Black T. Elastic crack growth in finite elements with minimal remeshing. Int J Numer Methods Eng. 1999;45(5):601–20.
23. Dugdale DS. Yielding of steel sheets containing slits. J Mech Phys Solids. 1960;8(2):100–4.
24. Hillerborg A, Modéer M, Petersson PE. Analysis of crack formation and crack growth in concrete by means of fracture mechanics and finite elements. Cem Concr Res. 1976;6(6): 773–81.
25. Barenblatt GI. The mathematical theory of equilibrium cracks in brittle fracture. In: Adv Appl Mech, vol. 7; 1962. p. 55–129.
26. Needleman A. A continuum model for void nucleation by inclusion debonding. J Appl Mech. 1987;54(3):525–31.
27. Tvergaard V, Hutchinson JW. The relation between crack growth resistance and fracture process parameters in elastic-plastic solids. J Mech Phys Solids. 1992;40(6):1377–97.
28. Camacho GT, Ortiz M. Computational modelling of impact damage in brittle materials. Int J Solids Struct. 1996;33(20–22):2899–938.
29. Geubelle PH, Baylor JS. Impact-induced delamination of composites: a 2D simulation. Compos Part B. 1998;29(5):589–602.
30. Mi Y, Crisfield MA, Davies GAO, Hellweg HB. Progressive delamination using interface elements. J Compos Mater. 1998;32(14):1246–72.
31. Camanho PP, Davila CG, De Moura MF. Numerical simulation of mixed-mode progressive delamination in composite materials. J Compos Mater. 2003;37(16):1415–38.
32. Kumar RS. Crack-growth resistance behavior of mode-I delamination in ceramic matrix composites. Acta Mater. 2017;131:511–22.
33. Li X, Liang L, Xie J, Chen L, Wei Y. Thickness-dependent fracture characteristics of ceramic coatings bonded on the alloy substrates. Surf Coat Technol. 2014;258:1039–47.
34. Song SH, Paulino GH, Buttlar WG. A bilinear cohesive zone model tailored for fracture of asphalt concrete considering viscoelastic bulk material. Eng Fract Mech. 2006;73(18):2829–48.
35. Zhou QC, Ju YT, Wei Z, Han B, Zhou CS. Cohesive zone modeling of propellant and insulation interface debonding. J Adhes. 2014;90(3):230–51.
36. Marzi S, Hesebeck O, Brede M, Kleiner F. A rate-dependent cohesive zone model for adhesively bonded joints loaded in mode I. J Adhes Sci Technol. 2009;23(6):881–98.
37. Valoroso N, Debruyne G, Laverne J. A cohesive zone model with rate-sensitivity for fast crack propagation. Mech Res Commun. 2014;58:82–7.

38. May M, Hesebeck O, Marzi S, Böhme W, Lienhard J, Kilchert S, et al. Rate dependent behavior of crash-optimized adhesives–experimental characterization, model development, and simulation. Eng Fract Mech. 2015;133:112–37.
39. Zhao G, Xu J, Feng Y, Tang J, Chen Y, Xin S, et al. A rate-dependent cohesive zone model with the effects of interfacial viscoelasticity and progressive damage. Eng Fract Mech. 2021;248: 107695.
40. Yang Z, Zhu Z, Xia Y, Yang F, Sun Y, Jiang H. Modified cohesive zone model for soft adhesive layer considering rate dependence of intrinsic fracture energy. Eng Fract Mech. 2021;258: 108089.
41. Zhou W, Tang L, Liu X, Ma G, Chen M. Mesoscopic simulation of the dynamic tensile behaviour of concrete based on a rate-dependent cohesive model. Int J Impact Eng. 2016;95: 165–75.
42. Destrade M, Gilchrist MD, Motherway JA, Murphy JG. Bimodular rubber buckles early in bending. Mech Mater. 2010;42(4):469–76.
43. Coleman BD, Gurtin ME. Thermodynamics with internal state variables. J Chem Phys. 1967;47 (2):597–613.
44. Chaudhury MK. Rate-dependent fracture at adhesive interface. J Phys Chem B. 1999;103(31): 6562–6.
45. Kachanov LM. Time to failure under creep condition. Izvestia Akademii Nauk SSSR Tech Nauk. 1958;8:26–31.
46. Giraldo-Londoño O, Spring DW, Paulino GH, Buttlar WG. An efficient mixed-mode rate-dependent cohesive fracture model using sigmoidal functions. Eng Fract Mech. 2018;192:307–27.

Stress Analysis of Adhesive in a Cracked Steel Plate Repaired with CFRP

Nomenclature

a	Crack length
b	Width of steel-plate
F	Longitudinal load
L_a	Length of adhesive
L_f	Length of CFRP
L_s	Length of steel-plate
M_s, M_f	Bending moment of steel-plate and CFRP, respectively
N_s, N_f	Longitudinal tension of steel-plate and CFRP, respectively
Q_s, Q_f	Shear force of steel-plate and CFRP, respectively
t_a	Thickness of adhesive
t_f	Thickness of CFRP
t_s	Thickness of steel-plate
u_f	Total displacements at the interfaces of adhesive-CFRP
u_s	Total displacements at the interfaces of adhesive-steel
u_{2f}^M	Displacements caused by moment at the surface of adhesive-CFRP
u_{1s}^M	Displacements caused by moment at the surface of adhesive-steel
u_f^T	Longitudinal displacement functions for CFRP
u_s^T	Longitudinal displacement functions for steel-plate
u_{1f}^T	Displacement at the top surface of CFRP
u_{1s}^T	Displacement at the surface of adhesion-steel
u_{2f}^T	Displacement at the surface of adhesive-CFRP
γ_a	Adhesive shear strain
σ	Peel stress
σ_f	Longitudinal stresses at CFRP
σ_s	Longitudinal stresses at the steel-plate and CFRP
$\sigma_{FE-\max}^b$	Maximum peel stresses obtained by FE method
$\sigma_{\max}^b$	Maximum peel stresses obtained by the theoretical method
τ	Shear stress

Y. Zhang, *Interfacial Mechanics of Offshore Engineering Equipment*,
https://doi.org/10.1007/978-981-95-3200-1_8

τ_f	Shear stress at the CFRP layer
τ_s	Shear stress at the steel-plate
$\tau^b_{FE-\max}$	Maximum shear stresses obtained by FE method
$\tau^b_{\max}$	Maximum shear stresses obtained by the theoretical method

1 Introduction

Carbon fiber reinforced polymer (CFRP) patch-repair technology has been used to strengthen steel structures and repair the surface defects of aircraft [1, 2]. This technology has also been used to repair the surface defects of petroleum pipelines and offshore platforms due to its many advantages, such as requiring no welding and allowing for quick repairs [3–7]. Compared with traditional repair technology like welding and bolt fastening, CFRP patch-repair technology does not destroy the destruction of original structure and stress concentration. CFRP repair also offers corrosion-resistance and better fatigue properties than other methods [8].

There are many rules regarding the repair of pipeline defects in the oil and gas industry, including ASME PCC-2-2015, DNV-RP-F113, and ISO 24817-2015 [9–11]. These rules offer material parameters for CFRP used to repair pipeline defects like cracks and corrosion, including formulas for the necessary thickness and length of the CFRP layer. However, the initiation and propagation of fatigue cracks often occur in the service period of offshore platforms and boats, and no guidelines or rules existed for their repair until recently [12].

Many researchers have studied the model of a cracked plate repaired by CFRP or composite patches. Tsouvalis and Mirisiotis [13] analyzed patch strain and crack propagation rates under cyclic loading, using experimental and FE models to study the repair of cracked steel-plate. Albedah and Khan [14] confirmed that composite patches increase the fatigue life of cracked structures, but the level of improvement depends strongly on the adhesive layer, and adhesives greatly decrease the fatigue life of repaired structures. Nateche and Meliani [15] created a procedure to determine a patch-repairing index to measure the residual harm of a crack-like defect after repair. Okafor and Singh [16] performed linear and nonlinear static finite element analysis (FEA) of bonded composite patch repair, and the results indicated that the region of maximum skin stress shifts from the crack front to the patch edges. Klug and Sun [17] used large deflection theory to study the effect of different variables, including thermal residual stresses and host and repair plate thickness, on the stress of a single-sided repair model. Naboulsi and Mall [18, 19] introduced an analytical procedure to characterize the stress intensity factor (SIF) in a panel repaired with a bonded composite patch. Rezgani and Madani [20] proved that repair by patch composite increases the rigidity of the damaged structure and retards crack propagation. Ramji and Srilakshmi [21] used six 3D FEMs to obtain an optimum composite patch shape for application to an inclined cracked panel. Chung [22] used experimental and FEM models to study the optimal patch length, whose results

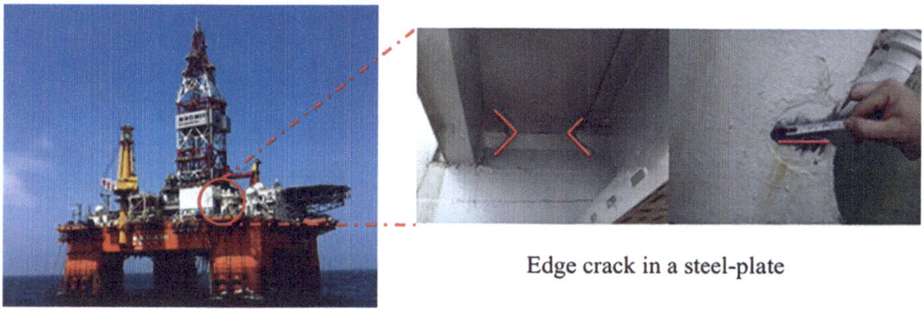

Fig. 1 Edge crack at an offshore platform

indicated that performance is better when the patch length is 1.5 times the panel length. Osnes [23] used experimental and theoretical methods to predict the failure loads of adhesively bonded lap-shear joints.

The distribution of the interface stress in cracked steel-plate repaired with CFRP can be similarly analyzed by a model of a double- or single-bonded joint. Tsai [24] proposed theoretical solutions for adhesively bonded lap joints based on the assumption of linear shear stress distributions through the thickness of the adherends. Zhang [25] analyzed the stresses of an adhesive joint under tension when taking the thickness of the adhesive layer into account. Deng [26] analyzed the interface stress of a beam reinforced with CFRP, and two conditions were considered in the theoretical and numerical calculation. Miller et al. [27] performed experimental and analytical studies to quantify the force transfer between CFRP and a steel substrate. The results showed that approximately 98% of the total force transfer occurs within the first 100 mm from the end of the CFRP plate. M. Ramji et al. [28] used three approaches to investigate the mechanical behavior of a beam reinforced with CFRP. Their research showed that increasing the reinforcement length could increase the failure loads in the repaired model as the interfacial stress peak was reduced.

This paper analyzes the stress distribution of edge cracks in steel-plates repaired with CFRP (the cracked steel-plate in a platform can be seen in Fig. 1). The theoretical method and FEM analyses were used to calculate the stresses in the repaired model under some rational assumptions, with good consistency. The relationships of the maximum shear stress and peel stress with the crack length were obtained, and these can be used to predict the adhesive failure in the repair model based on quadratic stress criteria.

2 Theoretical Analysis of a Cracked Steel-Plate Repaired by CFRP

The right of Fig. 1 shows a steel-plate in a platform with an edge crack. The CFRP repair model can be assumed to be shown in Fig. 2a, the CFRP double-patch model of a steel plate with an edge crack, which consists of steel-plate, adhesive, and

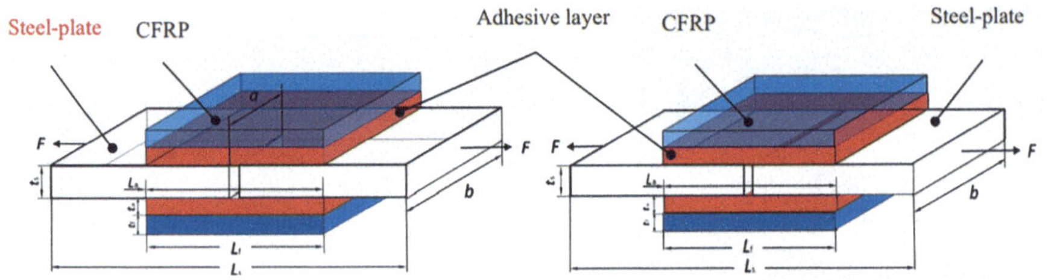

(a) Steel-plate with edge crack repaired by CFRP (b) The model with crack running through the steel plate

Fig. 2 A cracked steel plate repaired with CFRP. (**a**) Steel-plate with edge crack repaired by CFRP. (**b**) The model with crack running through the steel plate

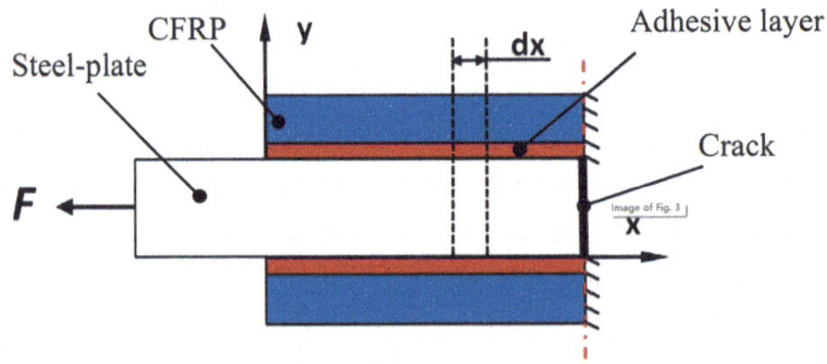

Fig. 3 2D model of half of a repaired steel plate

CFRP. The longitudinal tension force at the end of the steel-plate will partly be transferred from the adhesive layer to the CFRP layer.

In this paper, the limiting case when the crack runs through the steel-plate (as shown in Fig. 2b) was studied first with the theoretical method. The following assumptions were made to simplify the stress analysis in the CFRP double-patch model: (1) The shear and peel stresses in the adhesive layer do not vary through the thickness of the adhesive layer as it is very small. The longitudinal tension of the adhesive layer is neglected. (2) The shear stress in the steel and CFRP layer varies linearly with thickness. (3) The displacement function at the steel-adhesive, and CFRP-adhesive interfaces is continuous. As the 3D model (Fig. 2b) was symmetric along the longitudinal direction, it was simplified to the 2D model shown in Fig. 3.

Figure 4 shows an infinitesimal element of the 2D model. In the figure, τ and σ are respectively the shear stress and peel stress at the adhesion layer, and N, Q, and M are the longitudinal tension, shear force, and bending moment, respectively, at the CFRP and steel-plate.

Based on Fig. 4, the longitudinal, transverse, and moment equilibria can be respectively described as:

$$\frac{dN_s}{dx} - 2\tau b = 0, \quad \frac{dN_f}{dx} + \tau b = 0 \tag{1}$$

$$dQ_s = 0, \quad \frac{dQ_f}{dx} - \sigma b = 0 \tag{2}$$

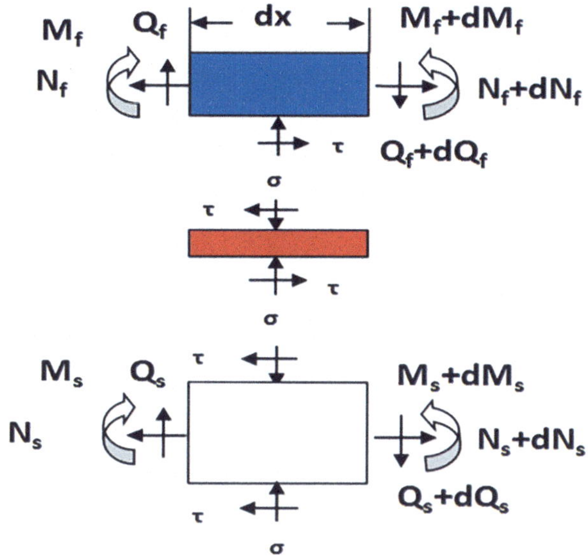

Fig. 4 The infinitesimal element

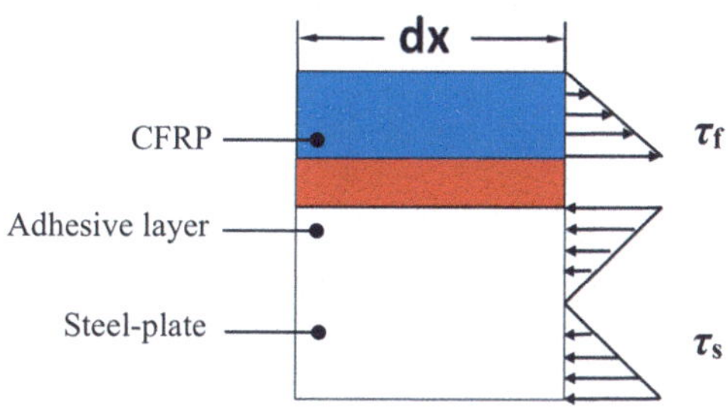

Fig. 5 The assumption of a linear shear stress distribution

$$\frac{dM_s}{dx} - Q_s = 0, \quad \frac{dM_f}{dx} - Q_f + \tau b \frac{t_f}{2} = 0 \tag{3}$$

The longitudinal force equilibrium of the repaired model can be described as:

$$F = N_s + 2N_f \tag{4}$$

2.1 Shear Stress Distribution in the CFRP Layer and Steel-Plate

Based on Fig. 5, the linear distribution of shear stress at the CFRP layer and steel-plate can be described by Eqs. (5a) and (5b). τ_s and τ_f are the shear stress at the steel-plate and CFRP layer, respectively. The shear stress in the free surface of the CFRP layer and the symmetry plane of the steel-plate is zero, while the shear stress in the interface of adhesive-CFRP and adhesive-steel plate is τ.

$$\tau_f = \tau \frac{y_1}{t_f}, \quad 0 \le y_1 \le t_f \tag{5a}$$

$$\tau_s = \tau \left| 1 - \frac{y_2}{t_s/2} \right|, \quad 0 \le y_2 \le t_s \tag{5b}$$

where τ is the shear stress at the adhesive layer, and y_1 and y_2 are the distance from respectively the top surface of the CFRP and steel. The longitudinal displacement functions u_f^T for the CFRP and u_s^T for the steel are defined as:

$$u_f^T(y_1) = u_{1f}^T + \int_0^{y_1} \gamma_f(y_1{}')dy_1{}' = u_{1f}^T + \frac{\tau y_1^2}{2G_f t_f} \tag{6a}$$

$$u_s^T(y_2) = u_{1s}^T + \int_0^{y_2} \gamma_s(y_2{}')dy_2{}' = u_{1s}^T + \frac{\tau}{G_s}\left(y_2 - \frac{y_2^2}{t_s}\right) \tag{6b}$$

where u_{1f}^T is the displacement at the top surface of CFRP, and u_{1s}^T is the displacement at the adhesive-steel interface. From assumption 3, the displacement at the adhesive-CFRP interface can be described as:

$$u_f^T(y_1 = t_f) = u_{1f}^T + \frac{\tau t_f}{2G_f} = u_{2f}^T \tag{7}$$

where u_{2f}^T is the displacement at the adhesive-CFRP interface. It can be rewritten as:

$$u_{1f}^T = u_{2f}^T - \frac{\tau t_f}{2G_f} \tag{8}$$

Substituting Eq. (8) into Eq. (6a), we can write the longitudinal displacement function of CFRP as:

$$u_f^T(y_1) = u_{2f}^T - \frac{\tau t_f}{2G_f} + \frac{\tau y_1^2}{2G_f t_f} \tag{9}$$

The longitudinal resultant forces N_s and N_f can be described as:

$$N_s/2 = \int_0^{t_s/2} \sigma_s b dy_2 \tag{10a}$$

$$N_f = \int_0^{t_f} \sigma_f b dy_1 \tag{10b}$$

where σ_s and σ_f are the longitudinal stresses at the steel-plate and CFRP, respectively. Differentiating Eqs. (6b) and (9) with respect to x and substituting them into Eqs. (10a) and (10b), respectively, we obtain:

$$N_s/2 = \int_0^{t_s/2} E_s b \frac{du_s^T}{dx} dy_2 = E_s t_s b \left(\frac{du_{1s}^T}{dx} + \frac{t_s}{6G_s} \frac{d\tau}{dx} \right) /2 \tag{11a}$$

$$N_f = \int_0^{t_f} E_f b \frac{du_f^T}{dx} dy_1 = E_f t_f b \left(\frac{du_2^T}{dx} - \frac{t_f}{3G_f} \frac{d\tau}{dx} \right) \tag{11b}$$

Then, Eqs. (11a) and (11b) can be rewritten as:

$$\frac{du_{1s}^T}{dx} = \frac{N_s}{E_s t_s b} - \frac{t_s}{6G_s} \frac{d\tau}{dx} \tag{12a}$$

$$\frac{du_{2f}^T}{dx} = \frac{N_f}{E_f t_f b} + \frac{t_f}{3G_f} \frac{d\tau}{dx} \tag{12b}$$

2.2 Shear Stress Distributions of Adhesive Layer

The adhesive shear strain γ_a is defined as:

$$\gamma_a = \left(u_f - u_s\right)/t_a \tag{13}$$

Differentiating Eq. (13) with respect to x gives:

$$-\frac{t_a}{G_a} \frac{d\tau}{dx} = \frac{du_f}{dx} - \frac{du_s}{dx} \tag{14}$$

where, u_f and u_s are respectively the total displacements at the adhesive-CFRP and adhesive-steel interfaces,

$$\frac{du_f}{dx} = \frac{du_{2f}^T}{dx} + \frac{du_{2f}^M}{dx} \tag{15a}$$

$$\frac{du_s}{dx} = \frac{du_{1s}^T}{dx} - \frac{du_{1s}^M}{dx} \tag{15b}$$

where, u_{2f}^M and u_{1s}^M are the respective displacements caused by moments at the adhesive–CFRP and adhesive–steel plate interfaces,

$$\frac{du_{2f}^M}{dx} = \frac{M_f}{E_f I_f} \frac{t_f}{2} \tag{16a}$$

$$\frac{du_{1s}^M}{dx} = \frac{M_s}{E_s I_s} \frac{t_s}{2} \tag{16b}$$

Substituting Eqs. (12a), (12b), (16a) and (16b) respectively into Eq. (15a) and (15b) gives

$$\frac{du_f}{dx} = \frac{N_f}{E_f t_f b} + \frac{t_f}{3G_f} \frac{d\tau}{dx} + \frac{M_f}{E_f I_f} \frac{t_f}{2} \tag{17a}$$

$$\frac{du_s}{dx} = \frac{N_s}{E_s t_s b} - \frac{t_s}{6G_s} \frac{d\tau}{dx} - \frac{M_s}{E_s I_s} \frac{t_s}{2} \tag{17b}$$

Substituting Eq. (17a), (17b), (5a) and (5b) into Eq. (14) gives

$$\left(\frac{t_a}{G_a} + \frac{t_s}{6G_s} + \frac{t_f}{3G_f}\right) \frac{d\tau}{dx} = -\left(\frac{N_f}{E_f A_f} - \frac{F - 2N_f}{E_s A_s} + \frac{M_f}{E_f I_f} \frac{t_f}{2} + \frac{M_s}{E_s I_s} \frac{t_s}{2}\right) \tag{18}$$

Differentiating Eq. (1) with respect to x and substituting it into Eq. (18) gives the governing equation of the repaired model:

$$\left(\frac{t_a}{G_a} + \frac{t_s}{6G_s} + \frac{t_f}{3G_f}\right) \frac{d^2 N_f}{dx^2} = \left(\frac{2}{E_s t_s} + \frac{1}{E_f t_f}\right) N_f - \frac{F}{E_s t_s} + \frac{bM_f}{E_f I_f} \frac{t_f}{2} + \frac{bM_s}{E_s I_s} \frac{t_s}{2} \tag{19}$$

Equation (19) can be rewritten as:

$$A_1 \frac{d^2 N_f}{dx^2} = B_1 N_f - C_1 \tag{20}$$

where,

$$A_1 = \frac{t_a}{G_a} + \frac{t_s}{6G_s} + \frac{t_f}{3G_f}$$

$$B_1 = \frac{2}{E_s t_s} + \frac{1}{E_f t_f}$$

$$C_1 = \frac{F}{E_s t_s} - \frac{bM_f t_f}{E_f I_f 2} - \frac{bM_s t_s}{E_s I_s 2}$$

The general solution of Eq. (20) is in the form:

$$N_f(x) = H_1 \cosh(\omega x) + H_2 \sinh(\omega x) + C_1/B_1 \tag{21}$$

where, $\omega = \sqrt{B_1/A_1}$

H_1 and H_2 are constant coefficients determined from the boundary conditions. The boundary conditions for the shear stress of the adhesive layer are:

$$N_\mathrm{f}(0) = M_\mathrm{f}(0) = M_\mathrm{s}(0) = 0 \tag{22a}$$

$$N_\mathrm{f}(L_\mathrm{f}) = F/2, \ \ M_\mathrm{f}(L_\mathrm{f}) = M_\mathrm{s}(L_\mathrm{f}) = 0 \tag{22b}$$

H_1 and H_2 can be obtained using:

$$H_1 = -\frac{C_1}{B_1} \tag{23a}$$

$$H_2 = \frac{\frac{F}{2} + \frac{C_1}{B_1}(\cosh(\omega L_\mathrm{f}) - 1)}{\sinh(\omega L_\mathrm{f})} \tag{23b}$$

Differentiating Eq. (21) with respect to x and substituting Eq. (1) into Eq. (21) gives the governing equation of shear stress:

$$\tau(x) = -(\omega H_1 \sinh(\omega x) + \omega H_2 \cosh(\omega x))/b \tag{24}$$

2.3 Peel Stress Distributions of Adhesive Layer

The peel stress at the adhesive layer is defined as:

$$\sigma = E_a(\nu_s - \nu_f)/t_a \tag{25}$$

where,

$$\frac{d^2\nu_f}{dx^2} = \frac{M_f}{E_f I_f}$$

$$\frac{d^2\nu_s}{dx^2} = \frac{M_s}{E_s I_s}$$

Differentiating Eq. (25) twice with respect to x gives

$$\frac{d^2\sigma}{dx^2} = \frac{E_a}{t_a}\left(\frac{M_s}{E_sI_s} - \frac{M_f}{E_fI_f}\right) \tag{26}$$

Differentiating Eq. (26) and substituting Eq. (3) into Eq. (24) gives

$$\frac{d^3\sigma}{dx^3} = \frac{E_a}{t_a}\left(\frac{1}{E_sI_s}Q_s - \frac{1}{E_fI_f}\left(Q_f - \tau b\frac{t_f}{2}\right)\right) \tag{27}$$

Differentiating Eq. (27) gives the governing equation

$$\frac{d^4\sigma}{dx^4} = \frac{E_a}{t_a}\left(\frac{1}{E_sI_s}\frac{dQ_s}{dx} - \frac{1}{E_fI_f}\frac{dQ_f}{dx} + \frac{1}{E_fI_f}\frac{bt_f}{2}\frac{d\tau}{dx}\right) \tag{28}$$

Substituting Eq. (2) into Eq. (28) gives:

$$\frac{d^4\sigma}{dx^4} + \frac{bE_a}{t_a}\frac{1}{E_fI_f}\sigma = \frac{bt_fE_a}{2t_a}\frac{1}{E_fI_f}\frac{d\tau}{dx} \tag{29}$$

Equation (29) can be rewritten as:

$$\frac{d^4\sigma}{dx^4} + A_2\sigma = B_2\frac{d\tau}{dx} \tag{30}$$

where,

$$A_2 = \frac{bE_a}{t_aE_fI_f}$$
$$B_2 = \frac{bt_fE_a}{2t_aE_fI_f}$$

Equation (30) has a general solution in the form:

$$\sigma(x) = e^{-\beta x}(S_1\cos(\beta x) + S_2\sin(\beta x)) + e^{\beta x}(S_3\cos(\beta x) + S_4\sin(\beta x)) + \frac{B_2}{A_2}\frac{d\tau}{dx} \tag{31}$$

where,

$$\beta = \sqrt[4]{A_2/4}$$

The peel stress in the adhesive layer approaches zero at a position far from the end of the CFRP layer and the symmetry plane. In this situation, $S_3 = 0$, and $S_4 = 0$, and Eq. (31) can be written as:

$$\sigma(x) = e^{-\beta x}(S_1 \cos(\beta x) + S_2 \sin(\beta x)) + \frac{B_2}{A_2}\frac{d\tau}{dx} \tag{32}$$

where S_1 and S_2 are constant coefficients determined from the boundary conditions. The boundary conditions for peel stress of the adhesive layer are:

$$M_f(0) = M_s(0) = Q_f(0) = Q_s(0) = 0 \tag{33}$$

Substituting the boundary conditions of Eq. (33) into Eq. (26) and Eq. (27), respectively, gives:

$$\frac{d^2\sigma(0)}{dx^2} = 0 \tag{34a}$$

$$\frac{d^3\sigma(0)}{dx^3} = B_2\tau(0) \tag{34b}$$

Differentiating Eq. (32) two and three times, respectively, with respect to x, we obtain

$$\frac{d^2\sigma(x)}{dx^2} = 2\beta^2 e^{-\beta x}(S_1 \sin(\beta x) - S_2 \cos(\beta x)) + \frac{B_2}{A_2}\frac{d^3\tau}{dx^3} \tag{35a}$$

$$\begin{aligned}\frac{d^3\sigma(x)}{dx^3} &= 2\beta^3 e^{-\beta x}(-S_1 \sin(\beta x) + S_2 \cos(\beta x) + S_1 \cos(\beta x) + S_2 \sin(\beta x)) \\ &+ \frac{B_2}{A_2}\frac{d^4\tau}{dx^4}\end{aligned} \tag{35b}$$

S_1 and S_2 of Eq. (32) can be obtained by substituting Eqs. (35a) and (35b) into Eqs. (34a) and (34b):

$$S_1 = \frac{B_2}{2\beta^3}\tau(0) - \frac{B_2}{2\beta^3 A_2}\frac{d^4\tau(0)}{dx^4} - \frac{B_2}{2\beta^2 A_2}\frac{d^3\tau(0)}{dx^3} \tag{36a}$$

$$S_2 = \frac{B_2}{2\beta^2 A_2}\frac{d^3\tau(0)}{dx^3} \tag{36b}$$

The equation of peel stress in the adhesive layer is

$$\sigma(x) = e^{-\beta x}(S_1 \cos(\beta x) + S_2 \sin(\beta x)) - \frac{B_2}{A_2}\frac{\omega^2}{b}(H_1 \cosh(\omega x) + H_2 \sinh(\omega x)) \tag{37}$$

3 Finite Element Analysis of a Cracked Steel-Plate Repaired by CFRP

The 3D meshing analytical model shown in Fig. 6 was compared with the theoretical analytical model. ABAQUS and MATLAB were used in the 3D meshing analytical model and theoretical analytical model, respectively. Material properties of the proposed CFRP repairing model are shown in Table 1. The length of steel was $L_s = 240$mm, the lengths of the adhesive layer and CFRP were $L_a = L_f = 200$mm, and the width of the entire model was $b = 80$mm. The thickness of steel plate, CFRP, and adhesive layer were $t_s = 6$mm, $t_f = 2.5$mm, $t_a = 0.2$mm, respectively.

The whole model would be symmetric. The symmetric boundary condition was employed at the symmetry plane of the repaired model. The crack length in steel-plate was ranged from 10 to 70 mm (10, 20, 30, 40, 50, 60, and 70 mm were considered). The longitudinal tension force at the end of the steel was $F = 0.48$ kN. To obtain accurate stress results at the end and the symmetry planes of the adhesive layer, a refined mesh was deployed in these areas, as shown in Fig. 6. All components in the repaired model were modeled with 8-node linear brick elements (element C3D8R), and the total number of elements was 513,322.

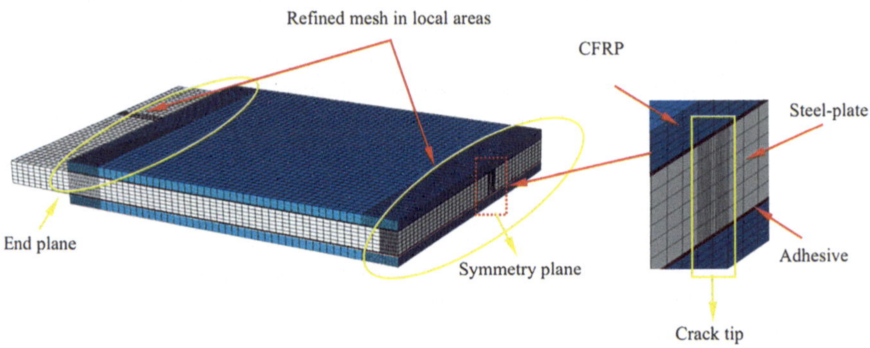

Fig. 6 FE mesh of repaired half-model

Table 1 Material properties of the repaired model

Material	Property	Value
Steel-plate	Density (ρ/kg/m^3)	7800
	Young's modulus (E_s/GPa)	211
	Poisson's ratio (μ_s)	0.3
CFRP	Young's modulus (E_f/GPa)	$E_1 = 134$, $E_2 = 10.3$, $E_3 = 10.3$
	Shear modulus (G_f/GPa)	$G_{12} = 5.5$, $G_{13} = 5.5$, $G_{23} = 3.2$
	Poisson's ratio (μ_f)	$\mu_{12} = 0.33$, $\mu_{13} = 0.33$, $\mu_{23} = 0.53$
Adhesive layer	Young's modulus (E_a/GPa)	2.8
	Poisson's ratio (μ_a)	0.4

4 Results and Discussion

4.1 Comparison of Results of Theoretical Analysis and FEM Analysis for the Through Crack Case

Figure 7 shows the shear stress and peel stress distributions of the adhesive layer when the crack length was equal to the steel-plate width, obtained using both the theoretical method and FEM. Stress distributions were almost the same except at the end plane and the symmetry plane, where singular stress fields existed. The maximum shear stress occurred at the symmetry plane. The absolute values of the maximum and minimum peel stress at the end and symmetry planes were the same. Figure 7 shows that the result of the FEM model was consistent with that of the theoretical model, demonstrating that this FEM model could be used to analyze stresses of the repair model.

4.2 Stress Distribution of Adhesive Layer Under Different Crack Lengths

To analyze the effect of different crack lengths on the stress distributions of the adhesive layer, different FEM models (crack lengths varies from 10 to 70 mm) were modeled. Because the maximum stresses in the adhesive occurred at the end plane and symmetry plane, we only discuss the values at the two planes to effectively analyze the effect of crack lengths on the stress distributions.

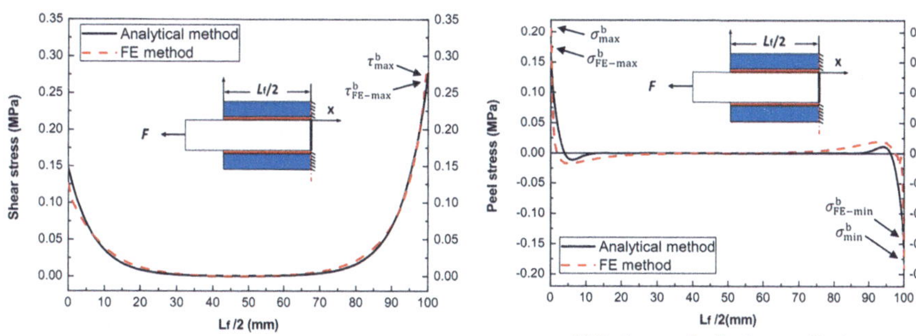

(a) Shear stress distribution on adhesive (b) Peel stress distribution on adhesive

Fig. 7 Comparison of results of theoretical method and FE method at the adhesive layer. ($\tau^{b}_{\max}$ and $\tau^{a}_{FE-\max}$ are the maximum shear stresses by the theoretical method and FE method, respectively. Maximum and minimum peel stresses uses the same subscripts). (**a**) Shear stress distribution on adhesive. (**b**) Peel stress distribution on adhesive

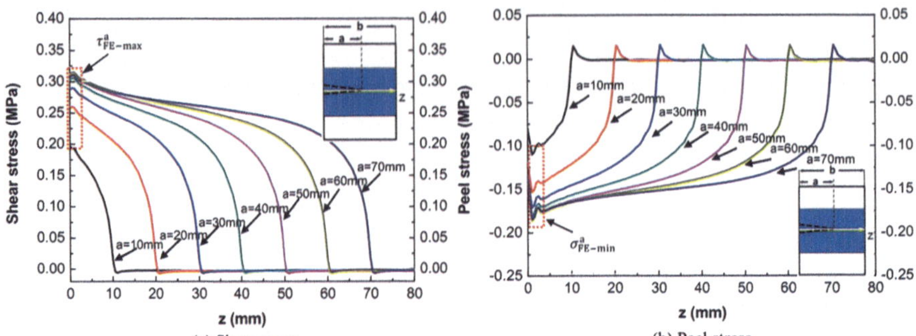

(a) Shear stress
(b) Peel stress

Fig. 8 Stress distributions of adhesive layer at symmetry plane under different crack lengths. (**a**) Shear stress. (**b**) Peel stress

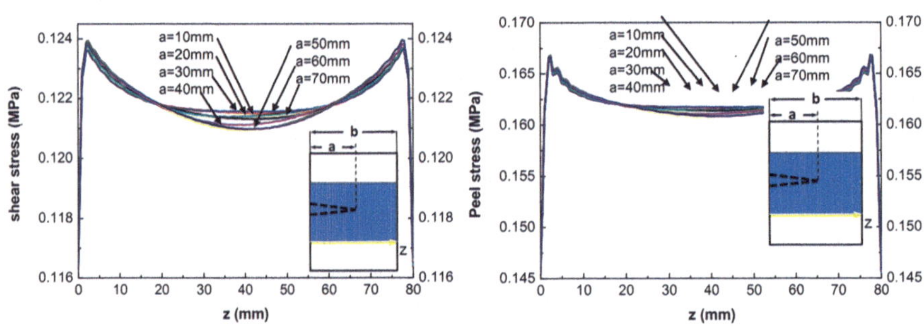

(a) Shear stress
(b) Peel stress

Fig. 9 Stress distributions of adhesive layer at end plane under different crack lengths. (**a**) Shear stress. (**b**) Peel stress

Figure 8 shows the stress distributions of the adhesive layer at the symmetry plane under different crack lengths. In Fig. 8a, it can be seen that the shear stress in the adhesive layer decreased along the direction of the crack until it reached the end of the crack, and $\tau^a_{FE-\max}$ in the adhesive layer increased with the increase of crack length. In Fig. 8b, we see that the absolute value of the peel stress in the adhesive layer also decreased along the direction of the crack until it reached the end of the crack, and $\sigma^a_{FE-\min}$ also increased with the increase of crack length. Moreover, $\tau^a_{FE-\max}$ and $\sigma^a_{FE-\min}$ were found at the same point and they did not appear on the end plane, but instead were seen a small distance from the end.

Figure 9 shows the stress distributions of the adhesive at the end plane with different crack lengths. The tendencies of the stress distributions for shear stress and peel stress were nearly the same, and the crack length had little effect on the stress distributions. Therefore, the effect of the crack length on the adhesive layer stress was neglected at the end plane in this paper.

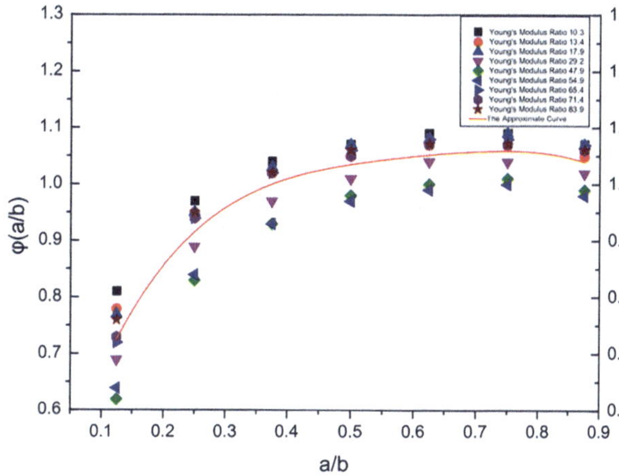

Fig. 10 Shear stress ratio $\tau^{a}_{FE-\max}/\tau^{b}_{\max}$ of adhesive versus crack length with different material combinations

4.3 Adhesive Damage Prediction Method

As shown in Fig. 8, the maximum stress at the adhesive layer increased as the crack length increased, indicating that the repaired construction will deboned from the adhesive when the stress reaches its maximum. The relationship between the maximum stress and crack length is discussed at the symmetry plane.

Figure 10 shows the curve of $\tau^{a}_{FE-\max}/\tau^{b}_{\max}$ versus the crack length under different materials' (CFRP and adhesive) Young's modulus ratios (shown in Table 2). The horizontal axis is the crack length ratio, the superscript "*a*" with values 10, 20, 30, 40, 50, 60, and 70 mm, represents the crack length, and the superscript "*b*" represents the crack length when it equals the steel width of 80 mm, which means $b = 80$ mm in this paper. The vertical axis is the ratio $\varphi(a/b)$ of the maximum shear stress between the two cases when the crack lengths equal *a* and *b*:

$$\varphi(a/b) = \frac{\tau^{a}_{FE-\max}}{\tau^{b}_{\max}} \tag{38}$$

Figure 10 shows that with different material combinations of CFRP and adhesive, the shear stress ratio $\varphi(a/b)$ can be approximated to a curve written as the equation

$$\varphi(a/b) = 0.3 + 4.0 \times (a/b) - 9.3 \times (a/b)^2 \tag{39}$$

Similarly, Fig. 11 shows the peel stress ratio versus the crack length under different materials' (CFRP and adhesive) Young's modulus ratios (shown in Table 2), with a curve expressed by

$$\psi(a/b) = 0.3 + 4.6 \times (a/b) - 10.6 \times (a/b)^2 \tag{40}$$

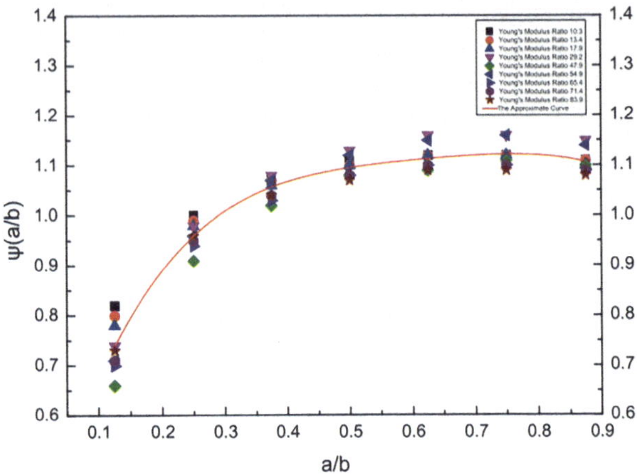

Fig. 11 Peel Stress ratio $O_{FE-max}^{-a}/O_{max}^{-b}$ of adhesive versus crack length with different material combinations

Table 2 Material combinations having different Young's modulus ratios

CFRP young's modulus (GPa)	Adhesive young's modulus (GPa)	Young's modulus ratios
134	13	10.3
134	10	13.4
134	7.5	17.9
134	4.59	29.2
134	2.8	47.9
153.8	2.8	54.9
183	2.8	65.4
200	2.8	71.4
235	2.8	83.9

Quadratic stress criterion was adopted to judge the failure of adhesive in the repaired model. This is given by

$$\left(\frac{\langle\sigma_n\rangle}{\sigma^{\max}}\right)^2+\left(\frac{\tau_s}{\tau_s^{\max}}\right)^2+\left(\frac{\tau_t}{\tau_t^{\max}}\right)^2=\begin{cases}<1\\ \geq 1\end{cases} \tag{41}$$

where $\langle\sigma_n\rangle$ is the peeling stress, and τ_s and τ_t are the shear stress with respect to the parallel and vertical directions of longitudinal tension, respectively. $\sigma^{\max}$, $\tau_s^{\max}$, and $\tau_t^{\max}$ are the limitary stresses of the adhesive. The adhesive will experience initial damage when the result of Eq. (41) is greater than one. As the shear stress in the parallel direction is much greater than in the vertical direction when the whole repaired model only experiences longitudinal force, the shear stress in the vertical direction τ_t can be considered to be zero. The Quadratic stress criterion can be rewritten as:

$$\left(\frac{\langle\sigma_n\rangle}{\sigma^{\max}}\right)^2 + \left(\frac{\tau_s}{\tau_s^{\max}}\right)^2 = \begin{cases} < 1 \\ \geq 1 \end{cases} \tag{42}$$

and $\langle\sigma_n\rangle$, τ_s in Eq. (41) can be obtained from Eqs. (43a) and (43b):

$$\tau_s = \varphi(a/b) \times \tau_{\max}^b \tag{43a}$$

$$\langle\sigma_n\rangle = \psi(a/b) \times \sigma_{\max}^b \tag{43b}$$

where $\tau_{\max}^b$ and $\sigma_{\max}^b$ are the shear stress and peel stress of the model with a through crack, and they could be derived from Eqs. (24) and (37). Substituting the values $\langle\sigma_n\rangle$, τ_s into the adhesive failure criteria Eq. (42), we obtain a simple method to predict the failure of the adhesive in the repaired steel plates with different crack lengths. Moreover, we can use this method to determine the critical stresses in the adhesive layer or the critical longitudinal load F.

5 Conclusions

In this paper, the model of a steel plate with an edge crack repaired by CRRP was analyzed. The 3D model with the crack running through the steel plate could be simplified to a 2D model, and the shear stress and peel stress distributions on the adhesive were analyzed with a theoretical method and a finite element method. It was demonstrated that the results obtained by the two methods were very similar. Using FEM, a repair model with different crack lengths was constructed and stress distributions on the adhesive were calculated on the end plane and symmetry plane; it was found that the crack length had little effect on the stresses on the end plane, whereas the maximum stresses on the symmetry plane varied with crack length. To describe the relationship between the maximum stresses and the crack, the curve of shear stress and peel stress ratios, $\tau_{\max}^a/\tau^b$ and $\sigma_{\max}^a/\sigma_{\max}^b$, with different crack lengths and material combinations were obtained, and the curve was expressed as a polynomial. Under a longitudinal load, adhesive debonding failure may occur, and a quadratic stress criterion was used to estimate the adhesive failure. Using the ratio expressions and the 2D theoretical results, a simple method was proposed to determine the critical stresses in the adhesive layer under different crack lengths and predict the adhesive failure in the repaired model.

References

1. Baker AA, Rose LRF, Jones R. Advances in the bonded composite repair of metallic aircraft structure. Elsevier Applied Science; 2002.
2. Baker AA. Bonded composite repair of fatigue-cracked primary aircraft structures. Compos Struct. 1999;47(1–4):431–43.

3. Bruce WA, Harwig DD, Harris ID, et al. Development of internal trenchless repair technology for gas transmission pipelines. In: GTI conference & exhibition, Phoenix, USA; 2004. p. 1–14.
4. Alexander C. Guideline for repairing damaged pipelines using composite materials. In: NACE international corrosion conference, Nashville, USA; 2007. p. 071441.
5. Duell JM, Wilson JM, Kessler MR. Analysis of a carbon composite overwrap pipeline repair system. Int J Press Vessel Pip. 2008;85(11):782–8.
6. Geoge DM, Echtermeyer AT, Leong KH, et al. Repair of floating offshore units using bonded fiber composite materials. Compos Part A Appl Sci Manuf. 2009;40(9):1364–80.
7. Turton TJ, Dalzel JJ, Livingstone F. Oil platforms, destroyers and frigates case studies of QinetiQ's marine composite patch repairs. Compos Part A Appl Sci Manuf. 2005;36(8):1066–72.
8. Cheng JQ, Wu XX, Li GQ, et al. Design and analysis of a smart composite pipe joint system integrated with piezoelectric layers under bending. Int J Solids Struct. 2007;44(1):298–319.
9. American Society of Mechanical Engineers. Repair of pressure equipment and piping. USA; 2015.
10. Veritas DN. Pipeline subsea repair. NOR; 2010.
11. International Organization for Standardization. Petroleum, petrochemical and natural gas industries. 2015.
12. Osnes H, Mcgeorge D, Weitzenbock JR, et al. Predicting failure of bonded patches using a fracture mechanics approach. Int J Adhes Adhes. 2012;37(37):102–11.
13. Tsouvalis NG, Mirisiotis LS, Dimou DN. Experimental and numerical study of the fatigue behavior of composite patch reinforced cracked steel plates. Int J Fatigue. 2009;31(10):1613–27.
14. Albedah A, Khan MAS, Benyahia F, et al. Experimental analysis of the fatigue life of repaired cracked plate in aluminum alloy 7075 with bonded composite patch. Eng Fract Mech. 2015;145:210–20.
15. Nateche T, Meliani MH, Khan SMA, et al. Residual harmfulness of a defect after repairing by a composite patch. Eng Fail Anal. 2015;48:166–73.
16. Okafor AC, Singh N, Enemuoh UE, et al. Design, analysis and performance of adhesively bonded composite patch repair of cracked aluminum aircraft panels. Compos Struct. 2005;71(2):258–70.
17. Klug JC, Sun CT. Large deflection effects of cracked aluminum plates repaired with bonded composite patches. Compos Struct. 1998;42(3):291–6.
18. Naboulsi S, Mall S. Characterization of fatigue crack growth in aluminum panels with a bonded composite patch. Compos Struct. 1997;37(3):321–34.
19. Naboulsi S, Mall S. Nonlinear analysis of bonded composite patch repair of cracked aluminum panels. Compos Struct. 1998;41(3–4):303–13.
20. Rezgni L, Madani K, Feaugas X. Influence of water ingress onto the crack propagation rate in a AA2024-T3 plate repaired by a carbon/epoxy patch. Aerosp Sci Technol. 2016;55:359–65.
21. Ramji M, Srilakshmi R, Prakash BM. Towards optimization of patch shape on the performance of bonded composite repair using FEM. Compos Part B. 2013;45(1):710–20.
22. Chung KH, Yang WH. A study on the fatigue crack growth behavior of thick aluminum panels repaired with a composite patch. Compos Struct. 2003;60(1):1–7.
23. Osnes H, Georgr MD, Weitzenbock JR, et al. Predicating failure of bonded patches using a fracture mechanics approach. Int J Adhes Adhes. 2012;37:102–11.
24. Tsai MY, Oplinger DW, Morton J. Improved theoretical solutions for adhesive lap joints. Int J Solids Struct. 1998;35(12):1163–85.
25. Zhang Y, Duan ML, Wang Y. Analytical study of the strength of adhesive joints of riser pipes, ships and offshore. Structure. 2015;10(5):545–53.
26. Deng J, Lee MMK, Moy SSJ. Stress analysis of steel beams reinforced with a bonded CFRP plate. Compos Struct. 2004;65(2):205–15.
27. Miller TC, Chajes MJ, Mertz DR, et al. Strengthening of a steel bridge girder using CFRP plates. J Bridg Eng. 2001;6(6):514–22.
28. De Jesus AMP, Pinto JMT, Morais JJL. Analysis of solid wood beams strengthened with CFRP laminates of distinct lengths. Constr Build Mater. 2012;35:817–28.

Rigorous SIF-Based Prediction of Fatigue Life Improvement for Prestressed-CFRP-Repaired Cracked Steel Plates

Nomenclature

a	Crack length (mm)
C	Material constants
F_{1-un}^{FEM}	Dimensionless SIFs of the unrepaired steel plate obtained by FEM
F_{1-0}^{FEM}	Dimensionless SIFs of the steel plate repaired with normal CFRP by FEM
$F_{1-\sigma_p}^{FEM}$	Dimensionless SIFs of the steel plate repaired with prestressed CFRP by FEM
F_{1-un}^{EXACT}	The exact dimensionless SIF of the unrepaired steel plate
F_{1-0}^{EXACT}	The exact dimensionless SIF of the steel plate repaired with normal CFRP
$F_{1-\sigma_p}^{EXACT}$	The exact dimensionless SIF of the steel plate repaired with prestressed CFRP
K_{fc}	Fracture toughness (MPa mm$^{1/2}$)
K_{1-un}^{FEM}	SIFs of the unrepaired steel plate obtained by FEM (MPa mm$^{1/2}$)
K_{1-0}^{FEM}	SIFs of the steel plate repaired with normal CFRP by FEM (MPa mm$^{1/2}$)
$K_{1-\sigma_p}^{FEM}$	SIFs of the steel plate repaired with prestressed CFRP by FEM (MPa mm$^{1/2}$)
K_{1-un}^{EXACT}	The exact SIF of the unrepaired steel plate (MPa mm$^{1/2}$)
K_{1-0}^{EXACT}	The exact SIF of the steel plate repaired with normal CFRP (MPa mm$^{1/2}$)
$K_{1-\sigma_p}^{EXACT}$	The exact SIF of the steel plate repaired with prestressed CFRP (MPa mm$^{1/2}$)
m	Material constants
N	Number of cyclic
$N_{c\text{-}un}$	The number of cycles to failure for unrepaired steel plates
$N_{c\text{-}\sigma p}$	The number of cycles to failure for repaired steel plates
R	Stress ratio
W	Width of steel plate (mm)
σ	Stresses loaded at both ends of the plate (MPa)
σ_p	Prestress (MPa)
σ_{z-un}^{FEM}	The FEM stress at the crack tip of the unrepaired specimen (MPa)

Y. Zhang, *Interfacial Mechanics of Offshore Engineering Equipment*,
https://doi.org/10.1007/978-981-95-3200-1_9

σ_{z-0}^{FEM} The FEM stress at the crack tip of the specimen repaired with normal CFRP (MPa)

$\sigma_{z-\sigma_p}^{FEM}$ The FEM stress at the crack tip of the specimen repaired with prestressed CFRP (MPa)

Subscripts

max Maximum
min Minimum

1 Introduction

Steel construction, which is utilized in offshore engineering equipment, has comprehensive advantages of factory manufacturing, fast installation, short construction period, good seismic performance, and fast investment recovery. In offshore engineering equipment, steel plates are the most common and widely utilized in steel structures. Those marine structures and ocean engineering are subjected to fatigue loads, such as wind, wave and current, which could cause cracks in these steel structures, and the cracks may propagate even fracture. Typical repair methods for cracked plates are grinding, welding, fixture, and composite repair [1–3]. Among them, the composite repair is widely used not only for repair but for maintenance because it is low cost, relatively easy to process, and can be completed in a short time. Composite repair techniques were originally used to reinforce concrete structures. With the advancement of technology, they are becoming more and more diverse. The repairing of damaged marine structures and equipment with composite material is widely used as shown in Figs. 1, 2, and 3, such as repairing cracked steel plates on FPSO, repairing corroded deck on FOU, repairing semi-submersible platform ballast tank [4].

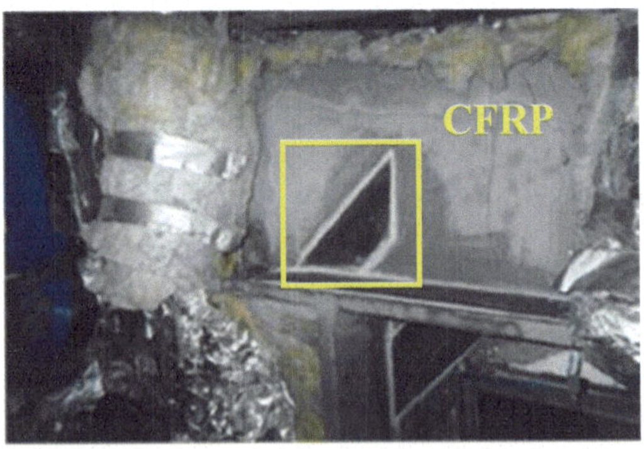

Fig. 1 Repairing cracked steel plates on FPSO with CFRP

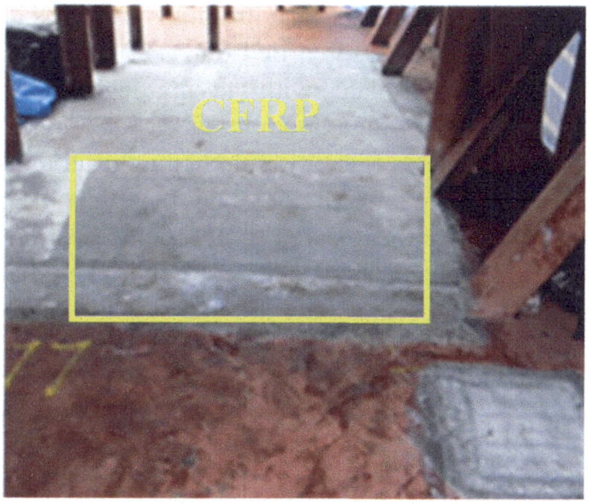

Fig. 2 Repairing corroded deck on FOU with CFRP

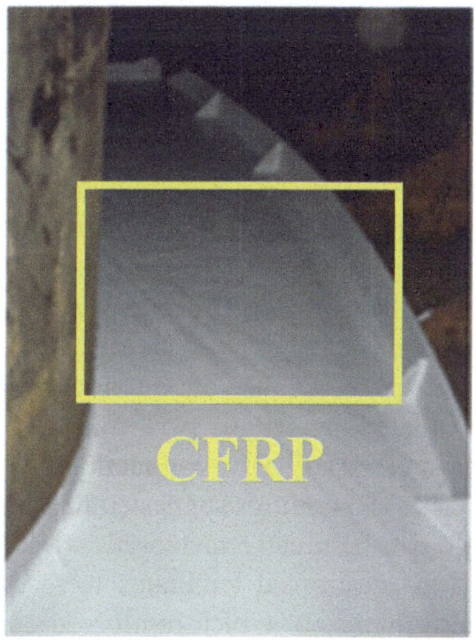

Fig. 3 Repairing semi-submersible platform ballast tank with CFRP

In the previous studies, Liu et al. [5] studied the parameters of thickness, length, and structural configuration of the patch, had a significant impact on fatigue life extension of cracked steel plate for repairs utilizing single-sided and double-sided CFRP. For reinforced concrete beams strengthened with prestressed CFRP, Xue et al. [6] noted out that compression damage, tensile damage, and debonding damage may be the potential bending damage modes, and estimated the flexural strength of the beams under the respective damage modes. In addition, the presence of prestressed CFRP was considered in the prediction of cracking moment, crack width, and deflection of reinforced beams. Wu et al. [7] investigated the fatigue performance of cracked steel plates fixed using CFRP with an ultra-high modulus. The impacts of CFRP bond length, bond width, and bond placement were considered. A comparison of the strengthening effect of normal Young's modulus, high

Young's modulus and ultra-high modulus CFRP plates with and without prestressing was also carried out. Yu et al. [8] evaluated the fatigue performance of cracked steel plates with varying degrees of initial damage and discovered that employing ultra-high modulus CFRP laminates and fixing at an incipient stage resulted in more efficient strengthening. Kafodya et al. [9] investigated the durability of pultruded unidirectional CFRP plates submerged in water and saltwater at room temperature and subjected to prolonged bending strains of various ultimate strain, and the mechanisms of CFRP strength degradation was studied. Al-Mosawe et al. [10] found a considerable in bond strength reinforcement under impact load, while the effective bond length was largely unaffected by high loading rates. Batuwitage et al. [11] proposed a parametric study to explore the effect of surface preparation process for steel plate, primer, number of CFRP layers and exposure duration in corrosion conditions on the failure modes and tensile capacity of double strap joints. Pang et al. [12] conducted an experiment to investigate the bonding behavior of the CFRP-steel interface under quasi-static cyclic loading, and researched the effects of bond length, adhesive thickness, adhesive type and loading protocols on it. Liu et al. [13] presented a novel CFRP-steel tube composite member to prevent debonding failure, and studied the strengthening impact of length of the thick-walled tube and the number of CFRP layers on the peak bearing capacity and the maximum ductile displacement.

The safety and reliability of the repaired structure depends on their fatigue performance. Ineffective fatigue performance can lead to wasted resources and continued failure of refurbished steel plate, substantial financial losses and serious safety hazards [14, 15]. Therefore, it is of great significance to investigate the fatigue performance of cracked steel plates repaired with composite materials during fatigue loadings. Several studies are focusing on the fatigue performance. Täljsten et al. [16] evaluated the fatigue performance of structures repaired with non-prestressed and prestressed CFRP by conducting fatigue tensile tests on aged steel plates with central crack. The findings suggested that when non-prestressed CFRP was utilized, fatigue life can be greatly increased, while crack extension appears to be totally stopped when prestressed laminates were employed. Feng et al. [17] studied the fatigue performance of CFRP reinforced cracked steel plates at various temperatures. The present analytical approach for forecasting the fatigue life of composite structure when temperature effects were considered has been enhanced in this research, and it was also confirmed by the corresponding experiments. Li et al. [18] demonstrated that double-sided repair with CFRP was more efficient than single-sided repair. When utilized with prestressed CFRP, the effect was further enhanced. A larger prestress level resulted in a bigger improvement in fatigue life of central inclined cracked steel plate. Liu et al. [19] devised a useful SIF method for forecasting the fatigue life of center cracked steel plate through a series of finite element case analyses, which results were confirmed by comparison with test results in various papers. Cabrilo et al. [20] used the SIF formula and the Paris-Erdogan formula to evaluate the fatigue crack propagation rate in the base metal, heat affected zone, and welded metal zone to investigate the fatigue performance of welded armor steel. Chen et al. [21] looked at the effectiveness of CFRP for improving the fatigue performance of porous crack-containing steel plates, as well as the influence of hole size and placement on repair efficiency. CFRP repair may enhance the fatigue life of porous broken steel plates to the same level as centrally cracked plates, according to the findings. Considering the effects of initial crack depth, CFRP modulus, CFRP thickness and residual stress of welding, Jie et al. [22] examined the range of SIFs for unrepaired and cruciform welded joints repaired with CFRP, fitted a new equation to calculate the range of SIFs based on numerical results, and predicted the fatigue life using a modified Paris equation, which correlated well with the experimental results. Hassan et al. [23] fabricated several sets of standard fatigue test specimens to test the fatigue performance of cracked steel plates with different reinforcement configurations, repair schemes and initial crack lengths, and also observed the development of debonding in composite structures. Lepretre et al. [24] evaluated SIF for metal plates with a crack emanating from the central hole using the virtual crack closure technique, taking account the effects of different reinforcing configurations and CFRP types. Experiments were used to verify the correctness of the crack extension FEA, and a parametric analysis of the fatigue performance impacted by prestressing on CFRP was carried out. The results show that prestressing is a powerful reinforcing choice. Tan et al. [25] created a local model of steel joists with integral nodes and a FEM of steel truss bridges to assess the fatigue life. Combining the SIF equation and the Paris-Erdogan equation allowed them to analyze the initial crack propagation and forecast the remaining life. To obtain a model for the corrosion-fatigue life prediction of reinforced concrete square piles in marine environments, Shao et al. [26] developed a crack growth model based on the Paris-Erdogan equation. He utilized an example to demonstrate the model's validity and performed a parametric analysis of the key influencing factors. Wang et al. [27] created a novel prestressing system using CFRP for the reinforcement of damaged steel beams. The effects of bonding layer, prestressing level and CFRP plate area on the reinforcement efficiency were investigated and subsequently parametrically analyzed by using finite element modeling, which showed that the reinforcement efficiency increased with the increase of the prestressing level. Liu et al. [28] proposed an extended finite element analysis model considering various stages of crack propagation and verified its validity with comparative analysis with experimental data, which showed that the model can reasonably predict the fatigue performance of center-cracked steel plates reinforced with partially covered CFRP. Deng et al. [29] used a combination of shape memory alloy (SMA) and CFRP to achieve hybrid self-prestressing reinforcement for fatigue cracks. Static and fatigue tensile tests revealed that SMA/CFRP composite reinforcement can effectively retard steel cracking and provide better strengthening than CFRP alone.

As mentioned above, several previous studies discussed the fatigue life by fatigue testing and the SIF analysis for unrepaired/repaired structures. However, there was no detail discussion for the accuracy of the SIF formula, and the useful method was not proposed to predict the fatigue life of these structures. Since the steel plates repaired with CFRP is now widely used because of their advantages, the fatigue life prediction has been requested to evaluate the safety. This paper experimentally investigates the fatigue performance of cracked steel plates repaired with conventional CFRP firstly. Referring to test specimens, a series of FEMs, including unrepaired and repaired steel plates, were constructed to analyze the SIF. The errors due to the FEM mesh were evaluated by comparing with the numerical solution. The exact SIF of repaired plates were obtained through proportional method. Based on the rigorous SIF, the significant parameters C and m of crack propagation were gained. The fatigue life prediction equations for cracked steel plates with different repair configurations were built. The numerical method and experimental results were used to evaluate the fatigue life of repaired plates, with good consistency. This numerical method can also be utilized to predict the fatigue life of steel plates repaired with CFRP considering the effect of prestress.

2 Fatigue Experiment of Cracked Plate Repaired by CFRP

2.1 Fatigue Experiment Model

To investigate the efficiency of CFRP repair on prolonging fatigue life, fatigue tests were conducted on cracked steel plates unrepaired and repaired with CFRP. It is crucial to obtain the real mechanical property parameters of the steel plate. Figure 4 shows the geometry and dimensions of the tensile specimen used to obtain mechanical properties based on National Standard of China (GB/T 228.1-2021, 2022) [30]. Figure 5 shows the nominal stress versus strain curves of Q235. Table 1

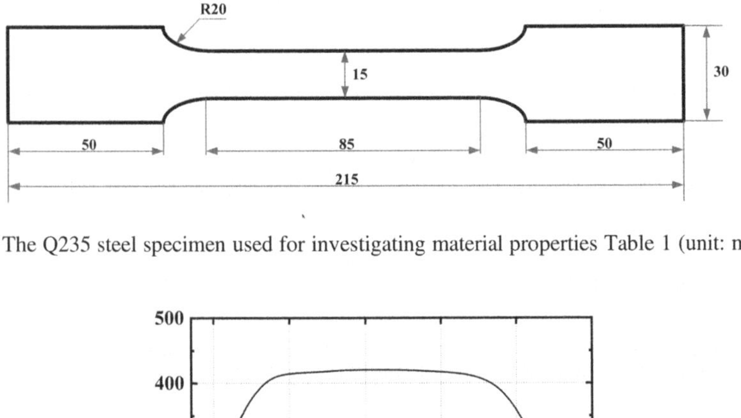

Fig. 4 The Q235 steel specimen used for investigating material properties Table 1 (unit: mm)

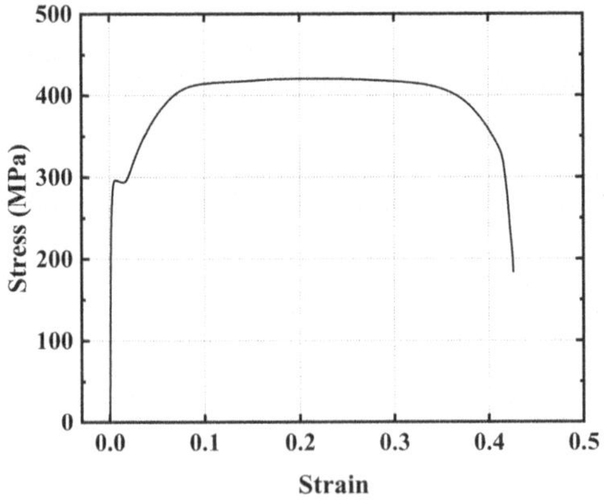

Fig. 5 Nominal stress versus strain curves

Table 1 The main mechanical properties of Q235

Young's modulus (GPa)	Yield strength (MPa)	Tensile strength (MPa)	Poisson's ratio	Fracture toughness (MPa m$^{1/2}$)	Fatigue crack growth threshold (MPa m$^{1/2}$)
206	297	420	0.3	83.16	6.424

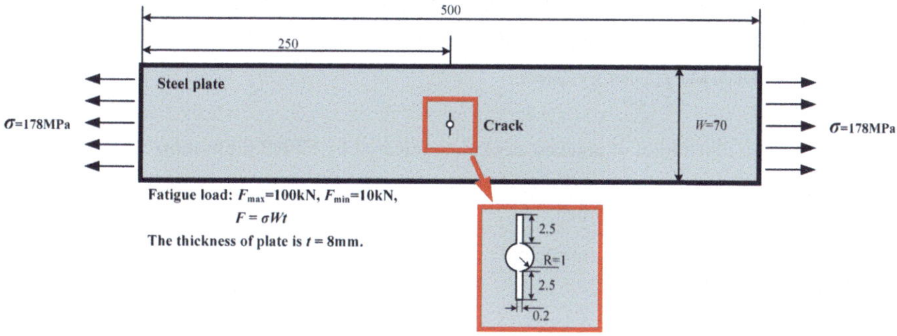

Fig. 6 Schematic illustration of cracked steel plate without repair (unit: mm)

presents the mechanical properties of Q235 obtained according to Fig. 4, fracture toughness [31] and fatigue crack growth threshold [32].

In this study, the fatigue crack propagation pattern will be investigated to clarify the effect of repaired by CFRP. Figure 6 shows the cracked steel plate specimen subjected to cyclic tensile loading without repairing. The specimens were designed considering the stress intensity factors where the steel plate width W is ten times larger than the initial crack length $2a$.

Liu et al. [5] investigated the effective repair length and the effective number of layers of CFRP reinforcement, the results showed that the length of CFRP exceeding 50% of the total length of specimen was sufficient and that five layers of CFRP were sufficient to meet the repair expectations. Aljabar et al. [33] evaluated the fatigue life enhancement of single- and double-sided CFRP repairs; the results showed that the double-sided CFRP repair is much more useful than the single-sided CFRP repair. According to these references, the length of the CFRP in the fatigue test was determined to be 250 mm with a width of 70 mm, and five layers were stacked with a thickness of 0.22 mm per layer and pasted on the upper and lower planes of the cracked steel plate, as illustrated in Fig. 7. The laying angles of the fibers are 0°, 45°, 90°, −45°, 0°. The dimensions of the specimens are shown in Table 2. Table 3 shows material properties of CFRP T700 provided by the manufacturers.

The cracked steel plate specimen shown in the upper panel of Fig. 8. For the repairment, the surface was firstly descaled and the CFRP bonded region was sanded, the five CFRP layers were bonded together, and cured together at room temperature. The cured CFRP sheets were then bonded to the surface of the steel plate. Finally, the bonded specimens were cured for 7 days at room temperature

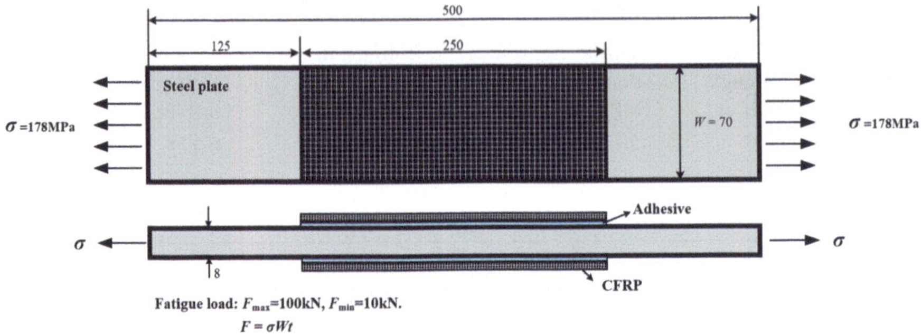

Fig. 7 Schematic illustration of cracked steel plate repaired by CFRP (unit: mm)

Table 2 Dimensions of the specimen

Specimen	Length L (mm)	Width W (mm)	Thickness t (mm)
Steel plate	500	70	8
CFRP	250	70	1.1 (0.22 × 5)

Table 3 Material properties of CFRP

Density (g/cm^3)	Young's modulus (GPa)	Shear modulus (GPa)	Poisson's ratio	Thermal expansion coefficient (/°C)
1.76	$E_1 = 233$, $E_2 = 23.1$, $E_3 = 23.1$	$G_{12} = 27.58$, $G_{13} = 27.58$, $G_{23} = 5.89$	$\mu_{12} = 0.34$, $\mu_{13} = 0.28$, $\mu_{23} = 0.28$	$\alpha_{11} = -0.7 \times 10^{-6}$

Fig. 8 Test specimens, C1 (up) and D1 (down)

about 20 °C. The repaired specimen is shown in Fig. 8. Table 4 shows the mechanical properties of adhesive Araldites® 2015 provided by manufacturers.

The remote tensile stress σ is applied so that the maximum stress $\sigma_{\max} = 178$ MPa, which does not exceed 60% of the steel yield strength. The stress ratio R is fixed as $R = 0.1$ at a frequency of 10 Hz. The experiments were conducted on the MTS LFV500-T5000 Nm fatigue testing machine until fracture occurred and each unrepaired and repaired fatigue specimen is tested two times. Figure 9 depicts the experimental set-up. To track the crack propagation, so-called beach marking method is used by inserting smaller stress amplitude cycles with limited number of

Table 4 Mechanical properties of adhesive

Young's modulus (GPa)	Shear modulus (GPa)	Poisson's ratio	Tensile yield strength (MPa)	Tensile failure strength (MPa)	Shear yield strength (MPa)	Shear failure strength (MPa)
1.80	0.68	0.33	12.63	21.63	14.6	17.9

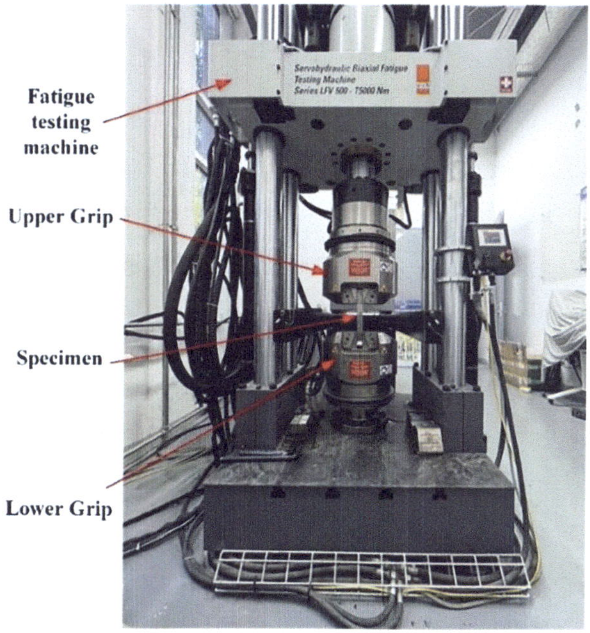

Fig. 9 Experimental setup

cycles to the original applied cyclic loading. Figure 10 illustrates the beach mark loading with the stress ratio $R = 0.55$ with a frequency of 10 Hz. For unrepaired plates, after the 30,000 cycles of fatigue load, the 15,000 cycles of beach marking load are added. While for repaired plates cycles of fatigue load are 60,000, and the cycles of marking load was 30,000, the other load setting stayed the same.

2.2 *Experimental Results*

The loading of the fatigue testing machine makes the specimen undergo crack propagation until fracture. The fractured central cross section of the unrepaired specimen is shown as Fig. 11. In smooth surface regions, beach marks are seen as extensions of stable fatigue cracks. The rough region indicates final unstable fracture. From Fig. 11, the critical crack length where unstable fracture occurs can be determined as $2a_{c\text{-}un} = 44.10$ mm for the unrepaired specimen.

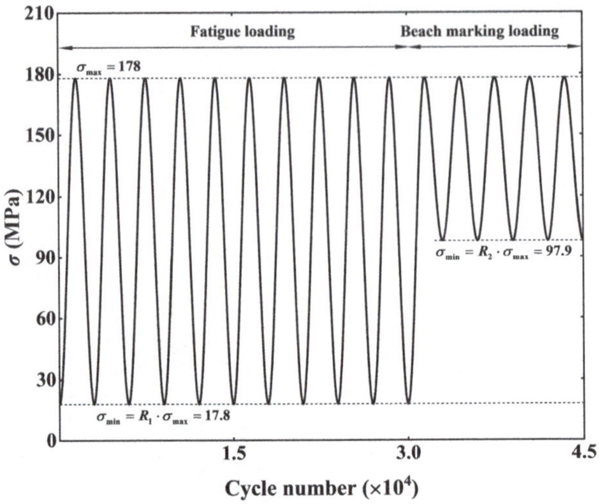

Fig. 10 Test stress range of "Beach Marking" technique

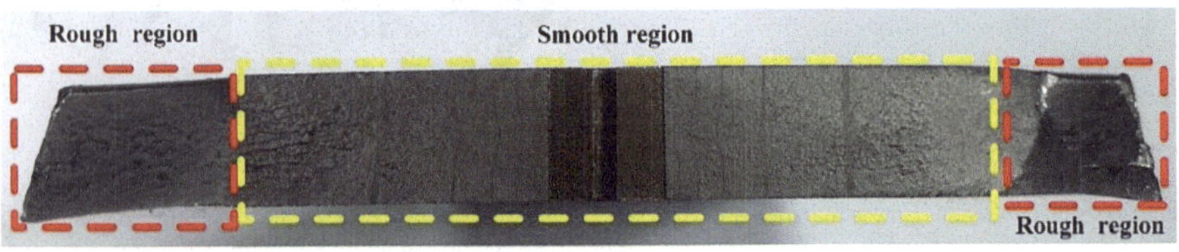

Fig. 11 Fractured central cross section of the unrepaired specimen

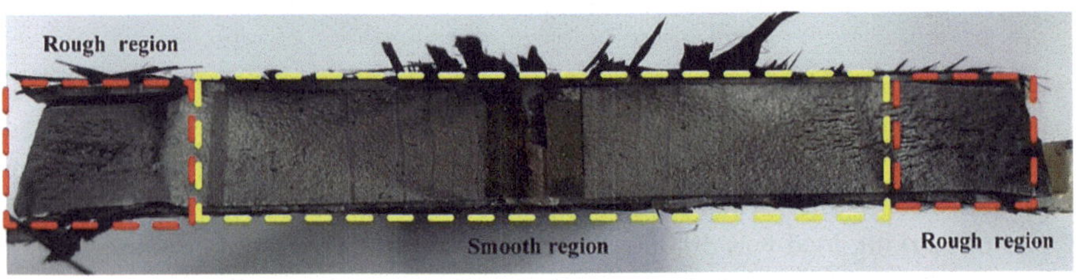

Fig. 12 Fractured central cross section of the repaired specimen

The fractures central cross section of the repaired specimen is shown as Fig. 12. In smooth surface regions, beach marks can be also observed as the fatigue crack extension. Compared to Fig. 11, the rough region where the final unstable fracture happened is comparatively smaller and the stable fatigue crack extension region is larger due to the CFRP reinforcement. From Fig. 12, the critical crack length can be determined as $2a_{c\text{-}0} = 53.00$ mm for the repaired specimen with zero prestress.

The relation between the crack length and the number of cycles experimentally obtained by using beach marking method, as presented in Fig. 13. The labels C1and D1 in the figure indicate the unrepaired and steel plates repaired with CFRP,

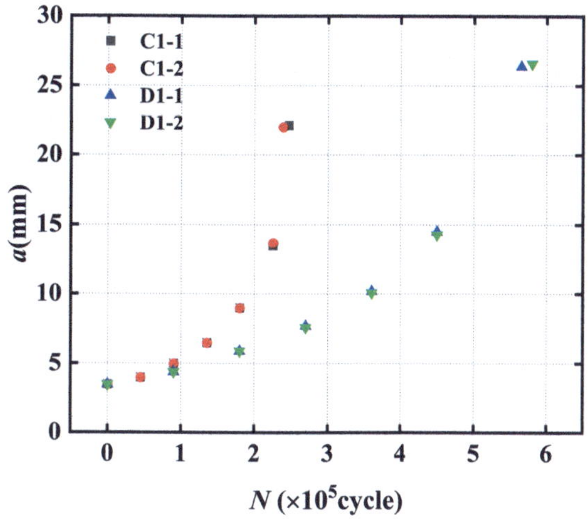

Fig. 13 Crack length *a* versus number of cyclic *N* experimentally obtained by using beach marking method

Table 5 Critical crack length, number of cycles to failure and fatigue fracture toughness of specimens in Figs. 11 and 12

Specimen number	Final crack length $2a$ (mm)	Average value $2a$ (mm)	Fatigue fracture toughness K_{fc} calculated by Eqs. (4) and (5) (MPa $m^{1/2}$)	Number of cycles to failure N (10^6)	Average values of N (10^6)
C 1–1	44.20	44.10	63.02	0.246	0.242
C 1–2	44.00			0.238	
D 1–1	52.80	53.00	55.31	0.565	0.5725
D 1–2	53.20			0.580	

respectively, consistent with Fig. 8. −1 and −2 represent the first and the second specimen of each type to have the mean experimental values. The labels in the subsequent figures represent the same meaning. Focusing on the end point where final unstable fracture occurs, the critical crack length is extended by 1.2 times by the CFRP repair and the number of cycles to failure is extended by 2.37 times by the CFRP repair. The experiments have demonstrated that CFRP repair is effective in improving the fatigue life of cracked steel plates. The experiments have demonstrated that CFRP repair is effective in improving the fatigue life of cracked steel plates, which is consistent the fatigue experiments conducted by [24], and it was proved that the presence of CFRP inhibited crack opening and thus reduced fatigue crack propagation rate.

The final crack length and number of load cycles to failure and the fracture toughness for the unrepaired/repaired specimens by using SIF equation discussed later are summarize in Table 5. The fatigue fracture toughness of the unrepaired steel plate can be obtained as $K_{fc} = 63.02$ MPa $m^{1/2}$ from the final crack length $2a_{c\text{-}un} = 44.10$ mm and the SIF solution for Fig. 6. Also, the fatigue fracture toughness of the repaired steel plate can be obtained as $K_{fc} = 55.31$ MPa $m^{1/2}$

from the final crack length $2a_{c\text{-}0}$ = 53.00 mm and the SIF results analytically obtained for the repaired plate in Fig. 7. Compared to the average fracture toughness K_{fc} of unrepaired steel plates, the relative error of the fracture toughness K_{fc} obtained through repaired steel plates was 12.24%. This difference can be explained from the extension of the delamination area of the CFRP repair plate. This difference can be explained from the extension of the delamination area of the CFRP repair plate [24].

2.3 Finite Element Model

Finite element software, Abaqus [34], was applied for investigating the SIF of unrepaired/repaired specimens and thus analyzing the correlation between the crack extension length and the number of load cycles (*a-N*). A series of FEMs were constructed to obtain the SIF for different crack lengths. Figure 14 shows as example of FEM mesh constructing a contour integral region at the tip of the crack. The element type is C3D6 (6-node tetrahedral 3D stress element type) used for contour integral region are indicated by yellow circles, and the element type is C3D8 (8-node hexahedral 3D stress element type) are used for other regions. The element type used for CFRP in the repair structure is also C3D8. In addition, the cohesive model was applied to model the interfacial adhesion region. The bilinear model was

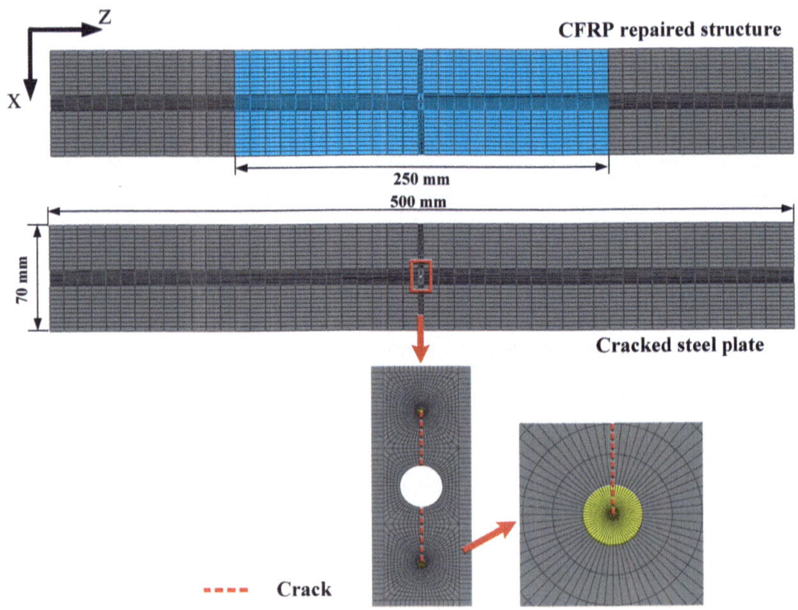

Fig. 14 FEM mesh to calculate SIF whose yellow circle shows a path for the contour integral method installed in Abaqus

Table 6 Numerical test matrix for SIF analysis

Configuration	Half crack length (a/mm)	Prestress (σ_p/MPa)
Unrepaired	3.5, 4, 5, 7, 9, 11, 13, 15, 17, 19, 21, 23, 25, 27	N/A
Repaired with CFRP	3.5, 4, 5, 7, 9, 11, 13, 15, 17, 19, 21, 23, 25, 27	0, 10, 20, 30, 40, 50, 80, 100

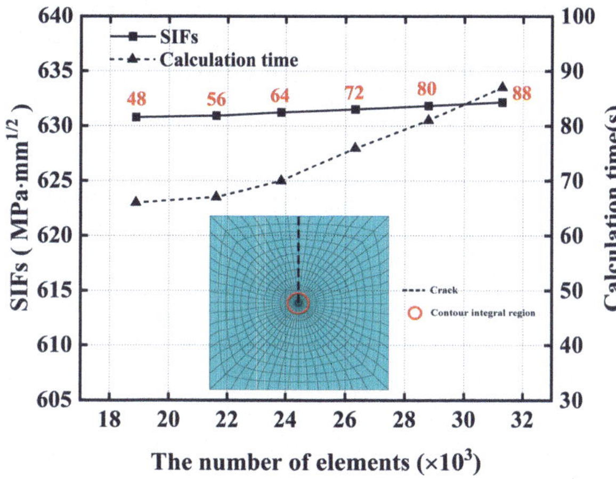

Fig. 15 Influence of the element numbers on the integration path on SIF and calculation time

employed to characterize the cohesive model, in which the key parameters are shown in Table 4.

In previous studies, it was discovered that the configuration of prestress was employed as a technique to improve repair performance of CFRP. In this study, several analyses were carried out to examine the effect of prestress on the mechanical and fatigue characteristics of repaired structures. Under a load of 60% yield strength, the axial displacement of a cracked steel plate is determined to be 0.22 mm. Furthermore, the existence or absence of prestress distinguishes the condition of the CFRP, with prestress $\sigma_p = 10, 20, 30, 40, 50, 80$ and 100 MPa. The techniques for introducing prestress on CFRP are often directly applying and temperature effect. The temperature rise approach is employed to produce the prestress in this investigation. The coefficients of thermal expansion are shown in Table 3. The dimensions of the FE models are same with Table 2, where the boundary conditions are added to the model using a setup with loading ($\sigma = 178$ MPa) at one end and fixed constraints at the other. The numerical matrix is presented in Table 6.

The variation of the SIF with increasing the total number of the mesh by varying element size is depicted in Fig. 15 when the crack length $a = 3.5$ mm ($2a/W = 0.1$). By averaging the values of the six element nodes in the direction of the plate's thickness, the specific value of SIF for crack tip is determined. In Fig. 15, the CPU time is also indicated as a dotted line with the red figure showing the number of elements on the integral path. As shown in Fig. 15, when the total element number is less than 32,000 and the element number on the integral path is less than 90, the SIF is still slightly increasing with increasing the mesh. However, when the total number of the mesh is larger than 32,000, the real calculation times needs more than a few days by using the computer currently used (DESKTOP-3O0CATT, CPU: AMD Ryzen 5 3500X 6-Core Processor 3.59 GHz, Memory: 16 GB). Such calculation time is not suitable for systematic calculation for various crack lengths and various repair conditions. Therefore, the following systematic calculations will be performed by using the mesh when the total number of the mesh is about 24,000 with the element number 64 on the integral path. As will be discussed later, the values of the SIF may have some errors less than 10%, the correction method will be shown in Sect. 4 based on the idea of proportional stress fields.

2.4 The Proportional Method for Correcting SIF

The dimensionless SIFs $F_1^{FEM} = K_1^{FEM}/(\sigma\sqrt{\pi a})$ obtained by the contour integral method installed in Abaqus is exhibited in Table 7 when the relative crack length $2a/W = 0.1, 0.2, 0.6, 0.7714$. F_{1-un}^{FEM} denotes the dimensionless SIF of the unrepaired specimen. And, notations F_{1-0}^{FEM}, F_{1-50}^{FEM}, F_{1-100}^{FEM} denote the dimensionless SIFs of the repaired specimens when the prestress $\sigma_p = 0$, 50, 100 MPa, respectively. The SIF ratios $F_{1-0}^{FEM}/F_{1-un}^{FEM}$, $F_{1-50}^{FEM}/F_{1-un}^{FEM}$, $F_{1-100}^{FEM}/F_{1-un}^{FEM}$ are also indicated in the parenthesis just below the value of F_{1-0}^{FEM}, F_{1-50}^{FEM}, F_{1-100}^{FEM}, respectively.

Similarly, the FEM stress value at the crack tip (see Fig. 14) of the unrepaired specimen is indicated as σ_{z-un}^{FEM}. Also, σ_{z-0}^{FEM}, σ_{z-50}^{FEM}, σ_{z-100}^{FEM} denote the stress at the crack tip of the repaired specimen obtained by the FEM when the prestress $\sigma_p = 0$, 50, 100 MPa, respectively. All values are obtained from the same mesh node of the crack tip of different models. Then, the FEM stress ratios $\sigma_{z-0}^{FEM}/\sigma_{z-un}^{FEM}$, $\sigma_{z-50}^{FEM}/\sigma_{z-un}^{FEM}$, $\sigma_{z-100}^{FEM}/\sigma_{z-un}^{FEM}$ are also in the parenthesis just below the value of σ_{z-0}^{FEM}, σ_{z-50}^{FEM}, σ_{z-100}^{FEM}, respectively.

The SIF ratio ≈ the FEM stress can be seen as $\frac{F_{1-0}^{FEM}}{F_{1-un}^{FEM}} \approx \frac{\sigma_{z-0}^{FEM}}{\sigma_{z-un}^{FEM}}$, $\frac{F_{1-50}^{FEM}}{F_{1-un}^{FEM}} \approx \frac{\sigma_{z-50}^{FEM}}{\sigma_{z-un}^{FEM}}$, $\frac{F_{1-100}^{FEM}}{F_{1-un}^{FEM}} \approx \frac{\sigma_{z-100}^{FEM}}{\sigma_{z-un}^{FEM}}$.

Table 7 The dimensionless SIF and the FEM stress value at the crack tip of the unrepaired/repaired specimens

$2a/W$	F_{1-un}^{FEM}	σ_{z-un}^{FEM}/MPa	F_{1-0}^{FEM} $\left(\frac{F_{1-0}^{FEM}}{F_{1-un}^{FEM}}\right)$	σ_{z-0}^{FEM} $\left(\frac{\sigma_{z-0}^{FEM}}{\sigma_{z-un}^{FEM}}\right)$	F_{1-50}^{FEM} $\left(\frac{F_{1-50}^{FEM}}{F_{1-un}^{FEM}}\right)$	σ_{z-50}^{FEM} $\left(\frac{\sigma_{z-50}^{FEM}}{\sigma_{z-un}^{FEM}}\right)$	F_{1-100}^{FEM} $\left(\frac{F_{1-100}^{FEM}}{F_{1-un}^{FEM}}\right)$	σ_{z-100}^{FEM} $\left(\frac{\sigma_{z-100}^{FEM}}{\sigma_{z-un}^{FEM}}\right)$
0.1000	1.070	975.5	0.9724 (0.9092)	890.9 (0.9132)	0.9475 (0.8859)	868.3 (0.8901)	0.9226 (0.8627)	845.7 (0.8670)
0.2000	1.075	1389	0.9325 (0.8676)	1210 (0.8714)	0.9083 (0.8451)	1179 (0.8489)	0.8841 (0.8226)	1148 (0.8264)
0.6000	1.247	2801	0.8861 (0.7103)	1998 (0.7133)	0.8618 (0.6908)	1944 (0.6938)	0.8378 (0.6716)	1890 (0.6746)
0.7714	1.474	3753	0.9500 (0.6447)	2429 (0.6472)	0.9243 (0.6273)	2364 (0.6299)	0.8986 (0.6098)	2290 (0.6102)

From Table 7, the following relation in Eq. (1) can be seen.

$$\frac{\sigma_{z-\sigma_p}^{FEM}}{\sigma_{z-un}^{FEM}} = \frac{F_{1-\sigma_p}^{FEM}}{F_{1-un}^{FEM}} \left(= \frac{K_{1-\sigma_p}^{FEM}}{K_{1-un}^{FEM}} \right) \tag{1}$$

where $\sigma_{z-\sigma_p}^{FEM}$ denotes the stress at the crack tip (see Fig. 14) in the FEM calculation and $F_{1-\sigma_p}^{FEM}$, $K_{1-\sigma_p}^{FEM}$ denote the SIFs of the specimens repaired with CFRP under prestress σ_p. Equation (1) means that the FEM stress ratio at the crack tip agrees well with the SIF ratio obtained by the contour integral method. Equation (1) suggests that the FEM stress ratio is useful for expressing the SIF ratio since only the stress value at the crack tip is focused without considering the integral path.

In the previous studies, the relation between the FEM stress at the crack tip (for example σ_{z-un}^{FEM} in Eq. (1)) and the exact SIF (for example K_{1-un}^{EXACT} in Eq. (2)) was investigated for various crack problems when the same FEM mesh is applied. Then, it was found that the FEM stress ratio is mesh-independent and coincide with the ratio of the exact SIF of repaired plate to the exact solution of unrepaired plate, as shown in Eq. (2) if the same FEM mesh is applied [35, 36].

$$\frac{\sigma_{z-\sigma_p}^{FEM}}{\sigma_{z-un}^{FEM}} = \frac{F_{1-\sigma_p}^{EXACT}}{F_{I-un}^{EXACT}} \tag{2}$$

The relation (1) was found to be useful for calculating SIFs accurately and conveniently since we can just focus the FEM crack tip stress without considering the contour integral. From Eq. (1) and Eq. (2), we have Eq. (3).

$$\frac{F_{1-\sigma_p}^{FEM}}{F_{1-un}^{FEM}} = \frac{F_{1-\sigma_p}^{EXACT}}{F_{I-un}^{EXACT}} \tag{3}$$

In this way, based on the proportional stress filed relations (2), (3) between the FEM and real ones, the exact SIF of cracked steel plates repaired with prestressed CFRP can be obtained as $F_{1-\sigma_p}^{EXACT} = F_{1-un}^{EXACT}\left(\sigma_{z-\sigma_p}^{FEM}/\sigma_{z-un}^{FEM}\right)$ or $F_{1-\sigma_p}^{EXACT} = F_{1-un}^{EXACT}\left(F_{1-\sigma_p}^{FEM}/F_{1-un}^{FEM}\right)$.

The SIF F_{1-un}^{FEM} in Fig. 6 and the SIF F_{1-0}^{FEM} in Fig. 7 obtained by the contour integral method is shown in Table 8. The central crack plate specimen is often used for fatigue experiment and the SIF can be provided as the following Eq. (4). The previous studies showed that the Eq. (4) can be used as the exact reference solution especially in the range $a/W \leq 0.7714$ in Table 8 considered in this study [37–39].

Table 8 The exact SIF F_{1-un}^{EXACT} obtained from Eq. (4) and the exact SIF F_{1-0}^{EXACT}

$2a/W$	F_{1-un}^{FEM}	F_{1-un}^{EXACT} obtained from Eq. (4)	F_{1-0}^{FEM} $(F_{1-0}^{FEM}/F_{1-un}^{FEM})$	F_{1-0}^{EXACT} obtained from $=F_{1-un}^{EXACT}\cdot(F_{1-0}^{FEM}/F_{1-un}^{FEM})$
0.1000	1.070	1.006	0.9724 (0.9088)	0.9142
0.1143	1.069	1.008	0.9655 (0.9032)	0.9104
0.1429	1.069	1.012	0.9532 (0.8917)	0.9024
0.2000	1.075	1.024	0.9325 (0.8674)	0.8883
0.2571	1.085	1.041	0.9149 (0.8432)	0.8778
0.3143	1.099	1.064	0.8997 (0.8187)	0.8710
0.3714	1.118	1.092	0.8875 (0.7938)	0.8669
0.4286	1.140	1.128	0.8792 (0.7712)	0.8699
0.4857	1.169	1.173	0.8764 (0.7497)	0.8794
0.5429	1.204	1.230	0.8787 (0.7298)	0.8977
0.6000	1.247	1.303	0.8861 (0.7106)	0.9259
0.6571	1.303	1.397	0.8993 (0.6902)	0.9642
0.7143	1.375	1.522	0.9194 (0.6687)	1.018
0.7714	1.474	1.698	0.9500 (0.6445)	1.095

$$
\begin{aligned}
K_{1-un}^{EXACT} &= \sigma\sqrt{\pi a}F_{1-un}^{EXACT} \\
F_{1-un}^{EXACT} &= \left[1-0.025\left(\frac{2a}{W}\right)^2+0.06\left(\frac{2a}{W}\right)^4\right]\sqrt{\sec\left(\frac{\pi a}{W}\right)} \qquad 0\leq\frac{2a}{W}<1
\end{aligned}
\tag{4}
$$

The difference of F_{1-un}^{EXACT} and F_{1-un}^{FEM} can be regarded as the FEM error caused by the mesh in Fig. 14 as discussed in Fig. 15. Since the same FEM mesh is applied to analyzing the specimen repaired by CFRP, the same error can be expected for the value of F_{1-0}^{FEM}. Therefore, by applying Eq. (3) based on the proportional stress fields [40, 41], the exact vale of the SIF for repaired specimen can be obtained and indicated as F_{1-0}^{EXACT} in Table 8.

3 Rigorous SIF-Based Fatigue Life Prediction of Cracked Plate Repaired by Prestressed CFRP

3.1 *The Fitting Equation for SIF*

Based on the joint utilization of finite element analysis and the proportionality method, the SIFs of the repaired structures in different prestressing states were obtained. The exact SIF for various prestress σ_p, F_{1-0}^{EXACT}, F_{1-10}^{EXACT}, F_{1-20}^{EXACT}, F_{1-30}^{EXACT}, F_{1-40}^{EXACT}, F_{1-50}^{EXACT}, F_{1-80}^{EXACT}, F_{1-100}^{EXACT} obtained from $F_{1-\sigma_p}^{EXACT}=F_{1-un}^{EXACT}\left(F_{1-\sigma_p}^{FEM}/F_{1-un}^{FEM}\right)$ are summed up in Table 9. From Table 9, it is seen that the SIFs of the repaired plate with the prestress $\sigma_p = 0$, 10, 20, 30, 40, 50, 80 and 100 MPa are much smaller than

Table 9 The exact SIF F_{1-0}^{EXACT}, F_{1-10}^{EXACT}, F_{1-20}^{EXACT}, F_{1-30}^{EXACT}, F_{1-40}^{EXACT}, F_{1-50}^{EXACT}, F_{1-80}^{EXACT}, F_{1-100}^{EXACT} for various prestress σ_p

$2a/W$	F_{1-un}^{EXACT}	F_{1-0}^{EXACT}	F_{1-10}^{EXACT}	F_{1-20}^{EXACT}	F_{1-30}^{EXACT}	F_{1-40}^{EXACT}	F_{1-50}^{EXACT}	F_{1-80}^{EXACT}	F_{1-100}^{EXACT}
0.1000	1.006	0.9142	0.9099	0.9053	0.9006	0.8959	0.8912	0.8772	0.8678
0.1143	1.008	0.9104	0.9059	0.9012	0.8965	0.8918	0.8872	0.8731	0.8637
0.1429	1.012	0.9024	0.8981	0.8934	0.8887	0.8840	0.8794	0.8653	0.8560
0.2000	1.024	0.8883	0.8843	0.8797	0.8750	0.8704	0.8658	0.8520	0.8428
0.2571	1.041	0.8778	0.8733	0.8688	0.8642	0.8597	0.8552	0.8415	0.8324
0.3143	1.064	0.8710	0.8660	0.8615	0.8570	0.8525	0.8480	0.8346	0.8257
0.3714	1.092	0.8669	0.8629	0.8585	0.8540	0.8495	0.8450	0.8316	0.8227
0.4286	1.128	0.8699	0.8651	0.8606	0.8560	0.8515	0.8469	0.8334	0.8244
0.4857	1.173	0.8794	0.8750	0.8702	0.8654	0.8608	0.8561	0.8420	0.8327
0.5429	1.230	0.8977	0.8933	0.8884	0.8835	0.8788	0.8739	0.8592	0.8496
0.6000	1.303	0.9259	0.9203	0.9152	0.9100	0.9050	0.8999	0.8848	0.8749
0.6571	1.397	0.9642	0.9581	0.9528	0.9476	0.9423	0.9370	0.9214	0.9109
0.7143	1.522	1.018	1.012	1.007	1.001	0.9958	0.9902	0.9738	0.9625
0.7714	1.698	1.095	1.089	1.083	1.077	1.071	1.065	1.047	1.035

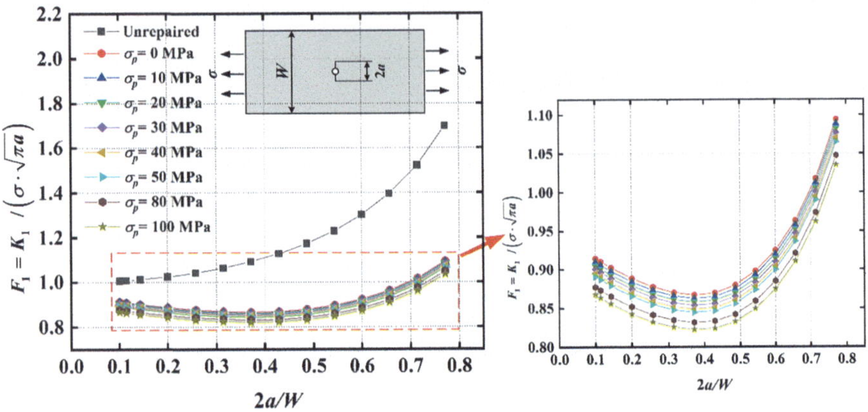

Fig. 16 The exact SIF of the unrepaired plate F_{1-un}^{EXACT} and the repaired plate $F_{1-\sigma_p}^{EXACT}$ with various prestress $\sigma_p = 0$, 10, 20, 30, 40, 50, 80 and 100 MPa

the SIF of the unrepaired plate. For example, when $2a/W = 0.7714$, the SIF with $\sigma_p = 100$ MPa is reduced to about 60% of the unrepaired SIF as $F_{1-\sigma_p}^{EXACT} \approx 0.6F_{1-un}^{EXACT}$.

The exact SIF of the unrepaired plate F_{1-un}^{EXACT} can be expressed as shown in Eq. (4), that is, $F_{1-un}^{EXACT} = \left[1 - 0.025(2a/W)^2 + 0.06(2a/W)^4\right]\sqrt{\sec(2a/W)}$, when $2a/W \to 0$, $F_{1-un}^{EXACT} \to 1$ and when $2a/W \to 1$, $F_{1-un}^{EXACT} \to \infty$. As shown in Table 8, F_{1-un}^{EXACT} increases monotonically with increasing the crack length $2a/W$. Instead, the exact value F_{1-0}^{EXACT} of the repaired plate is in the range $0.9142 \le F_{1-0}^{EXACT} \le 1.095$ when $0.1 \le 2a/W \le 0.7714$. In other words, due to the repairment, the repaired plate specimen can be regarded as a central crack in an infinite plate with $2a/W \to 0$, which can be expressed as $K_{1-\sigma_p}^{EXACT} \approx \sigma\sqrt{\pi a}$. Those results shows that the fatigue life can be improved significantly by the repair as shown in Fig. 13.

With increasing the prestress σ_p, the SIF decreases irrespective of the crack length $2a/W$ are plotted in Fig. 16. Compared to the SIF with $\sigma_p = 0$, the SIF with $\sigma_p = 100$ MPa can be reduced by several percent. To clarify the prestress effect on the SIF on the SIF, Fig. 17 shows the SIF ratio $F_{1-\sigma_p}^{EXACT}/F_{1-un}^{EXACT}$ versus $2a/W$ for various prestress σ_p. From Fig. 17, it is seen the SIF ratio $F_{1-\sigma_p}^{EXACT}/F_{1-un}^{EXACT}$ decreases linearly with increasing the prestress ratio $\sigma_p/\sigma_{p\max}$. The SIF ratio $F_{1-\sigma_p}^{EXACT}/F_{1-un}^{EXACT}$ also decreases linearly with increasing the relative crack length $2a/W$ in the range $0.1 \le 2a/W \le 0.7714$ as shown in Eq. (5). The coefficient of determination (COD) of the fitting expression was very close to 1, indicating that the fitting quality is very high.

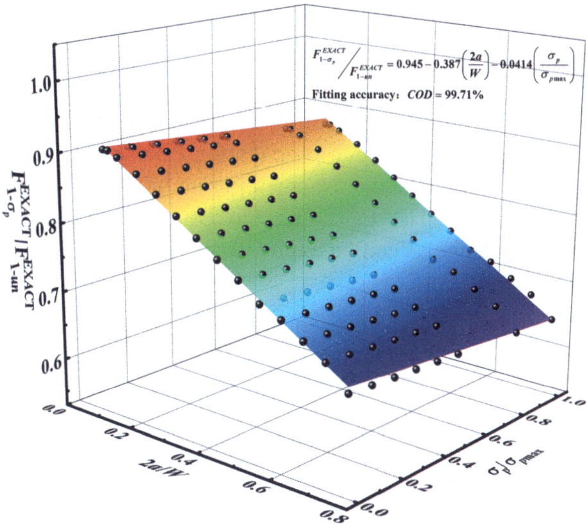

Fig. 17 The exact SIF ratio $F^{EXACT}_{1-\sigma_p}/F^{EXACT}_{1-un}$ for various prestress $\sigma_p = 0$, 10, 20, 30, 40, 50, 80 and 100 MPa

$$F^{EXACT}_{1-\sigma_p}/F^{EXACT}_{1-un} = 0.945 - 0.387\left(\frac{2a}{W}\right) - 0.0414\left(\frac{\sigma_p}{\sigma_{p\max}}\right) \tag{5}$$

In other words, $F^{EXACT}_{1-\sigma_p}$ can be expressed as follows.

$$\begin{aligned}
K^{EXACT}_{1-\sigma_P} &= F^{EXACT}_{1-\sigma_p}\sigma\sqrt{\pi a} \\
F^{EXACT}_{1-\sigma_P} &= \left[0.945 - 0.387\left(\frac{2a}{W}\right) - 0.0414\left(\frac{\sigma_p}{\sigma_{p\max}}\right)\right] \\
&\quad \times \left[1 - 0.025\left(\frac{2a}{W}\right)^2 + 0.06\left(\frac{2a}{W}\right)^2\right]\sqrt{\sec\left(\frac{\pi a}{W}\right)} \\
0.1 \le \frac{2a}{W} &\le 0.7714, \quad 0 \le \frac{\sigma_p}{\sigma_{p\max}} \le 1, \quad \sigma_{p\max} = 100\text{MPa}
\end{aligned} \tag{6}$$

where $K^{EXACT}_{1-\sigma_P}$ is the SIFs of steel plate repaired with prestressed CFRP (Unit: MPa m$^{1/2}$), σ_P is the prestress (Unit: MPa), $\sigma_{p\max}$ is the maximum of cycle load (Unit: MPa), a is the half crack length (Unit: mm), W is the width of steel plate (Unit: mm).

3.2 The Prediction of Fatigue Life

Obtaining the equation for calculating SIF of cracked steel plates repaired with prestressed CFRP is the first step in gaining an equation for fatigue life prediction. The difference between $F^{EXACT}_{1-\sigma_p}$ and F^{EXACT}_{1-un} is quite large up to about 60% at most, as shown in Table 9 and Fig. 16. However, Fig. 17 shows that the ratio $F_{1-\sigma_p}^{EXACT}/F_{1-un}^{EXACT}$ can be used conveniently to express $F_{1-\sigma_p}^{EXACT}$. Since the SIF of the cracked plate repaired by prestressed CFRP can be obtained in a convenient form as shown in Eq. (5), the fatigue life improvement can be predicted in the following way. It is known that fatigue crack propagation can be classified into three stages based on fracture mechanics, that is, crack initiation period, stable crack growth and unstable crack growth. The Pairs-Erdogan equation can be used to describe the stable extension stage as can be expressed in Eq. (6) [42].

$$\frac{da}{dN} = C(\Delta K)^m \tag{7}$$

$$\Delta K = K_{\max} - K_{\min} = (1 - R)K_{\max}, \quad R = K_{\min}/K_{\max} \tag{8}$$

where *da/dN* is the fatigue crack growth rate, ΔK is the stress intensity factor range. $K_{\max}$, $K_{\min}$ are maximum and minimum stress intensity factors in a load cycle, R is the stress ratio. In Eq. (6), C and m are material constants depending on R, obtained from the log-log pot of Eq. (6).

The beach marking method was employed to record fatigue crack growth in experiments, the ASTM E647-15 (2016) [43] indicated Secant method can determine the crack growth rate at each section recorded in testing. The formula is as follows:

$$\left(\frac{da}{dN}\right)_i = \frac{a_{i+1} - a_i}{N_{i+1} - N_i} \tag{9}$$

where a_i and a_{i+1} is the experimentally determined ith and i + 1th crack lengths; N_i and N_{i+1} is the number of load cycles corresponding to a_i and a_{i+1}, respectively. Since there were two specimens of cracked steel in both repair stages in the tests, both a and N were calculated as the average of the two specimens' observations. The average crack growth rate is determined by Eq. (8).

The log-log plot of $(da/dN)_i - (\Delta K)i$ relations for $i = 1$~6 is plotted in Fig. 18, and the linear relation can be seen except the last plots $i = 6$ of unrepaired plate. Although the crack propagation rates *da/dN* of unrepaired plate is slightly larger than *da/dN* of repaired plate, they can be expressed by the same equation within ±5.01% except the last plots $i = 6$ of unrepaired plate that is not on the linear relation. The equation to express the relation can be determined through intercepting and fitting the experimental data of the stable crack extension process. Then, $C = 6.670 \times 10^{-14}$, $m = 3.03$ are obtained.

Since Eq. (6) can be expressed as $\int dN = \int 1/[C(\Delta K)^m]da$, the number of cycles N can be expressed in Eq. (9) and Eq. (10) by substituting C and m into Eq. (6) using Eq. (4) and Eq. (5).

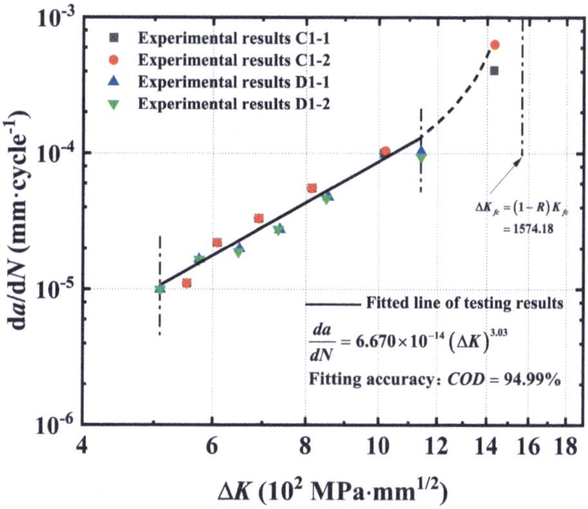

Fig. 18 $(da/dN) - (\Delta K)$ plots expressed as $da/dN = C(\Delta K)^m$, $C = 6.670 \times 10^{-14}$, $m = 3.03$ for the stable crack extension stage

$$N_{c-un} = \int_{a_0}^{a_{c-un}} \frac{da}{C\left[(1-R)K_{1-un}^{EXACT}\right]^m}, C = 6.670 \times 10^{-14}, \quad m = 3.03 \tag{10}$$

$$K_{1-un}^{EXACT} = \left[1 - 0.025\left(\frac{2a}{W}\right)^2 + 0.06\left(\frac{2a}{W}\right)^4\right]\sqrt{\sec\left(\frac{\pi a}{W}\right)}\sigma\sqrt{\pi a} \tag{11}$$

$$N_{c-\sigma_P} = \int_{a_0}^{a_{c-0}} \frac{da}{C\left[(1-R)K_{1-\sigma_P}^{EXACT}\right]^m}, C = 6.670 \times 10^{-14}, \quad m = 3.03 \tag{12}$$

$$\begin{aligned} K_{1-\sigma_P}^{EXACT} = & \left[0.945 - 0.387\left(\frac{2a}{W}\right) - 0.0414\left(\frac{\sigma_P}{\sigma_{P\max}}\right)\right]\left[1 - 0.025\left(\frac{2a}{W}\right)^2 \right. \\ & \left. + 0.06\left(\frac{2a}{W}\right)^4\right]\sqrt{\sec\left(\frac{\pi a}{W}\right)}\sigma\sqrt{\pi a} \end{aligned} \tag{13}$$

This equation can be used to rapidly calculate the fatigue life of Q235 steel plates with central cracks repaired by CFRP without or with different prestress, and this method can be applied to CRPP repairing plates with different materials, but the critical crack length should be known in advance to obtain fatigue life. Equation (9) is defined up to the final crack length $2a_{c\text{-}un} = 44.10$ mm and Eq. (10) is defined up to $2a_{c\text{-}0} = 53.0$ mm (see Table 5). From these final fatigue crack lengths where unstable fracture happens, the fatigue fracture toughness can be calculated for the unrepaired steel plate as $K_{fc} = 63.02$ MPa m$^{1/2}$ and for the repaired steel plate as $K_{fc} = 55.31$ MPa m$^{1/2}$ (see Table 5). Those values nearly coincide with each other within 12.24% difference. They are obtained under cyclic load and smaller than the fracture toughness obtained by a monotonic load $K_{fc} < K_c = 83.16$ MPa m$^{1/2}$ in

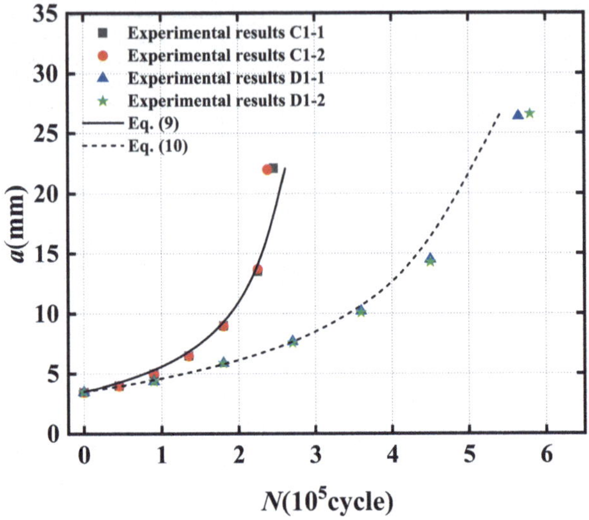

Fig. 19 Fatigue crack length versus number of cycles (*a-N* relation) obtained from Eqs. (9) and (10) compared to the experimental results in Fig. 13

Table 1. The difference of K_{fc} between the unrepaired and repaired plates can be explained by the spread of the peeling area of the CFRP repaired plate. This is because FEA always assumes a perfect bond until the final fracture yet crack opening and closing during cyclic loading may cause the peeling.

The basic form of the fatigue life prediction equation of unrepaired and repaired steel plates can be obtained by jointly using the rigorous SIF equation and the Paris-Erdogan equation. Equations (9) and (10) are defined from the initial crack length $2a_0 = 7.0$ mm to the final crack lengths $2a_{c\text{-}un} = 44.10$ mm, $2a_{c\text{-}0} = 53.0$ mm; therefore, $N_{c\text{-}un}$, $N_{c\text{-}\sigma p}$ are regarded as the number of cycles to failure for unrepaired and repaired steel plates. In this way, the fatigue crack length versus number of cycle (*a-N* relation) can be provided. Figure 19 shows the results of Eqs. (9) and (10) with $C = 6.670 \times 10^{-14}$, $m = 3.03$ in comparison with the experimental results. Equation (9) provides relatively larger value of fatigue life for the unrepaired plates; its difference is 7.81% compared to the experimental results. Equation (10) predicts the fatigue life of the CFRP repaired plate with less than 5.50% difference.

The predicted crack length (a-N relation) obtained from Eqs. (9) and (10) to clarify the effect of prestress are presented in Fig. 20. It should be mentioned that experiments are required for obtaining the final crack length of the repaired steel with prestressed CFRP. Nevertheless, the experiment was not carried out in this investigation.

To evaluate how prestress affects fatigue life, the experimental ultimate crack length of steel plates repaired with common CFRP was employed as the referenced value. Figure 20 shows that the fatigue life N_c can be extended by 2.07~2.41 times by repairing with the prestressed CFRP.

Fig. 20 Fatigue crack length versus number of cycles (*a*-*N* relation) obtained from Eqs. (9) and (10)

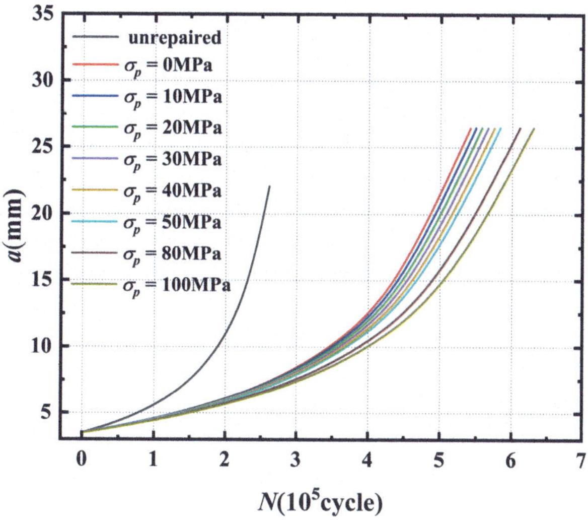

4 Conclusions

Cracked plates in steel structures are now widely repaired by prestressed CFRP. The safety and reliability of the repair are depending on their fatigue performance. In this study, first, the fatigue experiment was conducted for the unrepaired/repaired plates. Second, the SIF of the cracked plate with/without repair was calculated by using a standard FEM software. Third, the error due to the FEM mesh was evaluated by comparing with the exact solution for the unrepaired plate. Finally, the fatigue life extension was predicted for the cracked steel plate repaired by various prestressed CFRP. The conclusions can be summarized in the following way.

1. It was found that the error of the SIFs included in the FEA using a commercial software can be eliminated on the idea of the proportional method. The crack tip FEM stress ratio which was focused in the previous proportional method coincides with the FEM SIF ratio obtained by the finite element software when the same FEM mesh is applied to the unknown and reference problems. Since by taking the ratio the FEM mesh error is eliminated, the exact SIF of repaired plate can be provided from the exact solution of unrepaired plate and the FEM ratio.
2. It was found that the SIF ratio $F_{1-\sigma_p}^{EXACT}/F_{1-un}^{EXACT}$ of the repaired plate to the unrepaired plate decreases linearly with increasing the prestress $\sigma_p/\sigma_{p\max}$ in CFRP. The SIF ratio $F_{1-\sigma_p}^{EXACT}/F_{1-un}^{EXACT}$ also decreases linearly with increasing the relative crack length $2a/W$ as can be expressed in Eq. (5). Although the difference between $F_{1-\sigma_p}^{EXACT}$ and F_{1-un}^{EXACT} is quite large up to 60% at most it was found that the ratio $F_{1-\sigma_p}^{EXACT}/F_{1-un}^{EXACT}$ can be used conveniently to express $F_{1-\sigma_p}^{EXACT}$. The exact SIF expression (5) combined with Paris-Erdogan's law may predict the fatigue life of the repaired plate within 5.50% difference as shown in Fig. 19. The fatigue life N_c

can be extended by about two times longer by repairing the cracked steel plate with the prestressed CFRP.

3. The crack propagation rate experimentally obtained for unrepaired/repaired plates can be expressed by almost the same equation (see Fig. 19). Also, the fatigue fracture toughness of the unrepaired steel plate nearly coincides with the fatigue fracture toughness of repaired steel plate within 12.24% difference (see Table 5). The difference of K_{fc} between the unrepaired and repaired plates can be explained from that the possible spread of the peeling area of the CFRP repair plate. This is because FEA always assumes a perfect bond until the final fracture, yet crack opening and closing occurs during cyclic loading.

References

1. Al-Karawi H, Polach RUF, Al-Emrani M. Fatigue crack repair in welded structures via tungsten inert gas remelting and high frequency mechanical impact. J Constr Steel Res. 2020;172:12.
2. Mazaheri P, Asgarian B, Gholami H. Assessment of strengthening, modification, and repair techniques for aging fixed offshore steel platforms. Appl Ocean Res. 2021;110:102612.
3. Russian O, Belarbi A, Dawood M. Effect of surface preparation technique on fatigue performance of steel structures repaired with self-stressing SMA/CFRP patch. Compos Struct. 2022;280:114968.
4. Weitzenboeck JR, Mcgeorge D. A cold repair method for FPSOs. Presented at the offshore technology conference, Houston, TX, USA; 2012.
5. Liu H, Al-Mahaidi R, Zhao X. Experimental study of fatigue crack growth behaviour in adhesively reinforced steel structures. Compos Struct. 2009;90:12–20.
6. Xue WC, Tan Y, Zeng L. Flexural response predictions of reinforced concrete beams strengthened with prestressed CFRP plates. Compos Struct. 2010;92:612–22.
7. Wu C, Zhao XL, Al-Mahaidi R, Emdad MR, Duan WH. Fatigue tests of cracked steel plates strengthened with UHM CFRP plates. Adv Struct Eng. 2012;15:1801–15.
8. Yu QQ, Zhao XL, Al-Mahaidi R, Xiao ZG. Tests on cracked steel plates with different damage levels strengthened by CFRP laminates. Int J Struct Stab Dyn. 2014;14:1450018.
9. Kafodya I, Xian GJ, Li H. Durability study of pultruded CFRP plates immersed in water and seawater under sustained bending: water uptake and effects on the mechanical properties. Compos Part B Eng. 2015;70:138–48.
10. Al-Mosawe A, Al-Mahaidi R, Zhao XL. Bond behaviour between CFRP laminates and steel members under different loading rates. Compos Struct. 2016;148:236–51.
11. Batuwitage C, Fawzia S, Thambiratnam D, Al-Mahaidi R. Durability of CFRP strengthened steel plate double-strap joints in accelerated corrosion environments. Compos Struct. 2017;160: 1287–98.
12. Pang YY, Wu G, Wang HT, Su ZL, He XY. Experimental study on the bond behavior of CFRP-steel interfaces under quasi-static cyclic loading. Thin-Walled Struct. 2019;140:426–37.
13. Liu ZQ, Luo B, Wang Q, Feng B. Experimental and numerical investigation of the anti-debonding performance for novel CFRP-steel tube composite member under tension. J Build Eng. 2021;35:102004.
14. Colombi P, Fava G. Fatigue crack growth in steel beams strengthened by CFRP strips. Theor Appl Fract Mech. 2016;85:173–82.
15. Orcesi AD, Feraille A, Chataigner S. Fatigue strengthening of steel structures using high modulus CFRP plates: development of a life-cycle analysis approach. Constr Build Mater. 2019;227:116628.

16. Täljsten B, Hansen CS, Schmidt JW. Strengthening of old metallic structures in fatigue with prestressed and non-prestressed CFRP laminates. Constr Build Mater. 2009;23:1665–77.
17. Feng P, Hu LL, Zhao XL, Cheng L, Xu SH. Study on thermal effects on fatigue behavior of cracked steel plates strengthened by CFRP sheets. Thin-Walled Structures. 2014;82:311–20.
18. Li LZ, Chen T, Zhang NX, Hidekuma Y. Test on fatigue repair of central inclined cracked steel plates using different adhesives and CFRP, prestressed and non-prestressed. Compos Struct. 2019;216:350–9.
19. Liu RY, Chen T, Li LZ, Tateishi K. A practical stress intensity factor formula for CFRP-repaired steel plates with a central crack. J Constr Steel Res. 2019;162:105755.
20. Cabrilo A, Sedmak A, Burzic Z, Perkovic S. Fracture mechanics and fatigue crack propagation in armor steel welds. Eng Fail Anal. 2019;106:104155.
21. Chen T, Yao JX, Liu RY, Li LZ. Fatigue behavior of steel plates with multi-holes repaired by CFRP. Compos Struct. 2020;242:112163.
22. Jie ZY, Wang WJ, Fang RF, Zhuge P, Ding Y. Stress intensity factor and fatigue analysis of cracked cruciform welded joints strengthened by CFRP sheets considering the welding residual stress. Thin-Walled Struct. 2020;154:106818.
23. Hassan MM, Shafiq MA, Mourad SA. Experimental study on cracked steel plates with different damage levels strengthened by CFRP laminates. Int J Fatigue. 2021;142:105914.
24. Lepretre E, Chataigner S, Dieng L, Gaillet L. Stress intensity factor assessment for the reinforcement of cracked steel plates using prestressed or non-prestressed adhesively bonded CFRP. Materials. 2021;14:1625. https://doi.org/10.3390/ma14071625.
25. Tan H, Hu X, Wu X, Zeng Y, Tu X, Xu X, Qian J. Initial crack propagation of integral joint in steel truss arch bridges and its fatigue life accession. Eng Fail Anal. 2021;130:105777.
26. Shao W, Wang Y, Shi D. Corrosion-fatigue life prediction of reinforced concrete square piles in marine environments. Eng Fail Anal. 2022;138:106324.
27. Wang HT, Bian ZN, Chen MS, Hu L, Wu Q. Flexural strengthening of damaged steel beams with prestressed CFRP plates using a novel prestressing system. Eng Struct. 2023;284:115953.
28. Liu J, Zhao X-L, Xin H, Zhang Y. Prediction of fatigue crack propagation in center cracked steel plate strengthened with partially covered CFRP strip. Thin-Walled Struct. 2023;189: 110917.
29. Deng J, Fei Z, Wu Z, Li J, Huang W. Integrating SMA and CFRP for fatigue strengthening of edge-cracked steel plates. J Constr Steel Res. 2023;206:107931.
30. Standardization Administration of China. Metallic materials-tensile testing-part 1: method of test at room temperature. GB/T 228.1–2021, Standards Press of China, Beijing; 2022. (in Chinese).
31. Li Y, Lu P, Wang B, Hu X, Li D, Xiang Q. Evaluation of the fracture toughness of butt-welded joints using the boundary effect model. Eng Fract Mech. 2022;274:108777.
32. Dong DS, Mei X. Testing for Paris coefficients C, m and ΔK_{th} of Q235. J Mec Strength. 2003;2: 215–8.
33. Aljabar NJ, Zhao XL, Al-Mahaidi R, Ghafoori E, Motavalli M, Koay YC. Experimental investigation on the CFRP strengthening efficiency of steel plates with inclined cracks under fatigue loading. Eng Struct. 2018;172:877–90.
34. Abaqus. ABAQUS Version 2020. Dessault: Syst; 2020.
35. Nisitani H, Kawamura T, Fujisaki W, Fukuda T. Determination of highly accurate values of stress intensity factor or stress concentration factor of plate specimen by FEM. Trans Jpn Soc Mech Eng Seri A. 1999;65:26–31.
36. Nisitani H, Teranishi T. KI of a circumferential crack emanating from an ellipsoidal cavity obtained by the crack tip stress method in FEM. Eng Fract Mech Fract Damage Mech. 2004;71: 579–85.
37. Brown WF, Srawley Jr J. Plane strain crack toughness testing of high strength metallic materials, ASTM STP-410; 1966. p. 11.
38. Feddersen RE. Discussion to the above paper. ASTM STP-410; 1966. p. 77.

39. Isida M. Analysis of stress intensity factors for the tension of a centrally cracked strip with stiffened edges. Eng Fract Mech. 1973;5:647–65.
40. Oda K, Kamisugi K, Noda N. Analysis of stress intensity factor for interface cracks based on proportional method. Nihon Kikai Gakkai Ronbunshu, J-STAGE. 2009;A75:476–82.
41. Oda K, Lan X, Noda N, Michinaka K. Effect of arbitrary bi-material combination and bending loading conditions on stress intensity factors of an edge interface crack. Int J Struct Integr. 2012;3:457–75.
42. Paris P, Erdogan F. A critical analysis of crack propagation Laws. J Basic Eng. 1963;85:528–33.
43. ASTM E647-15. Standard test method for measurement of fatigue crack growth rates. West Conshohocken: ASTM International; 2016.

The Debonding Failure Analysis of Corroded Pipes Repaired with CFRP Under Tension and the Bending Moment

Nomenclature

A, I	Section area and polar moment of inertia
G, E	Shear modulus and elastic modulus
l	Adhesive layer length
l_1	Corrosion length
M, N, V	Cross-section bending moment, tensile load, and shear load
r, t	Radius and thickness
t_1	Corrosion depth
u, v, w	Radial, circumferential, and axial displacements in the cylindrical coordinate
θ_1	Corrosion angle
μ	Poisson's ratio
σ, τ	Normal stress and shear stress
Subscripts	r, θ, z are for the radial, the circumferential, and the axial in the cylindrical coordinate, respectively
Subscripts	s, a, c are for the metal pipe layer, the bonding layer, and the composite material layer, respectively

1 Introduction

The composite material are often used as an important repair method for corroded marine pipes, and it has attracted an increasing amount of attention in recent years. Fiber reinforced polymers (FRP) are typically composed of a polymer or plastic matrix and reinforced fiber. The most commonly used are carbon fiber reinforced polymers (CFRPs) and glass fiber reinforced polymers (GFRPs). By using experiments and numerical analyses, the effectiveness of composite materials for repairing cylindrical pipe structures has been verified [1, 2]. Due to the layered structural

Y. Zhang, *Interfacial Mechanics of Offshore Engineering Equipment*,
https://doi.org/10.1007/978-981-95-3200-1_10

characteristics of FRP bonded restorations, the bonding state between layers and the performance of the bonding interface are critical, but these are difficult issues in the research. There are few standards for repairing corroded pipes with composite materials. What does exist primarily includes PRCI code proposed by the Pipe Research Council International in 2006, the ASME PCC 2 code proposed by the American Society of Mechanical Engineers in 2006, the DNV-RP-F113 code proposed by the Det Norske Veritas in 2007, and the ISO 24817 code proposed in 2015 [3–5]. These standards specify partial requirements for composite materials to repair corroded pipes. There are still many problems to be solved in this new repair method.

According to the published literature [6, 7], it can be understood that no matter what material the repair piece belongs to, the failure mode of a composite reinforced repair piece under various limit load conditions is primarily a debonding failure. Zarrinzadeh [8] used a cohesive element to simulate the adhesive layer with damage parameters that are defined using the FORTRAN subroutine in ABAQUS for pipe crack propagation under a cyclic load. In addition, the debonding behavior of a cracked aluminum tube repaired using GFRP was obtained. Liu [9] used the staining method to study the patch debonding of CFRP wrapped cracked aluminum alloy tubes in an experiment. He visualized the debonding process and obtained relatively accurate experimental results.

Regarding interfacial stress, most scholars have directly applied the CFRP layer on the steel plate for the shear test of the CFRP laminate by placing two steel plates together for the tensile test of the double-end bond to study the bond failure mechanism. Based on the double CFRP strap-to-steel test, Yang [10] proposed a tri-linear bond-slip model and its closed-form solution, which included a strength and bond stress of the CFRP-to-steel interface. Zhang [11] proposed a new approximation numerical model to analyze the debonding stress and shear stress of CFRP patch reinforced steel plates. Zhang [12] used theoretical and finite element method to obtain the stress distribution in the adhesive layer of a cracked steel plate repaired with CFRP. The relationships between the maximum shear stress, the peel stress, and the crack length are also obtained. Based on this, he proposed a certain critical stress in the adhesive layer under different crack lengths and a simple method repair model of the adhesive failure. He [13] proposed new bond–slip models for the linear and nonlinear adhesive of the CFRP debonding failure mode according to experimental results.

Some scholars have studied the repairing effect of metal tubular structure repaired with CFRP under multiple loads. Xu and Li [14] solved theoretically the 3D stress state in an adhesively bonded fiber-reinforced composite tubular joint subjected to a torsional load through finite difference method. Parashar and Mertiny [15] studied various joining techniques for FRP piping through adhesive bonding and examined damage mechanisms for different loading conditions. Casalegno et al. [16] studied the joining properties of hypereutectic Al–Si alloy heat pipes to CRFP skins for radiator panel applications under three different epoxy adhesives. Also, they found thermal cycling did not significantly affect the mechanical properties of the joints or the failure propagation mode. Budhe [17] obtained the general derivation process of the failure pressure of perfect pipe reinforced by CFRP, on the basis of which the residual strength factor was used to obtain the failure pressure formula suitable for corroded pipes. Using an analytical method, Guz [18] studied the failure for different lay-ups and loading conditions. The stress distributions through the pipe thickness for various lay-ups were computed for fibre-reinforced pipes under different outer pressure magnitudes. Take into account the anisotropy of the material, stress and failure analysis was performed for the pipes with different lay-ups subjected to aouter pressure of various magnitudes analytically by Bouhafs [19]. In order to study the effect of Engagement Ratio on the shear strength of an epoxy adhesive applied to steel adherents, based on Volkersen theory, a stress distribution model was established to analyze its failure mechanism by Croccolo [20]. Considering orthotropy, Croccolo [21] deeply studied the axial push out force in adhesively bonded and press fitted connections made by a carbon epoxy bush press fitted into a steel housing by combining analytic and experimental methods. Okamoto [22] studied the mechanical properties of CFRP pressure vessels by internal pressure loading experiment and a finite element analysis using an equivalent method to consider the anisotropy CFRP. Ahmad [23] calculated the bending stiffness of the composite thick-walled tube using equivalent elastic modulus [24] and anisotropic constitutive model [25] separately, and comparison between two results shown that when the winding angle is from 50 to 90°, the results of the two are basically the same.

Due to the influence of seawater, low temperature, and other harsh environmental factors, the attenuation characteristics of the performance of the bond layer of repaired pipes have also attracted the attention of many scholars. Pang [26] investigated interfacial debonding between the CFRP plate and the steel plate under freeze-thaw cycles using experiments, and degradation models for the bond-slip parameters were formulated. Yang [27] tested the bond response of CFRP-to-steel joints after salt fog exposure using an experimental test.

Although many papers studied CFRP and structures repaired with CFRP considering material anisotropy, the initial debonding failure mechanism has not been studied theoretically. Debonding is a serious defect in CFRP repaired pipes, especially for marine pipes. This is because when the initial debonding occurs in a repaired marine pipe, the bonding layer will be quickly destroyed due to the effect of seawater. In this study, the stress distribution theory of the bonding layer of pipes repaired using CFRP under tension and bending is presented considering the anisotropic CFRP using an equivalent method [22]. Then, the Quads Damage criterion is used to calculate the initial debonding load, and the results are verified using the experiment and the finite element method. Finally, the attenuation law of the initial debonding load under long-term seawater immersion is examined.

2 The Theoretical Analysis of Interfacial Stresses in a Pipe Reinforced with CFRP

In order to obtain the interfacial stresses of a corroded pipe repaired with CFRP, the mechanical model of a perfect pipe reinforced with CFRP is established firstly, then they are modified with a residual strength factor considering the corrosion geometry. The cross section structure of a pipe reinforced with CFRP is shown in Fig. 1. It is primarily composed of an inner steel pipe layer, a middle adhesive layer, and an outer CFRP layer. The stress distribution of the bonding layer is obtained using a stress analysis. Since the focus of this paper is the interfaces of the CFRP and pipe, the composite layer is considered as a whole, and the equivalent modules and Poison's ratio are obtained [22]. As the adhesive layer is very thin, the stress is independent of the radial position. In addition, the tensile load is an axisymmetric force, so the stress is independent of the circumferential position. The $dz{*}d\theta$ infinitesimal segment shown in Fig. 2 is extracted for mechanical analysis. The free body diagram of the structure under tension and bending load is shown in Fig. 3.

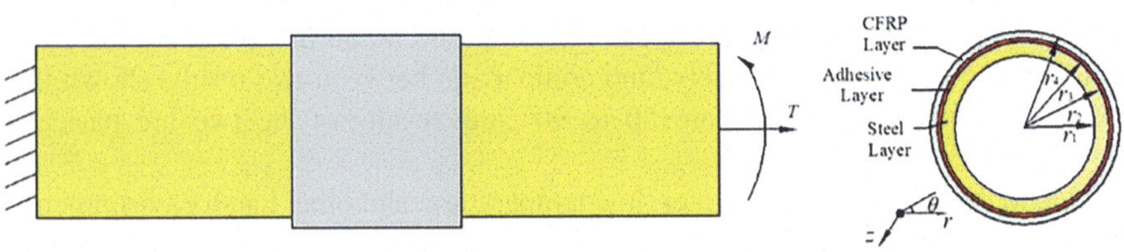

Fig. 1 The section structural sketch of pipeline reinforced with CFRP

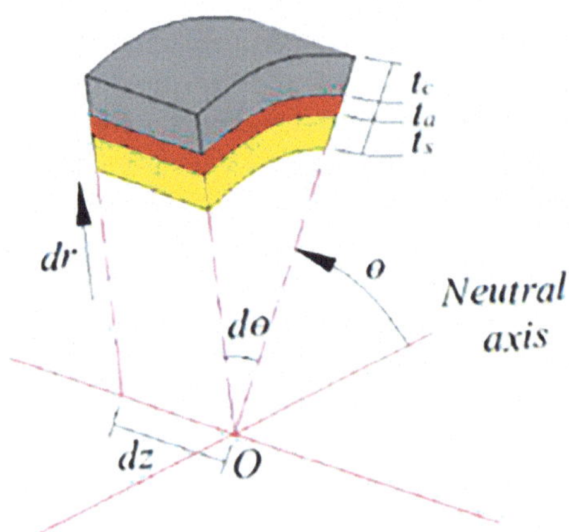

Fig. 2 The $dz{*}d\theta$ infinitesimal segment extracted for mechanical analysis

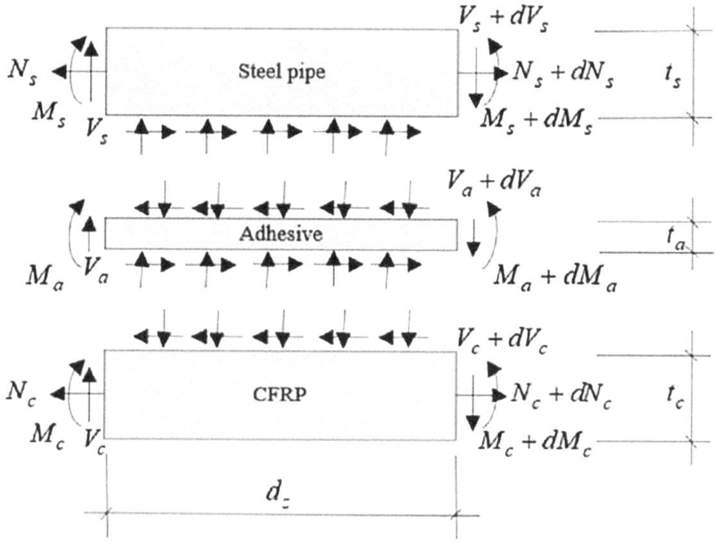

Fig. 3 The free body diagram of infinitesimal segment under tensile load

2.1 Assumptions

1. The structure satisfies the hypothesis of a plane section, and the circumferential displacement is 0; that is, the radial, circumferential, and axial displacement functions of any layer are respectively $u = u(r, z)$, $v = 0$, $w = w(r, z)$.
2. Ignoring the axial tension of the bonding layer, the bonding layer is regarded as the plane stress problem of the circumferential section, and the stress assumption of the bonding layer is $\sigma_z = 0$, $\sigma_\theta = \tau_{\theta r} = \tau_{\theta z} = 0$.
3. The bonding layer is very thin, so the stresses of the upper and lower bonding surfaces of the layer are the same.
4. All materials are considered to be linear elastic.

2.2 Stresses on the Adhesive Layer Interface Under a Tensile Load

1. Interfacial shear stresses
   First, it is assumed that the shear stress at the upper and lower interfaces of the bonding layer is τ_{rz}, and the shear stress does not change with the radial position.

The function of the shear strain can be expressed by the displacement of the bonding layer:

$$\gamma_{arz} = \frac{\partial w_a(r,z)}{\partial r} + \frac{\partial u_a(r,z)}{\partial z} \tag{1}$$

The shear stress at the interfaces of the bonding layer is:

$$\tau_{rz} = G_a \gamma_{arz} = G_a \left(\frac{\partial w_a(r,z)}{\partial r} + \frac{\partial u_a(r,z)}{\partial z} \right) \tag{2}$$

The derivative of the interface shear stress of the bonding layer with respect to z is obtained:

$$\frac{d\tau_{rz}}{dz} = G_a \left(\frac{\partial^2 w_a(r,z)}{\partial r dz} + \frac{\partial^2 u_a(r,z)}{\partial z^2} \right) \tag{3}$$

For the second term in Eq. (3), by using the derivative element of the dθ segment and approximating it to the beam structure, the bending deflection can be expressed as in Eq. (4):

$$\frac{\partial^2 u_a(r,z)}{\partial z^2} = \frac{M_a}{E_a I_a} \tag{4}$$

The moment equilibrium equation is then presented for the bonding layer as the following:

$$dM_\alpha + \tau_{zr} r d\theta t_a dz - \tau_{rz} r d\theta dz t_\alpha = 0 \tag{5}$$

Since both ends of the bonding layer are free, there is no external bending moment on the end face, so $M_a = 0$, i. e:

$$\frac{\partial^2 u_a(r,z)}{\partial z^2} = 0 \tag{6}$$

As the shear stress is assumed to be constant independent of r, the axial displacement of the bonding layer should be linearly related to the radial position in Eq. (3), and the radial derivative of the axial displacement of the bonding layer can be obtained.

$$\frac{\partial w_a(r,z)}{\partial r} = \frac{w_c(r=r_3) - w_s(r=r_2)}{t_a} \tag{7}$$

$$\frac{\frac{\partial w_c(r=r_3)}{\partial z}-\frac{\partial r w_s(r=r_2)}{\partial z}}{t_a} \tag{8}$$

Both the bending moment and the tensile load cause axial strain. The axial strain can be expressed for the inner surface of the composite material layer as the following:

$$\frac{\partial w_c(r=r_3)}{\partial z}=\frac{N_c}{E_cA_c}-\frac{M_ct_c/2}{E_cI_c} \tag{9}$$

In addition, for the outer surface of the steel pipe, it can be expressed as:

$$\frac{\partial w_s(r=r_2)}{\partial z}=\frac{N_s}{E_sA_s}+\frac{M_st_s/2}{E_sI_s} \tag{10}$$

By substituting Eqs. (9) and (10) into Eq. (8), the following is obtained:

$$\frac{\partial^2 w_a(\boldsymbol{r},z)}{\partial r\partial z}=\frac{\frac{N_c}{E_cA_c}-\frac{M_ct_c/2}{E_cI_c}-\frac{N_s}{E_sA_s}-\frac{M_st_s/2}{E_sI_s}}{t_a} \tag{11}$$

By substituting Eq. (11) and Eq. (6) into Eq. (3) and taking the second derivative of the shear stress with respect to z, the following is obtained:

$$\frac{d^2\tau_{arz}}{dz^2}=G_a\frac{\frac{1}{E_cA_c}\frac{dN_c}{dz}-\frac{t_c/2}{E_cI_c}\frac{dM_c}{dz}-\frac{1}{E_sA_s}\frac{dN_s}{dz}-\frac{t_s/2}{E_sI_s}\frac{dM_s}{dz}}{t_a} \tag{12}$$

For each differential term in Eq. (12), the equilibrium equation is solved separately. For the composite material portion, the bending moment equilibrium equation and the axial equilibrium equation are listed respectively below (Eqs. 13 and 14), and the higher order terms in the equation are ignored.

$$dM_c-V_cdz+\tau_{rz}r_3d\theta dzt_c/2-\sigma_{cr}r_3d\theta dzdz/2=0 \tag{13}$$

$$(N_c+dN_c)-N_c-\tau_{rz}2\pi r_3dz=0 \tag{14}$$

$$\frac{dM_c}{dz}=V_c-\tau_{rz}r_3d\theta t_c/2 \tag{15}$$

$$\frac{dN_c}{dz}=2\pi r_3\tau_{rz} \tag{16}$$

Similarly, for the steel portion, the bending moment equilibrium equation and the axial equilibrium equation are listed respectively below (Eqs. 17 and 18), and the higher order terms in the equation are ignored.

$$\frac{dM_S}{dz} = V_s - \tau_{rz} r_2 d\theta t_s/2 \tag{17}$$

$$\frac{dN_S}{dz} = -2\pi r_2 \tau_{rz} \tag{18}$$

By substituting Eqs. (15)–(18) into Eq. (12), the following is obtained:

$$\frac{d^2\tau_{rz}}{dz^2} = \frac{G_a}{t_a}\left(\frac{2\pi r_3}{E_c A_c} + \frac{2\pi r_2}{E_s A_s} + \frac{r_3 (t_c/2)^2 d\theta}{E_c I_c} + \frac{r_2 (t_s/2)^2 d\theta}{E_s I_s}\right)\tau_{rz} - \frac{G_a}{t_a}\left(\frac{t_c/2}{E_c I_c} V_c + \frac{t_s/2}{E_s I_s} V_s\right) \tag{19}$$

where V_c and V_s are the shear force on the composite portion and steel pipe, respectively. It is assumed that the shear stress of the section changes linearly along the radial direction [28].

$$V_c = \int_{r_3}^{r_4} \frac{\tau_{r_2}}{t_c}(r_4 - r) r dr d\theta = \left(\frac{r_4^3}{6} - \frac{r_3^2 r_4}{2} + \frac{r_3^3}{3}\right)\frac{\tau_{r_z}}{t_c} d\theta \tag{20}$$

$$V_s = \int_{r_1}^{r_2} \frac{\tau_{rz}}{t_s} r^2 dr d\theta = \left(\frac{r_2^3 - r_1^3}{3}\right)\frac{\tau_{rz}}{t_s} d\theta \tag{21}$$

By substituting Eqs. (20) and (21) into Eq. (19), the following is obtained:

$$\frac{d^2\tau_{rz}}{dz^2} = \frac{G_a}{t_a}\left(\begin{array}{l} \frac{2\pi r_3}{E_c A_c} + \frac{2\pi r_2}{E_s A_s} + \frac{r_3 (t_c/2)^2 d\theta}{E_c I_c} + \frac{r_2 (t_s/2)^2 d\theta}{E_s I_s} \\ -\frac{1}{2E_c I_c}\left(\frac{r_4^3}{6} - \frac{r_3^2 r_4}{2} + \frac{r_3^3}{3}\right) d\theta - \frac{1}{2E_s I_s}\left(\frac{r_2^3 - r_1^3}{3}\right) d\theta \end{array}\right)\tau_{rz} \tag{22}$$

$$\frac{d^2\tau_{rz}}{dz^2} = \lambda^2 \tau_{rz} \tag{23}$$

The shear stress of the adhesive is obtained as expressed in Eq. (24):

$$\tau_{rz} = c_1 e^{\lambda z} + c_2 e^{-\lambda z} \tag{24}$$

According to the shear stress boundary condition of the adhesive, the shear stress is an odd function of the axial position with the bonding center as the origin. Since the bending moments of the sections at the two boundaries are 0, and the axial tension loads at the two ends of composite layer are 0, the first-order derivative values of the shear stress at the two boundaries of the bonding layer are constant.

$$\frac{d\tau_{rz}}{dz}\left(z = -\frac{l}{2}\right) = \frac{d\tau_{rz}}{dz}\left(z = \frac{l}{2}\right) = -G_a \frac{N_s}{E_s A_s t_a} \tag{25}$$

2. Interfacial normal stresses

The normal stress of the bonding layer is only related to the axial position and is independent of the radial position. It is assumed that the normal stresses of the upper and lower interfaces both are σ_r, and thc normal stress of the bonding layer is expressed as:

$$\sigma_r = -E_a \frac{u_c(r=r_3) - u_s(r=r_2)}{t_a} \tag{26}$$

The second derivative of the normal stress with respect to z is then obtained:

$$\frac{d^2\sigma_r}{dz^2} = -\frac{E_\alpha}{t_\alpha} \frac{d^2u_c(r=r_3) - d^2u_s(r=r_2)}{dz^2} \tag{27}$$

The composite layer and the steel pipe layer are analyzed separately, and the radial strain is primarily caused by the radial normal stress and the bending moment. For the composite layer, the radial deformation caused by bending is:

$$\frac{d^2u_c(r=r3)}{dz^2} = \frac{M_c}{E_c I_c} \tag{28}$$

The radial displacement caused by the radial normal stress is calculated according to the theory of a thick-walled cylinder [29].

$$u_c(r=r_3) = \frac{1-\mu_c}{E_c} \frac{\sigma_r r_3^3}{r_4^2 - r_3^2} + \frac{1+\mu_c}{E_c} \frac{\sigma_r r_3 r_4^2}{(r_4^2 - r_3^2)} \tag{29}$$

$$\frac{d^2u_c(r=r_3)}{dz^2} = \left(\frac{1-\mu_c}{E_c} \frac{r_3^3}{r_4^2 - r_3^2} + \frac{1+\mu_c}{E_c} \frac{r_3 r_4^2}{r_4^2 - r_3^2}\right) \frac{d^2\sigma_r}{dz^2} \tag{30}$$

Similarly, for the steel pipe portion, the radial deformation caused by the bending is:

$$\frac{d^2u_s(r=r_2)}{dz^2} = \frac{M_s}{E_s I_s} \tag{31}$$

The radial displacement caused by radial normal stress is:

$$\frac{d^2u_s(r=r_2)}{dz^2} = -\left(\frac{1-\mu_s}{E_s} \frac{r_2^3}{r_2^2 - r_1^2} + \frac{1+\mu_s}{E_s} \frac{r_2 r_1^2}{r_2^2 - r_1^2}\right) \frac{d^2\sigma_r}{dz^2} \tag{32}$$

By substituting Eq. (30), Eq. (32) and Eq. (27) into Eq. (32), the following is obtained:

$$\frac{d^2\sigma_r}{dz^2} = \frac{E_a}{t_a}\left[\begin{array}{c} -\left(\frac{1-\mu_c}{E_c}\frac{r_3^3}{r_4^2-r_3^2}+\frac{1+\mu_c}{E_c}\frac{r_3r_4^2}{r_4^2-r_3^2}\right)\frac{d^2\sigma_r}{dz^2}-\frac{M_c}{E_cI_c}- \\ \left(\frac{1-\mu_s}{E_s}\frac{r_2^3}{r_2^2-r_1^2}+\frac{1+\mu_s}{E_s}\frac{r_1^2r_2}{r_2^2-r_1^2}\right)\frac{d^2\sigma_r}{dz^2}+\frac{M_s}{E_sI_s} \end{array}\right] \tag{33}$$

$$\begin{gathered} \left[\begin{array}{c} 1+\frac{E_a}{t_a}\left(\frac{1-\mu_c}{E_c}\frac{r_3^3}{r_4^2-r_3^2}+\frac{1+\mu_c}{E_c}\frac{r_3r_4^2}{r_4^2-r_3^2}\right) \\ +\frac{E_a}{t_a}\left(\frac{1-\mu_s}{E_s}\frac{r_2^3}{r_2^2-r_1^2}+\frac{1+\mu_s}{E_s}\frac{r_1^2r_2}{(r_2^2-r_1^2)}\right) \end{array}\right]\frac{d^4\sigma_r}{dz^4} \\ = \frac{E_a}{t_a}\left[\frac{1}{E_sI_s}\frac{d(V_s-\tau_{r2}r_2d\theta t_s/2)}{dz}-\frac{1}{E_cI_c}\frac{d(V_c-\tau_{rz}r_3d\theta t_c/2)}{dz}\right] \end{gathered} \tag{34}$$

By substituting Eqs. (15) and (17) into Eq. (33) and seeking the first order on both sides of the equation, the following is obtained:

For the steel pipe and the composite material portions, the radial equilibrium equations are listed respectively as follows:

$$\sigma_r r_2 d\theta dz - dV_s = 0 \tag{35}$$

$$\sigma_r r_3 d\theta dz + dV_c = 0 \tag{36}$$

By substituting Eqs. (35) and (36) into Eq. (34), the following is obtained:

$$\begin{gathered} \left[\begin{array}{c} 1+\frac{E_a}{t_a}\left(\frac{1-\mu_c}{E_c}\frac{r_3^3}{r_4^2-r_3^2}+\frac{1+\mu_c}{E_c}\frac{r_3r_4^2}{r_4^2-r_3^2}\right) \\ +\frac{E_a}{t_a}\left(\frac{1-\mu_s}{E_s}\frac{r_2^3}{r_2^2-r_1^2}+\frac{1+\mu_s}{E_s}\frac{r_1^2r_2}{(r_2^2-r_1^2)}\right) \end{array}\right]\frac{d^4\sigma_r}{dz^4} \\ = \frac{E_a}{t_a}\left[\frac{1}{E_sI_s}\frac{d(V_s-\tau_r r_2d\theta t_s/2)}{dz}-\frac{1}{E_cI_c}\frac{d(V_c-\tau_r r_3d\theta t_c/2)}{dz}\right] \\ = \frac{E_a}{t_a}\left(\frac{1}{E_sI_s}r_2d\theta+\frac{1}{E_cI_c}r_3d\theta\right)\sigma_r+\frac{E_a}{t_a}\left(-\frac{1}{E_sI_s}r_2d\theta t_s/2+\frac{1}{E_cI_c}r_3d\theta t_s/2\right)\frac{d\tau_{rz}}{dz} \end{gathered} \tag{37}$$

$$\sigma_r^{(4)} = \gamma^4\sigma_r + \beta\frac{d\tau_{rz}}{dz} \tag{38}$$

The interface normal stress is then obtained using the following:

$$\begin{gathered} \sigma_r = e^{-\gamma z}(s_1\cos(\gamma z)+s_2\sin(\gamma z))+e^{\gamma z}(s_3\cos(\gamma z)+s_4\sin(\gamma z)) \\ +\frac{\beta}{\lambda^4-\gamma^4}\frac{d\tau_{rz}}{dz} \end{gathered} \tag{39}$$

where the coefficients are the following:

$$\gamma=\sqrt[4]{\frac{\frac{E_a}{t_a}\left(\frac{1}{E_sI_s}r_2d\theta+\frac{1}{E_cI_c}r_3d\theta\right)}{1+\frac{E_a}{t_a}\left(\frac{1-\mu_c}{E_c}\frac{r_3^3}{r_4^2-r_3^2}+\frac{1+\mu_c}{E_c}\frac{r_3r_4^2}{r_4^2-r_3^2}\right)+\frac{E_a}{t_a}\left(\frac{1-\mu_s}{E_s}\frac{r_2^3}{r_2^2-r_1^2}+\frac{1+\mu_s}{E_s}\frac{r_1^2r_2}{(r_2^2-r_1^2)}\right)}}$$

$$\beta=\frac{\frac{E_a}{t_a}\left(-\frac{1}{E_sI_s}r_2d\theta t_s/2+\frac{1}{E_cI_c}r_3d\theta t_c/2\right)}{1+\frac{E_a}{t_a}\left(\frac{1-\mu_c}{E_c}\frac{r_3^3}{r_4^2-r_3^2}+\frac{1+\mu_c}{E_c}\frac{r_3r_4^2}{r_4^2-r_3^2}\right)+\frac{E_a}{t_a}\left(\frac{1-\mu_s}{E_s}\frac{r_2^3}{r_2^2-r_1^2}+\frac{1+\mu_s}{E_s}\frac{r_1^2r_2}{(r_2^2-r_1^2)}\right)} \tag{40}$$

According to the normal stress boundary conditions of the bonding layer, the radial strain of the bonding layer at the boundary is caused by the radial normal stress. There is no bending moment and shear force at the boundary interface, and the second derivative of the normal stress at the boundaries are 0.

$$\frac{d^2\sigma_r\left(-\frac{l}{2}\right)}{dz^2}=\frac{d^2\sigma_r\left(\frac{l}{2}\right)}{dz^2}=0 \tag{41}$$

The third derivatives of the normal stress at the boundaries are:

$$\frac{d^3\sigma_r\left(-\frac{l}{2}\right)}{dz^3}=\frac{\frac{E_a}{t_a}\left[\begin{array}{l}\frac{1}{E_cl_c}\left(\tau_{rz}\left(-\frac{l}{2}\right)r_3d\theta t_c/2\right)\\-\frac{1}{E_sl_s}\left(\tau_{rz}\left(-\frac{l}{2}\right)r_2d\theta t_s/2\right)\end{array}\right]}{1+\frac{E_a}{t_a}\left[\frac{1-\mu_c}{E_c}\frac{r_3^3}{r_4^2-r_3^2}+\frac{1+\mu_c}{E_c}\frac{r_3r_4^2}{r_4^2-r_3^2}\right]+\frac{E_a}{t_a}\left[\frac{1-\mu_s}{E_s}\frac{r_2^3}{r_2^2-r_1^2}+\frac{1+\mu_s}{E_s}\frac{r_1^2r_2}{r_2^2-r_1^2}\right]} \tag{42}$$

$$\frac{d^3\sigma_r\left(\frac{l}{2}\right)}{dz^3}=\frac{\frac{E_a}{t_a}\left[\begin{array}{l}\frac{1}{E_cl_c}\left(\tau_{rz}\left(\frac{l}{2}\right)r_3d\theta t_c/2\right)\\-\frac{1}{E_sl_s}\left(\tau_{rz}\left(\frac{l}{2}\right)r_2d\theta t_s/2\right)\end{array}\right]}{1+\frac{E_a}{t_a}\left[\frac{1-\mu_c}{E_c}\frac{r_3^3}{r_4^2-r_3^2}+\frac{1+\mu_c}{E_c}\frac{r_3r_4^2}{r_4^2-r_3^2}\right]+\frac{E_a}{t_a}\left[\frac{1-\mu_s}{E_s}\frac{r_2^3}{r_2^2-r_1^2}+\frac{1+\mu_s}{E_s}\frac{r_1^2r_2}{r_2^2-r_1^2}\right]} \tag{43}$$

Then the coefficients s_1, s_2, s_3 and s_4 can be solved by simultaneous Eqs. (39) and (41)–(43).

2.3 Stresses on the Adhesive Layer Interface Under the Bending Moment

The free body diagram under the bending moment is shown in Fig. 3, assuming that the stress at the bonding layer interface is comprised of the radial normal stress and the shear stress.

For the entire steel pipe layer, the bending load on the steel pipe layer at the z position is $M(z)$, and the size is related to the z coordinate. By ignoring the influence of the interface shear stress on the steel pipe strain, the section shape of the steel pipe remains unchanged. The strain of the steel pipe layer is proportional to the distance from the neutral layer. When the structure cross section is circular, it is proportional to the product of the radius r and sin θ(θ is for the angle between the line from this point to the center of the circle and the neutral axis in the cross section).

The axial normal stress of steel pipe layer under bending is:

$$\begin{aligned} \sigma_{sz} &= \frac{M_s(z)}{I_s} y = \frac{M_s(z)}{I_s} r \sin\theta \\ I_s &= \frac{\pi\left(d_2^4 - d_1^4\right)}{64} \end{aligned} \tag{44}$$

$$\varepsilon_{sz} = \frac{1}{E_s}\sigma_{sz} = \frac{M_s(z)}{E_s I_s} r \sin\theta \tag{45}$$

The moment equilibrium equation is presented for a dz segment of the steel pipe layer.

$$\begin{aligned} dM_s(z) - \int_0^{2\pi} \tau(z,\theta) r_2 d\theta dz r_2 \sin\theta = 0 \\ \frac{dM_s(z)}{dz} = \int_0^{2\pi} \tau(z,\theta) r_2 r_2 \sin\theta d\theta \end{aligned} \tag{46}$$

For the bonding layer, the internal and external surfaces of the adhesive are affected by shear stress and normal stress. By considering a dθ*dz element, the axial strain of the adhesive is:

$$\varepsilon_{az} = \frac{-\mu_a}{E_a}\sigma_{ar} \tag{47}$$

For the contact surface within the adhesion, assume that the radial stress at the interface of the bonding layer is $\sigma(z, \theta)$. Hence, the strain at the inner interface of the adhesive is equal to that of the steel pipe layer on the outer surface. Therefore, the following is obtained:

$$\begin{aligned} \frac{-\mu_a}{E_a}\sigma(z,\theta) &= \frac{M_s(z)}{E_s I_s} r_2 \sin\theta \\ \sigma(z,\theta) &= \frac{E_a M_s(z)}{\mu_a E_s I_s} r_2 \sin\theta \end{aligned} \tag{48}$$

Take the derivative of $\sigma(z, \theta)$ with respect to z and substitute Eq. (46) into the following equation:

$$\frac{d\sigma(z,\theta)}{dz} = -\frac{E_a r_2 \sin\theta}{\mu_a E_s I_s}\frac{dM_s(z)}{dz} = -\frac{E_a r_2 \sin\theta}{\mu_a E_s I_s}\int_0^{2\pi} \tau(z,\theta) r_2 r_2 \sin\theta d\theta \tag{49}$$

Based on the equation $\tau(z, \theta) = f(z) \sin \theta$, the following is obtained:

$$\frac{d\sigma(z,\theta)}{dz} = -\frac{\pi E_a r_2^3 f(z)}{\mu_a E_s I_s}\sin\theta \tag{50}$$

The bending moment equilibrium equation for the bonded layer is as follows:

$$(M_a + dM_a) + \tau(z,\theta) r_2 d\theta t_a dz + \tau(z,\theta) r_2 d\theta dz t_a - M_a = 0 \tag{51}$$

The bending moment, M_a, of the bonding layer is 0 because the external bending moment at the bonding boundary is 0. Since the shear stress of the cross-section of the bonding layer does not change with the radial position, the shear strain generated by the shear stress is the same in different radial positions and is represented by the following:

$$\gamma_{arz} = \frac{\tau(z,\theta)}{G_a} \tag{52}$$

The axial displacement relation of the inner and outer bonding surface is:

$$w_a(r = r_3) = w_a(r = r_2) - \frac{\tau(z,\theta)}{G_a} t_a \tag{53}$$

The derivative of both sides with respect to z can be obtained as follows:

$$\varepsilon_a(r = r_3) = \varepsilon_a(r = r_2) - \frac{t_a}{G_a}\frac{d\tau(z,\theta)}{dz} = \frac{M_s(z)}{E_s I_s} r_2 \sin\theta - \frac{t_a}{G_a}\frac{d\tau(z,\theta)}{dz} \tag{54}$$

For the CFPR layer, the internal surface is affected by shear stress and normal stress. By considering a $d\theta$*dz element, the axial equilibrium equation of the segment is:

$$\begin{gathered} N_c + dN_c + \tau(z,\theta)r_3 d\theta dz - N_c = 0 \\ \frac{dN_c}{dz} = \tau(z,\theta)r_3 d\theta \\ \frac{d\sigma_{cz}}{dz} = \frac{\tau(z,\theta)}{t_c} \end{gathered} \tag{55}$$

The bending moment equilibrium equation is:

$$\begin{gathered} (M_c + dM_c) - V_c dz + \tau(z,\theta)r_3 d\theta dz\frac{t_c}{2} - M_c = 0 \\ \frac{dM_c}{dz} = V_c - \tau(z,\theta)r_3 d\theta\frac{t_c}{2} \end{gathered} \tag{56}$$

V_c is the shear force in the section, and it is assumed that the shear stress in the axial section is distributed linearly along the radial direction. Therefore, the following is true:

$$V_c = \int_{r_3}^{r_4} \frac{\tau(z,\theta)}{t_c}(r_4 - r)rdrd\theta = \left(\frac{r_4^3}{6} - \frac{r_3^2 r_4}{2} + \frac{r_3^3}{3}\right)\frac{\tau(z,\theta)}{t_c}d\theta \tag{57}$$

The axial displacement of the interfaces between the composite material layers caused by shear strain is the following:

$$w_c' = \int_{t_c/2}^{t_c} \frac{\tau(z,\theta)}{t_c G_c} ydy = \frac{3t_c\tau(z,\theta)}{8G_c} \tag{58}$$

The axial stress of the interfaces between the composite material layers is the following:

$$\begin{gathered} \varepsilon_{cz}(r = r_3) = \frac{1}{E_c}(\sigma_{cz} - \mu_c\sigma(z,\theta)) + \frac{M_c}{E_c I_c}\frac{t_c}{2} + \frac{3t_c}{8G_c}\frac{d\tau(z,\theta)}{dz} \\ \frac{d\varepsilon_{cz}}{dz} = \frac{1}{E_c t_c}\frac{d\tau(z,\theta)}{dz} + \frac{\mu_c}{E_c}\frac{d\sigma_{cr}}{dz} + \frac{dM_c}{dz}\frac{t_c}{2E_c I_c} \end{gathered} \tag{59}$$

According to the equivalence of the interface strain between the CFRP layer and the adhesive, that is $\varepsilon_{cz}(r = r_3) = \varepsilon_{az}(r = r_3)$, the following is obtained:

$$\frac{1}{E_c}(\sigma_{cz} - \mu_c\sigma(z,\theta)) + \frac{M_c}{E_c I_c}\frac{t_c}{2} + \frac{3t_c}{8G_c}\frac{d\tau(z,\theta)}{dz} = \frac{M_s(z)}{E_s I_s}r_2\sin\theta - \frac{t_a}{G_a}\frac{d\tau(z,\theta)}{dz} \tag{60}$$

The derivative of both sides with respect to z can be obtained as follows:

$$\begin{aligned}\frac{1}{E_c}\left(\frac{d\sigma_{cz}}{dz}-\mu_c\frac{d\sigma(z,\theta)}{dz}\right)+\frac{t_c}{2E_cI_c}\frac{dM_c}{dz}+\frac{3t_c}{8G_c}\frac{d^2\tau(z,\theta)}{dz^2}\\=\frac{r_2\sin\theta}{E_sI_s}\frac{dM_s(z)}{dz}-\frac{t_a}{G_a}\frac{d^2\tau(z,\theta)}{dz^2}\end{aligned} \tag{61}$$

By substituting Eqs. (47), (51), (56), and (57) into the above equation, the following is obtained:

$$\begin{aligned}\frac{1}{E_c}\frac{\tau(z,\theta)}{t_c}+\frac{E_a\mu_c r_2\sin\theta}{E_c\mu_a E_sI_s}\int_0^{2\pi}\tau(z,\theta)r_2r_2\sin\theta d\theta+\frac{t_c}{2E_cI_c}\left(V_c-\tau(z,\theta)r_3d\theta\frac{t_c}{2}\right)\\+\frac{3t_c}{8G_c}\frac{d^2\tau(z,\theta)}{dz^2}=\frac{r_2\sin\theta}{E_sI_s}\int_0^{2\pi}\tau(z,\theta)r_2r_2\sin\theta d\theta-\frac{t_a}{G_a}\frac{d^2\tau(z,\theta)}{dz^2}\end{aligned} \tag{62}$$

Based on $\tau(z,\theta)=f(z)\sin\theta$, the expression of $f(z)$ can be extracted from the above equation.

$$\begin{aligned}f''(z)-\eta^2f(z)=0\\f(z)=c_1e^{\eta x}+c_2e^{\eta x}\end{aligned} \tag{63}$$

The solution result is as follows:

$$\begin{aligned}&\tau(z,\theta)=\left(c_1e^{T_2}+c_2e^{\eta_2}\right)\sin\theta\\&\eta=\sqrt{\frac{\frac{1}{E_ct_c}-\frac{\pi E_a\mu_c r_2^3}{E_c\mu_aE_sI_s}+\frac{6}{E_cr_3t_c^2}\left(\left(\frac{r_4^3}{6}-\frac{r_3^2r_4}{2}+\frac{r_3^3}{3}\right)\frac{1}{t_c}-\frac{t_cr_3}{2}\right)-\frac{\pi r_2^3}{E_sI_s}}{\frac{3t_c}{8G_c}+\frac{t_a}{G_a}}}\end{aligned} \tag{64}$$

According to Eq. (51), the expression of the normal stress can be obtained as the following:

$$\begin{aligned}\frac{d\sigma(z,\theta)}{dz}&=-\frac{\pi E_ar_2^3\sin\theta}{\mu_aE_sI_s}\left(c_1e^{\eta z}+c_2e^{\eta z}\right)\\\sigma(z,\theta)&=-\frac{\pi E_ar_2^3\sin\theta}{\eta\mu_aE_sI_s}\left(c_1e^{\eta z}-c_2e^{\eta z}\right)\end{aligned} \tag{65}$$

The constants c_1 and c_2 are determined by the boundary conditions at the end of bonding layer obtained from Eq. (48):

$$\sigma\left(-\frac{l}{2},\theta\right)=\sigma\left(\frac{l}{2},\theta\right)=-\frac{E_aM}{\mu_aE_sI_s}r_2\sin\theta \tag{66}$$

3 The Finite Element Analysis of the Interfacial Stresses in the Pipe Repaired with CFRP

Stress distributions on the adhesive are also obtained using the FEM model, and they are compared with the theoretical results. Then the effects of the corrosion defects on the stress distribution are studied by establishing the CFRP repaired corroded pipe model. By using the cohesive element to simulate the adhesive layer in ABAQUS, the failure state and the debonding load of the bonding layer is obtained.

3.1 The Geometric Model and Material Properties of the Repaired Pipe

The FEM model of a perfect pipe reinforced with CFRP and a corroded pipe repaired with CFRP are constructed in this study. The operation of repairing corroded pipe with CFRP is briefly described as the following procedure. First, the defect in the steel pipe is restored to a level state using filler. Then, the adhesive is evenly applied on the circumference of the defect area, and the CFRP lamina is bound to it to strengthen pipe. The repaired pipe structure is shown in Fig. 4. The structure section diagram from the inside to the outside includes the pipe, the filler, the adhesive layer, and the CFRP layer.

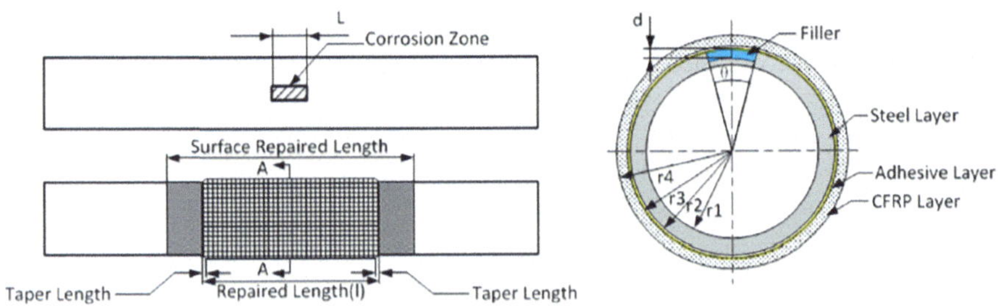

Fig. 4 The structural diagram of corroded pipe repaired with CFRP

Table 1 Material properties

Component	Density (g/cm^3)	Elastic module (GPa)	Shear module (GPa)	Poisson's ratio
Steel pipe	7.85	200	77	0.3
Filler	1.89	5.5	0.0246	0.43
Adhesive	1.05	2.86	1.04	0.38
CFRP lamina	1.9	82	0.32	31

Table 2 Material properties

Pipe		Corrosion defect		CFRP lamin	
OD (mm)	21.3	Angle (°)	15	Length (mm)	46
Thickness (mm)	1.6	Depth (mm)	1	Thickness (mm)	0.56
Length (mm)	300	Length (mm)	10	Number of plies	5

The type of pipe selected for this study is an API 5 L X60 [30] steel pipe, and the size of the steel pipe is the DN15 series. Rectangular corrosion defects commonly used by scholars are selected for this study [31]. A 3k-200 g high strength grade I bi-directional 0/90 plain weave carbon fiber fabric is selected as the reinforced composite material, which can enhance the bending and tensile resistance at the same time. The fillers used are the PH1120 steel repair agent, and the adhesive is TH960. The material properties are shown in Table 1, according to the manufacturer's information.

It is necessary to determine the thickness and length of the composite materials required for the repair of corroded pipes according to their properties and the size of the corrosion defect. Different calculation methods are used for different standards. Currently, the relevant standards primarily include PRCI, ASME PCC-2, DNV-RP-F113, and ISO 24817.

In this study, according to ASME PCC-2-2011, which is a repair standard for pressure equipment and pipes, the minimum repair thickness, $t_{\min}$ and length, $L_{\min}$ of CFRP laminate are determined: $t_{\min} = 0.56$ mm and $L_{\min} = 46$ mm. The geometric parameters of the repaired corroded pipeline are shown in Table 2.

In this study, the perfect pipe and the corroded pipe are assembled using the bonding layer and the CFRP layer, respectively (in Fig. 5). The pipe, filler, and CFRP layer are each set to the linear material using the equivalent elastic parameters and the equivalent Poisson's ratio [22]. In addition, the cohesive element is used to simulate the bonding layer [32]. The bilinear constitutive model is used as cohesive element model, and the critical fracture energy is 650 N/m calculated according to the method of Reference [33]. Static and implicit analysis steps are established in ABAQUS/Explicit to study the stress distributions.

The contact surfaces between the steel tube layer, the bonding layer, and the composite material layer are defined as the tie constraint. One end of the pipe is fixed, and the other end faces are coupled to a reference point. The axial force and bending moment are applied in sequence to the reference point, and the loaded surface of the system remains plane due to this loading method.

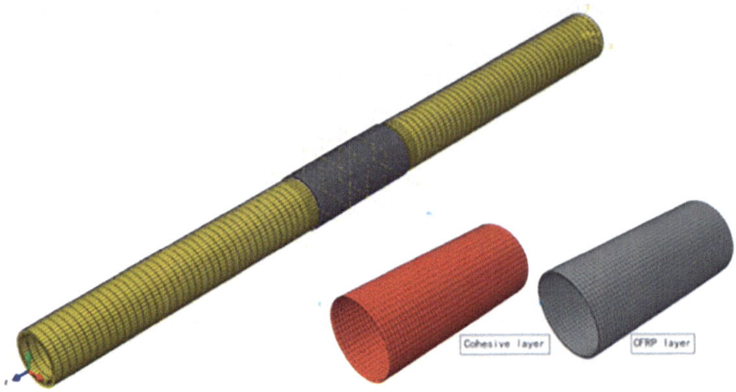

Fig. 5 Finite element model

The element type is C3D8R (linear reduced integration 3D stress element), the bonding layer mesh type is COH3D8 (3D 8-node cohesive element), and the maximum degradation value is set to 1 when the element is deleted. Specifically, when SDEG = 1, the element is considered to be invalid.

3.2 *Interfacial Stress Distributions in the Repaired Pipe*

The finite element stress results along the routing line shown in Figs. 6 and 7, including the stress of the intact pipe repaired and the stress of the corroded repaired pipe, are extracted. In addition, the specific parameters from the geometric modeling are carried into the interface stress formula to obtain the theoretical stress results, and then the two calculation results are compared and analyzed.

Figures 6 and 7 respectively show the theoretical and FE solution obtained when a 30 kN tensile load is applied and 150 Nm bending load is applied to the repaired perfect and corroded pipes. It can be seen from the figures that the stress distribution tendency of the FEM results is similar with that of the theoretical results, except that oscillation happens in FEM results, but not in analytical results, which is caused by simulation error. The normal stress shows an even function property along the axis, and the shear stress shows an odd function property along the axis. The maximum stress appears at the boundary. For the corroded pipe case, the stress concentration happens also at the end of the corrosion position, which is smaller than the end of the interface. As shown in Figs. 6 and 7, the stress distribution tendencies of the FE models are nearly the same with the theoretical results. Nevertheless, at the end of the bonding layer the peak values for the normal stress under tension have a difference of 10% between analytical and FE results, and it is close to the difference of normal stress under the bending load. As for shear stress, the difference of peak values under tension is a little larger than that under bending load.

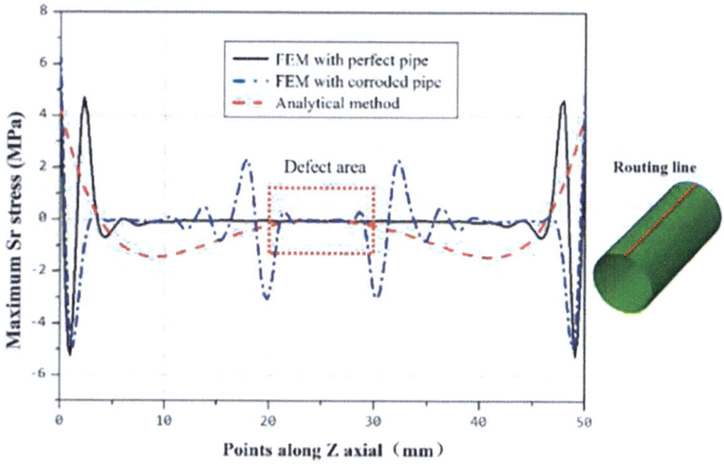

(a) Interfacial normal stress of bonding layer

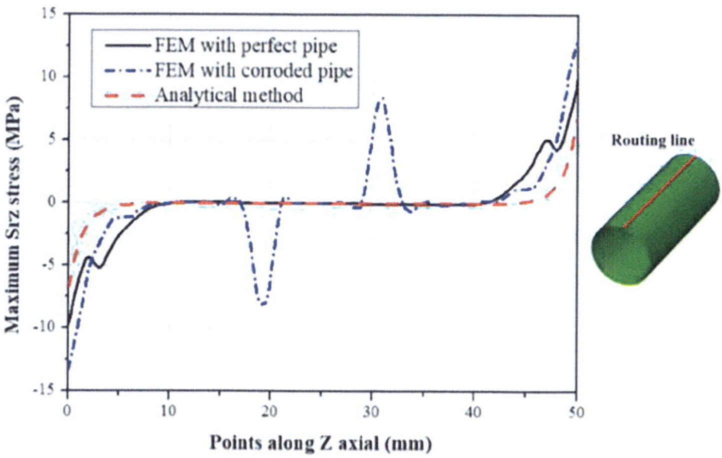

(b) Interfacial shear stress of bonding layer

Fig. 6 The theoretical results of interfacial stress of bond layer under 30 kN tensile load are compared with the finite element analysis results. (**a**) Interfacial normal stress of bonding layer. (**b**) Interfacial shear stress of bonding layer

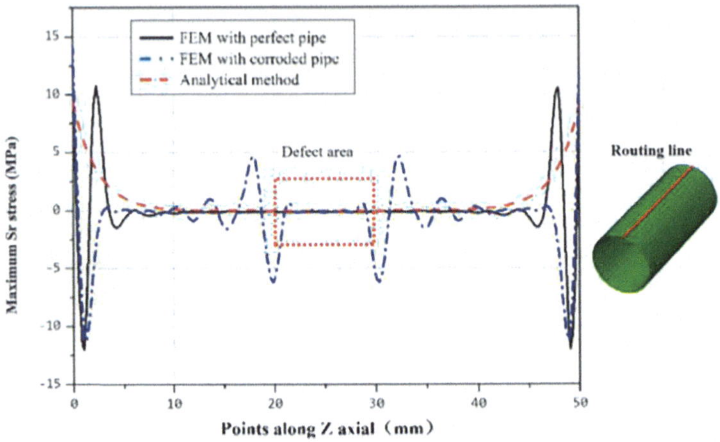

(a) Interfacial normal stress of bonding layer

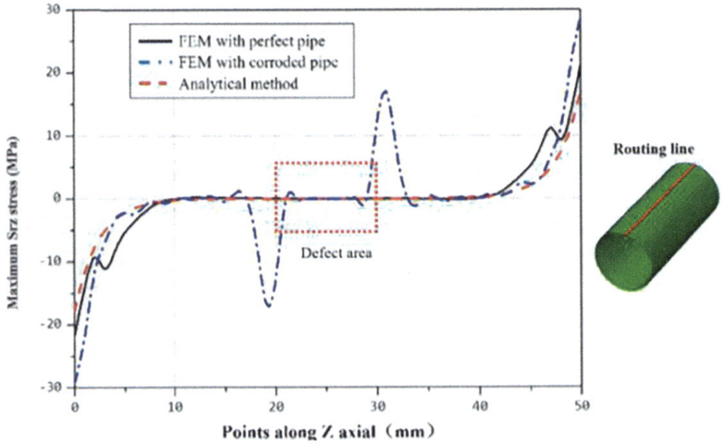

(b) Interfacial shear stress of bonding layer

Fig. 7 The theoretical results of interfacial stress of bond layer under 150 Nm bending moment are compared with the finite element analysis results. (**a**) Interfacial normal stress of bonding layer. (**b**) Interfacial shear stress of bonding layer

4 The Debonding Load Analysis of the Repaired Pipe

In the structure of a pipe repaired using adhesive and CFRP, adhesive debonding sometimes occurs due to a large load. Based on the stress distributions analysis from the above chapter and the adhesive failure criteria, the debonding load is now discussed.

The damage failure criteria of the adhesive layer primarily in ABAQUS software includes: Quade Damage [34], Maxe Damage [34], Quads Damage [35], Maxs Damage [36], Maxpe Damage [37], and Maxps Damage [38].

Among them, Quads Damage, which is square nominal stress criterion, is the most commonly used in computation. When the sum of squares of the nominal stress ratios in all directions is equal to 1, this judgment method indicates that the damage happens in the adhesive layer, and the mathematical expression can be expressed as:

$$\left\{\frac{t_n}{t_n^0}\right\}^2 + \left\{\frac{t_s}{t_s^0}\right\}^2 + \left\{\frac{t_t}{t_t^0}\right\}^2 = 1 \tag{67}$$

where t_n, t_s and t_t represent the stresses in the three directions on the bonding interface, namely the normal stress and the shear stresses in two directions. When the structure is subjected to the tensile load and bending load, the circumferential shear stress at the interface is 0, and the axial shear stress and radial normal stress can be obtained according to Eqs. (24), (39), (64), and (65) in Sect. 2. In addition, the stress distribution under constant load can be seen in Figs. 6 and 7. t_n^0, t_s^0 and t_t^0 represent the material strength in three directions, namely, the bonding peel strength and the bonding shear strengths in two directions, and typically the two shear strengths are the same.

By substituting the formulas of shear stress and normal stress in Sect. 3 into the Quads Damage debonding criterion, the initial debonding load of the perfect pipe reinforced using the CFRP under tension and bending is 48 kN and 132.35 Nm respectively, where the bonding peeling stress and shear stress of the adhesive are 22.52 MPa and 15.31 MPa, respectively. Based on the stress expressions of the bonding layer in a perfect pipe reinforced with CFRP, the residual strength factor α_θ is introduced to measure the effect of corrosion defect, and the stresses of a corroded pipe repaired with CFRP were obtained. The residual strength factor α_θ is calculated according to RSTRENG0.85 [39]:

$$M_f = 3.3 + 0.032\frac{L^2}{De} \tag{68}$$

$$\alpha_\theta = \frac{1 - 0.85\left(\frac{d}{e}\right)\left(\frac{1}{M_f}\right)}{1 - 0.85\left(\frac{d}{e}\right)} \tag{69}$$

$$\sigma_{\max} = \sigma_y + 69 \tag{70}$$

In Eqs. (68)–(70), σ_y is the ultimate strength of the pipeline material, M_f is the Fourier coefficient, d is the corrosion depth, L is the corrosion length, e is the wall thickness and D is the outside diameter of the pipeline.

Then $\alpha_\theta = 1.4575$ is obtained based on the geometric character of the corrosion damage. Then, the initial debonding loads of the pipe repaired with CFRP under tension and bending are obtained. The results are 34 kN and 94.53 Nm, respectively.

The finite element results of the initial debonding load corresponds to the initial failure load of cohesive element, which can be obtained directly based on the debonding failure expressed in Eq. (62). For the perfect pipe reinforced with CFRP, the initial loads are 44 kN and 126 Nm for the tension case and bending case, respectively, and for the corroded pipe repaired with CFRP, the initial loads are 36 kN and 105.52 Nm for the tension case and bending case, respectively.

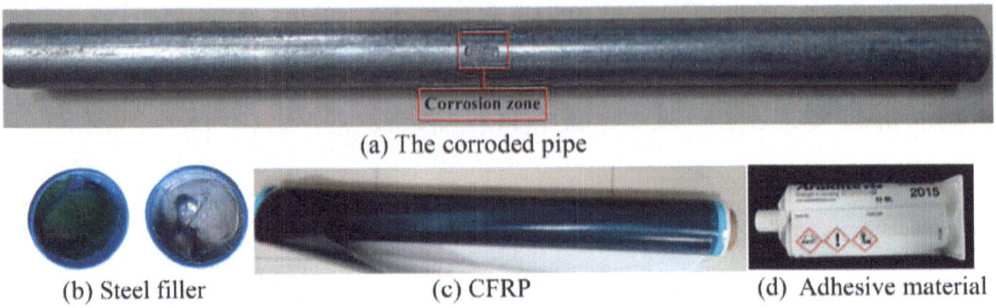

Fig. 8 The corroded pipe and repairing materials

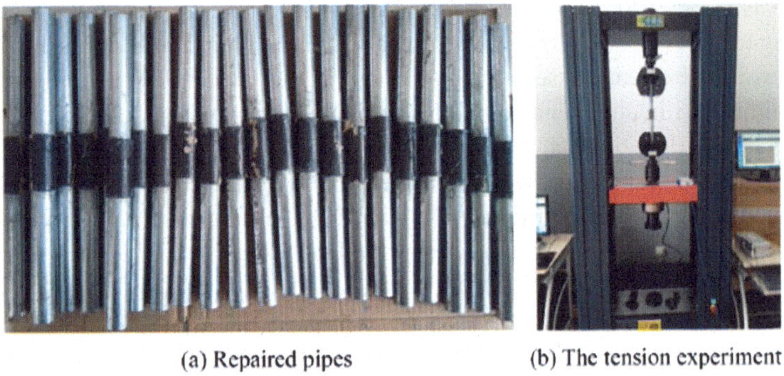

Fig. 9 Repaired pipes and tension experiment

To verify the theoretical results and finite elements results, the tension experiment is constructed for the perfect pipe reinforced with CFRP and the corroded pipe repaired with CFRP. The geometry of pipes and corrosion are the same with the FEM model. Firstly, the corroded pipe is filled with steel filler and then wrapped with adhesive impregnated CFRP (shown in Fig. 8). The perfect pipe is also wrapped with adhesive impregnated CFRP. In order to eliminate accidental errors, each group has three replicas, and they are put for 72 h for solidification. The repaired pipes and reinforced pipes are subjected to a tension load with speed of 0.01 mm/min in a 500 kN tensile testing machine (WDW-100 M) until fracture (Fig. 9), and the sampling frequency is 200 times/s. According to Fig. 10, after tensile fracture, it can be seen that the composite layer remains intact and the bonding layer is completely debonded. The relationship between loads and displacements using the replica mean values is shown in Fig. 11 and the standard deviation of each set of experimental data points does not exceed 0.37 kN. A sudden sound is heard during the experiment, which is caused by the debonding between the CFRP and the steel pipe [8], and the average initial debonding load is 45 kN for the perfect pipe reinforced with CFRP, and it is 38 kN for the corroded pipe repaired with CFRP.

Table 3 shows the experimental results, finite element results and theoretical results of the initial debonding load under tension and bending moment. It can be seen that the differences of initial debonding tension load and initial debonding bending moment are almost within 12% among the theoretical results, simulation

Fig. 10 Fracture graph of CFRP repair corroded pipe

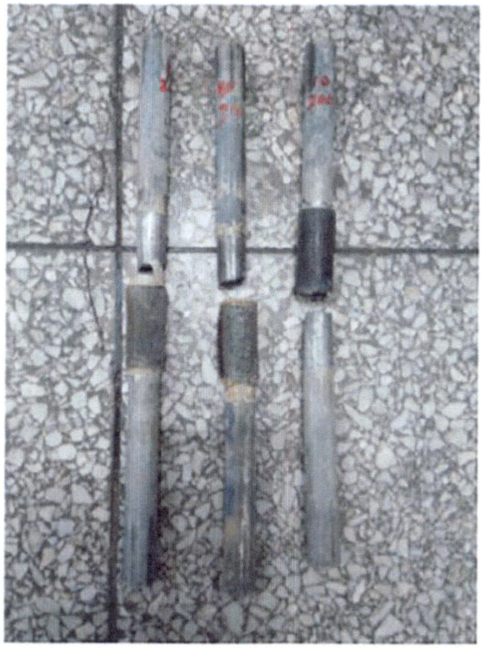

Fig. 11 The relationship between tension load and displacernent obtained from exneriments

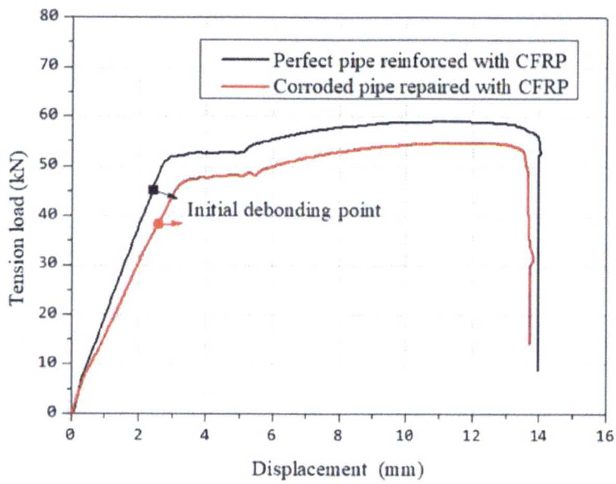

Table 3 The initial debonding loads

	Perfect pipe reinforced with CFRP			Corroded pipe repaired with CFRP		
	Theory (error)	FEM (error)	Experiment	Theory (error)	FEM (error)	Experiment
Tension	48 kN	44 kN (−8%)	45 kN (−6%)	34 kN	36 kN (+6%)	38 kN (+12%)
Bending	132.35 nm	126 nm (−5%)	–	94.53 nm	105.52 mm (+12%)	–

results and experimental results (tension only). The initial debonding loads for the corroded pipe repaired with CFRP could be obtained using the theoretical expressions derived in this paper.

5 The Debonding Load Considered the Durability

In this study, CFRP is used to repair offshore pipes that are immersed in the seawater and have long-time working loads. Hence, the adhesive property will be weakened to a certain extent due to long-time seawater immersion and working loads. Therefore, it is of great significance to study the adhesive property durability to accurately predict the working state of repaired pipes. Generally, the elastic modulus and Poisson's ratio of the bonding layer change little with time, but the bond strength changes greatly with the soaking time [40]. Peng [41] conducted an experiment on the bonding performance of CFRP to steel and obtained the law of the bonding strength durability of adhesive under seawater immersion. Peng also obtained the change rule of the shear strength and the peel strength of the adhesive with aging time, as shown in Fig. 12.

With the change in the ultimate bonding peel strength and shear strength of the bonding layer material obtained in Fig. 12, the attenuation of the initial debonding load of the corroded pipe repaired with CFRP at different aging times under tensile and bending loads is obtained, as shown in Fig. 13.

As it can be seen from Fig. 13, the initial debonding load is not a constant value, and it reduces with the aging time. In addition, the rate of decrease is large during the initial 1000 h. Therefore, if CFRP is used to repair the marine pipes, the failure loads should consider the material durability. Hence, the use of the theoretical method, the stresses, and the debonding loads can be obtained with good accuracy.

Fig. 12 The rule of CFRP sheet-sted shear strength varies with aging time

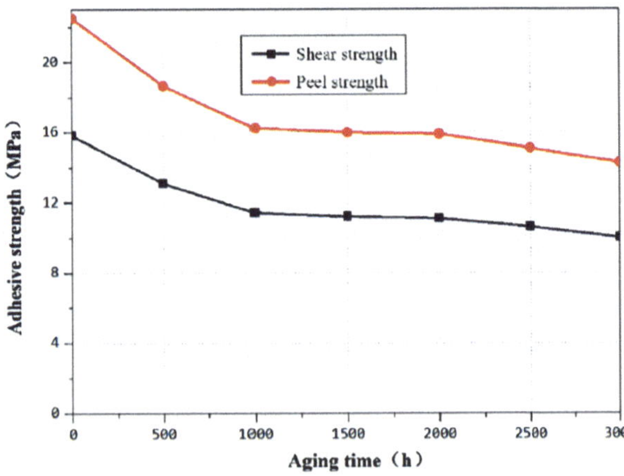

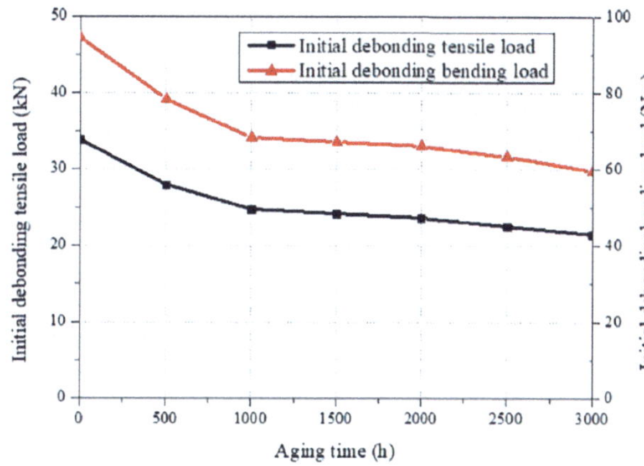

Fig. 13 The durability of initial debonding load

6 Conclusions

In this study, the debonding failure of the corroded pipe repaired with CFRP is analyzed using the theoretical derivation and a finite element calculation. The following conclusions are obtained.

1. The exact functional expressions of the interface normal stress and shear stress of the bonding layer of the reinforced pipe with CFRP are obtained. The stress distribution tendency of the FEM results is similar with that of the theoretical results, except that oscillation happens in FEM results, but not in analytical results.
2. According to the failure criterion of the adhesive layer, the debonding load is calculated using the theoretical method and the FEM, and they are compared with the experimental results. The experimental results and finite element results have good coherence with theoretical experimental results for the tensile case, and the errors are within 12%. It could be said that the debonding loads for the corroded pipe repaired with CFRP could be obtained using the theoretical expressions derived in this paper.

References

1. Patrick AJ. Composites—case studies of pipe repair applications. 2004.
2. Sen R, Mullins G. Application of FRP composites for underwater piles repair. Compos Part B Eng. 2007;38:751–8.
3. Jasket CE, Hart BO, Bruce WA. Pipe repair manual. In: Final Report to pipe Research Council International, Inc., Contract PR; 2006. p. 186–324.
4. Asme-PCC-2. Repair of pressure equipment and piping. America Petroleum Institute; 2008.
5. ISO-24817. Composite repairs for pipe work-qualifiacation and design installation testing and inspection. Petroleum petro chemical and natural gas industries; 2015.

6. Park S, Kim JW, Lee C, Park SK. Impedance-based wireless debonding condition monitoring of CFRP laminated concrete structures. NDT&E Int. 2011;44(2):232–8.
7. Sun W, He T, Liu SK. Developing an anchored CFRP reinforcement for efficiently and readily strengthening reinforced concrete structures. Compos Part B-Eng. 2019;176:107–99.
8. Zarrinzadeh H, Kabir MZ, Deylami A. Crack growth and debonding analysis of an aluminum pipe repaired by composite patch under fatigue loading. Thin-Walled Struct. 2017;112:140–8.
9. Liu JC, Qin MJ, Zhao QL, Chen L, Liu PF, Gao JG. Fatigue performances of the cracked aluminum-alloy pipe repaired with a shaped CFRP patch. Thin-Walled Struct. 2017;111:155–64.
10. Yang YM, Biscaia H, Chastre C, Silva MAG. Bond characteristics of CFRP-to-steel joints. J Constr Steel Res. 2017;138:401–19.
11. Zhang XM, Zhu LB, Hu MY, Liu Y. An analytical model and stress analysis of oneside bonded composite patch to metal reinforcement. Int J Adhes Adhes. 2015;58:63–9.
12. Zhang Y, Zhang K, Zhao HK, Xin JH, Duan ML. Stress analysis of adhesive in a cracked steel plate repaired with CFRP. J Constr Steel Res. 2018;145:210–7.
13. He J, Xian GJ. Debonding of CFRP-to-steel joints with CFRP delamination. Compos Struct. 2016;153:12–20.
14. Xu W, Li GQ. Finite difference three-dimensional solution of stresses in adhesively bonded composite tubular joint subjected to torsion. Int J Adhes Adhes. 2010;30(4):191–9.
15. Parashar A, Mertiny P. Adhesively bonded composite tubular joints: review. Int J Adhes Adhes. 2012;38:58–68.
16. Casalegno V, Salvo M, Rizzo S, et al. Joining of carbon fibre reinforced polymer to Al-Si alloy for space applications. Int J Adhes Adhes. 2018;82:146–52.
17. Budhe S, Banea MD, Barros S. Composite repair system for corroded metallic pipelines: an overview of recent developments and modelling. J Mar Sci Technol. 2020;25(2):379–96. Fig 10.13. The durability of initial debonding load.
18. Guz IA, Menshykova M, Paik JK. Thick-walled composite tubes for offshore applications: an example of stress and failure analysis for filament-wound multilayered pipes. Ships Offshore Struct. 2015;12(3):304–32.
19. Bouhafs M, Sereir Z, Chateauneuf A. Probabilistic analysis of the mechanical response of thick composite pipes under internal pressure. Int J Press Vessels Piping. 2012;95:7–15.
20. Croccolo D, Agostinis MD, Fini S, et al. Effect of the engagement ratio and of temperature on the shear strength of epoxy adhesive bonded aluminum alloy pinand-collar joints. J Adhes. 2018;94(11):932–50.
21. Croccolo D, Agostinis MD, Vincenzi N. Design of hybrid steel-composite interference fitted and adhesively bonded connections. Int J Adhes Adhes. 2012;37:19–25.
22. Okamoto S, Yamamoto M, Hosokawa K, et al. Mechanical properties of CFRP pressure vessels and identification of equivalent elastic moduli of CFRPs. In: 4th international conference on high performance structures and materials; 2008.
23. Ahmad MIG, Hoa SV. Flexural stiffness of thick walled composite tubes. Compos Struct. 2016;149:125–33.
24. Hyer M. Stress analysis of fiber-reinforced composite materials. Destech publications; 2009.
25. Jolicoeur C, Cardou A. Analytical solution for bending of coaxial orthotropic cylinders. J Eng Mech. 1994;120(12):56–74.
26. Pang YY, Wu G, Wang HT, Liu Y. Interfacial bond-slip degradation relationship between CFRP plate and steel plate under freeze-thaw cycles. Constr Build Mater. 2019;214:242–53.
27. Yang YM, Silva MAG, Silva RJC. Material degradation of CFRP-to-steel joints subjected to salt fog. Compos Part B Eng. 2019;173:106–884.
28. Zhang Y, Duan ML, Wang Y, et al. Analytical study of the strength of adhesive joints of riser pipes. Ships Offshore Struct. 2015;10(5):545–53.
29. Takagaki K, Minakuchi S, Takeda N. Thick-walled crack-free CFRP pipes: stress reduction usingatypical lay-up. Compos Struct. 2015;126:337–46.
30. API SPEC SL. Specification for line pipe. 2018.

31. Cabral HLD, Motta RS, Afonso SMB, et al. The development of a computational tool for generation of high quality FE models of pipelines with corrosion defects. J Braz Soc Mech Sci Eng. 2017;39(8):3137–50.
32. Zhou YW, Chen XL, Fan ZH, et al. Bond behaviors of FRP to concrete interface under the control of a novel end-anchorage system. Compos Struct. 2017;68(5):130–42.
33. Bocciarelli M, Colombi P, D'Antino T, et al. Intermediate crack induced debonding in steel beams reinforced with CFRP plates under fatigue loading. Eng Struct. 2018;171:883–93.
34. Huang J, Yao WX. High-temperature mechanical properties of strain isolation pad for thermal protection system. J Spacecr Rockets. 2018;55(4):848–55.
35. Xie ZH, Liu HH, Zhang ZL. Prediction on low velocity impact damage of interlayertoughened composite laminate. J Nanjing Univ Aeronaut Astronaut. 2013;45(5):686–92.
36. Pereira K, Bhatt Wahab MA. Prediction of fretting fatigue crack initiation location and direction using cohesive zone model. Tribol Int. 2018;127:245–54.
37. Enrico S, Fulvia T, Luca C, et al. Subject-specific finite element models implementing a maximum principal strain criterion are able to estimate failure risk and fracture location on human femurs tested in vitro. J Biomech. 2008;41(2):356–67.
38. Wang J, Ren L, Xie LZ, et al. Maximum mean principal stress criterion for three dimensional brittle fracture. Int J Solids Struct. 2016;102–103:142–54.
39. Ahammed M. Prediction of remaining strength of corroded pressurized pipelines. Int J Pres Ves Pip. 1997;122:655–9.
40. Zhang LL, Zhang L, Ma JX. Durability of carbon fiber reinforced polymer (CFRP) sheet under marine environment. J Build Mater. 2008;11(6):732–5.
41. Peng FM, Cai P, Zhang N, Zhang XS, Liu BX. Experimental research on the bond behavior of CFRP to steel. Ind Constr. 2009;39:112–6. 06.

The Attenuation Mechanism of CFRP Repaired Corroded Marine Pipelines Based on Experiments and FEM

1 Introduction

Marine pipelines play an important role in oil and gas development, and they are the lifelines of subsea systems. However, due to the harsh marine environment, marine pipelines are easy to damage, and corrosion is a common type of failure. Three methods are commonly used to repair marine pipelines: welding, fixture, and composite material repair methods [1–4]. The composite material repair method has been widely applied in the maintenance and repair of the oil and gas pipelines because of its low cost, relatively simple process, and short time consumption [5, 6]. However, with the increase of the service time, the repair effect will be affected by factors such as a long loading time and seawater immersion, and the performance of the composite repair will degrade. If the degradation is great, it may even cause the failure of the repaired pipeline, resulting in serious environmental pollution and huge economic losses [7–9]. Therefore, it is of great significance to study the durability of corroded pipelines repaired with composite materials during longtime seawater immersion and loadings.

The compound repair method was originally used for the reinforcement of bridge concrete structures. From the 1970s to the present, there have been many technologies for repairing pressure pipes and vessels with fiber composite materials, and many scholars have conducted a lot of research on the repair technology and repair effects of composite materials. Peck and Ramirez [10, 11] studied the mechanical properties of fiberglass and carbon fiber composites under different environmental conditions, it was found that the properties of carbon fibers were better. Abanilla [12] immersed the carbon fiber epoxy resin matrix composites at different temperatures, the results showed that the strength and elastic modulus of the composite decreased after immersing for a long time. Nguyen [13] studied the bonding properties of a CFRP-steel interface at different temperatures. It was proposed that the bonding failure time was the result of the triple actions of the temperature, time, and load. Almosawe [14] studied the bond behavior between CFRP laminates under impact

Y. Zhang, *Interfacial Mechanics of Offshore Engineering Equipment*,
https://doi.org/10.1007/978-981-95-3200-1_11

load. The results showed that there is a significant bonding strength enhancement under impact load. Batuwitage [15] carried out a parametric study to investigate the effect of surface preparation method, primer, number of CFRP layers and exposure duration on the failure modes and tensile capacity of double-strap joints. Feng [16] proposed an in-situ and efficient monitoring method for cable force by using smart twisted CFRP bar with embedded fiber Bragg grating (FBG) sensors. Based on a theoretical method and the FEM with different crack lengths, Zhang [17] studied the stress distributions of adhesive in a cracked steel plate repaired with CFRP. Gsab [18] studied the mechanical properties of steel pipes repaired with fiber reinforced polymer composites under bending load using experimental method and finite element method. Galvez [19] proposed a steel-CFRP adhesive joint that combined the good properties of both materials for verifying the suitability of this type of joints for structural applications in aggressive environments. Pang [20] manufactured a series of the CFRP-steel single-shear specimens to investigate the effects of bond length, adhesive thickness, adhesive type and loading procedures on the bond behavior of CFRP-steel interfaces. Fawad and Wang [21, 22] studied the effect of CFRP on the repair of concrete structure and damaged bridge structures. The research shows that CFRP can more effectively and feasibly increase the performance of concrete. Ying [23] proposed a surrogate-based optimization model of the external patch design for damaged CFRP laminates to improve the impact-resistance. Wu [24] investigated the effects of glass fiber sheet (GFS) on the corrosion behaviour and bond characteristics between steel and CFRP. Zql [25] proposed a novel CFRP-steel tube composite member (CFRP-STCM), which is strengthened with a thick-walled tube section at both ends to suppress premature debonding failure of the CFRP sheet. He and Lu [26, 27] systematically investigated the mechanical behavior of epoxy adhesives and CFRP-steel hybrid joints. Ar [28] studied to enhance the strength of low strength hybrid fiber reinforced concrete (HFRC) cylinders confined with different number of CFRP sheets by examining the behavior of low strength HFRC cylinders confined with CFRP sheets.

Kafodya [29] studied the durability behavior of pultruded unidirectional carbon fiber reinforced polymer (CFRP) plates immersed in water and seawater at room temperature, under sustained bending strain of 30% and 50% ultimate strain. Borrie [30] studied the durability of steel structures strengthened with CFRP laminates and sheets under seawater environment with increasing temperature. Correia [31] studied the durability of concrete (RC) slabs strengthened with prestressed CFRP laminate strips according the externally bonded reinforcement (EBR) technique under different environmental and loading conditions. Yang [32] studied on bond properties of double-strap CFRP-to-steel bonded joints with two different composite materials as well as adhesive coupons subjected to freeze-thaw cycles for 10,000 h. It was confirmed that the interfacial bond strength is critically related to the maximum shear stress and failure mode. Zhang [33] undertook an experimental to study the effect of specimen thicknesses on water absorptions and flexural strengths. Test results showed that the water uptake and flexural strength retention of CFRP laminates were significantly affected by the adopted specimen thickness. Miyano [34] reviewed an accelerated test method (ATM) developed by the team over the years for measuring the durability of CFRP laminates. Lu [35] studied the durability of RC beams externally strengthened with prestressed CFRP sheet in a chloride containing environment. Wma [36] presented an experimental study on the durability of adhesively bonded CFRP-aluminum alloy (CFRP-Al) joints subjected to a coupled temperature and alternating load.

However, there are few studies about CFRP repaired corroded marine pipelines. In recent years, the corroded marine pipelines are prone to be repaired with CFRP as its many advantages, but the durability are very important because the repaired pipelines are in complex load and environment conditions. For marine pipelines on the seabed, the bending moment is an important load. In this research, the durability of a corroded pipeline repaired with CFRP with longtime bending moment and longtime sea water immersion was studied, using an experimental method and the Finite Element Method (FEM), and the attenuation law and its mechanism were discussed.

2 Durability Study of CFRP Repaired Pipelines

First, the durability of CFRP repaired pipelines was studied using experiments. Figure 1 shows the geometric dimensions of the repaired corroded pipeline with a three-point bending moment. A corroded pipeline with a rectangular corrosion defect (the length was 10 mm and the depth was 1 mm with an angle of 15°) was repaired according to the ASME standard PCC-2-2011 [37]. A pipeline was filled with metal filler, and then it was wrapped with CFRP, which was bonded with the steel pipeline using adhesive. The length of the pipeline specimen was 300 mm and the outer radius was 21.3 mm with a thickness of 1.6 mm. The length of the CFRP was 50 mm and the thickness was 0.555 mm with five layers. The thickness of the adhesive was 0.1 mm.

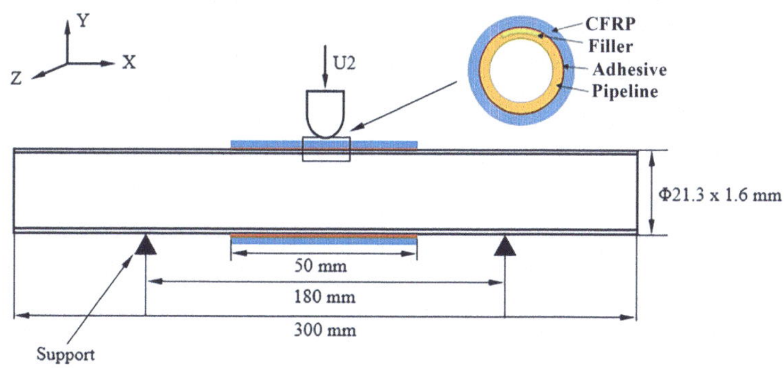

Fig. 1 The geometric dimensions of the repaired corroded pipeline under a three-point bending moment

Table 1 Material parameters of steel pipelines and CFRP

Material	Parameters	Value
API 5L X60	Yield strength (MPa)	485
	Tensile strength (MPa)	600
	Poisson's ratio	0.3
	Elastic modulus (GPa)	210
CFRP	Density (g/cm^3)	1.8
	Poisson's ratio	0.3
	Elastic modulus E_1(GPa)	114
	Elastic modulus E_2 (GPa)	114
	Elastic modulus E_3 (GPa)	8.61
	Poisson's ratio ν_{12}	0.3
	Poisson's ratio ν_{13}	0.45
	Poisson's ratio ν_{23}	0.45
	Shear modulus G_{12} (GPa)	3.0
	Shear modulus G_{13} (GPa)	4.1
	Shear modulus G_{23} (GPa)	4.1
	Elongation ratio	2.07%

The steel pipeline was API 5L X60, and the CFRP was 3K-200g with epoxy resin matrix and two-direction 0/90, five layers were wrapped outside the steel pipeline. The material parameters of the steel pipelines and the CFRP are shown in Table 1, and these data were provided by the manufacturers. The adhesive was Araldite 2015, and the elastic modulus of the adhesive was 2GPa, with a tensile strength of 22.3 MPa, and the ultimate elongation was 1.85%. The metal filler was double bond DB406, and its elastic modulus was 5.5GPa, with a Poisson's ratio of 0.4, and the shear strength was 15 MPa.

Figure 2a shows the specimens of the CFRP repaired corroded pipelines. These specimens were put in the three-point bending test machine, which is a 500kN Universal test machine (WDW-100 M) as shown in Fig. 2b, to study the ultimate loads. The loading speed is 0.05 mm/min, and loads and displacements data were measured by sensors in the loading device of test machine. Figure 2c shows the corroded pipeline and the repaired pipeline before and after bending failure. The crushing failure happened for the corroded pipeline and the repaired pipeline with the increase of the bending moment. CFRP wrinkles along the axial direction and partial interface slipping were found after the failure of the repaired pipelines. The results show that the debonding happened between the CFRP and the steel pipeline during the failure process.

To obtain accurate experimental data, each group of experiments was tested three times and the average value was used. The load-displacement curves for the corroded pipeline and the CFRP repaired pipeline are shown in Fig. 3. The ultimate load of the repaired pipelines was 6.85 kN, while the ultimate load of the corroded pipeline was 5.35 kN. It could be found that CFRP repair could improve the ultimate bending load, but the bending stiffness had no obvious change, as shown in Fig. 3.

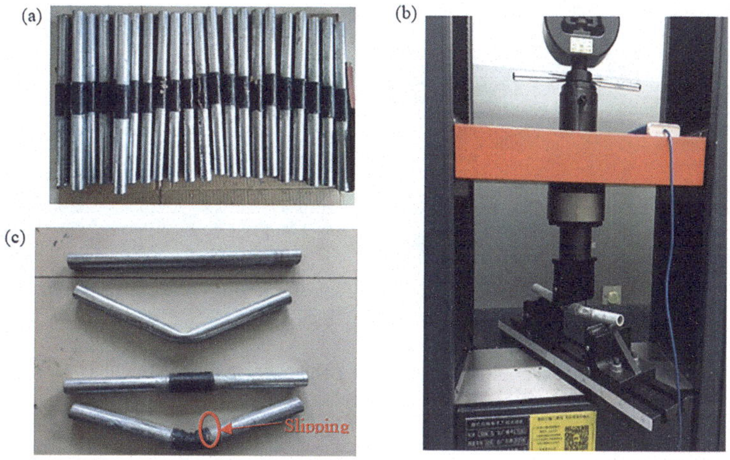

Fig. 2 CFRP repaired corroded pipelines. (**a**) The specimens of CFRP repaired corroded pipelines; (**b**) The specimen under a three-point bending test machine; (**c**) The corroded pipeline and the repaired pipeline before and after bending failure

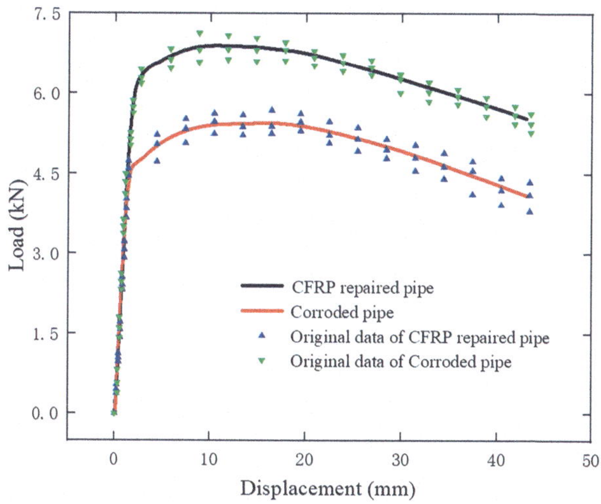

Fig. 3 The load-displacement curves of the corroded pipeline and the repaired pipeline

To study the durability of CFRP repaired corroded pipelines, two groups of experiments were constructed in this research. One group experienced the bending moment for 200, 400, 600, and 800 h as shown in Fig. 4a, and the bending moment was 30% of the ultimate bending moment of the original perfect pipelines. The other group experienced seawater immersion and the bending moment (0% and 30% of the ultimate bending moment) for 200, 400, 600, and 800 h, as shown in Fig. 4b. The seawater was configured according to the standard ASTM D1141-98 (2013), and five times concentration standard was used to do the accelerated testing. The

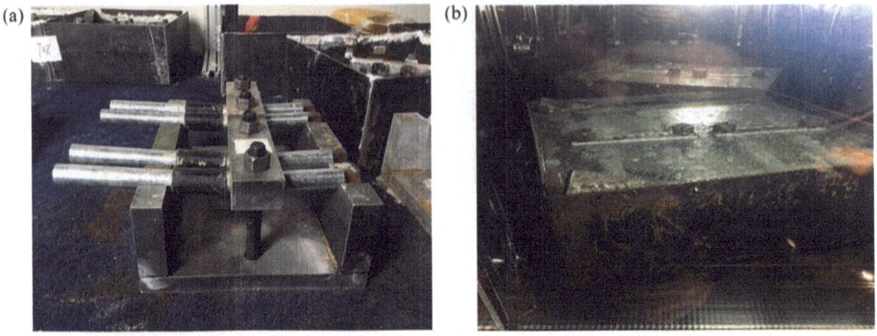

Fig. 4 Durability experiments of repaired pipelines. (a) Under bending moment; (b) under bending moment and seawater immersion

Table 2 Experimental conditions with bending moment

Type	Bending moment	Time (h)
CFRP repaired pipelines	30%	0
		200
		400
		600
		800

Table 3 Experimental conditions with seawater immersion and bending moment

Type	Bending moment	Time (h)	Seawater concentration	Temperature (°C)
CFRP repaired pipelines	0%	0	Five times ASTM standard	17
		200		
		400		
		600		
		800		
	30%	0		
		200		
		400		
		600		
		800		

seawater temperature is 17 °C. The experimental conditions are shown in Tables 2 and 3.

After the durability experiments, the ultimate loads were tested with a three-point bending experiment. The load-displacement relationships are shown in Fig. 5a–c for the CFRP repaired pipelines with different durability conditions. With the increase of the time, the ultimate load of the CFRP repaired pipelines for different conditions decreased gradually. For example, the ultimate loads were 6.7 kN, 6.4 kN, 6.2 kN, and 5.9 kN for seawater immersion and the bending moment for 200 h, 400 h, 600 h, and 800 h, respectively, but the stiffness of the repaired pipelines under durability

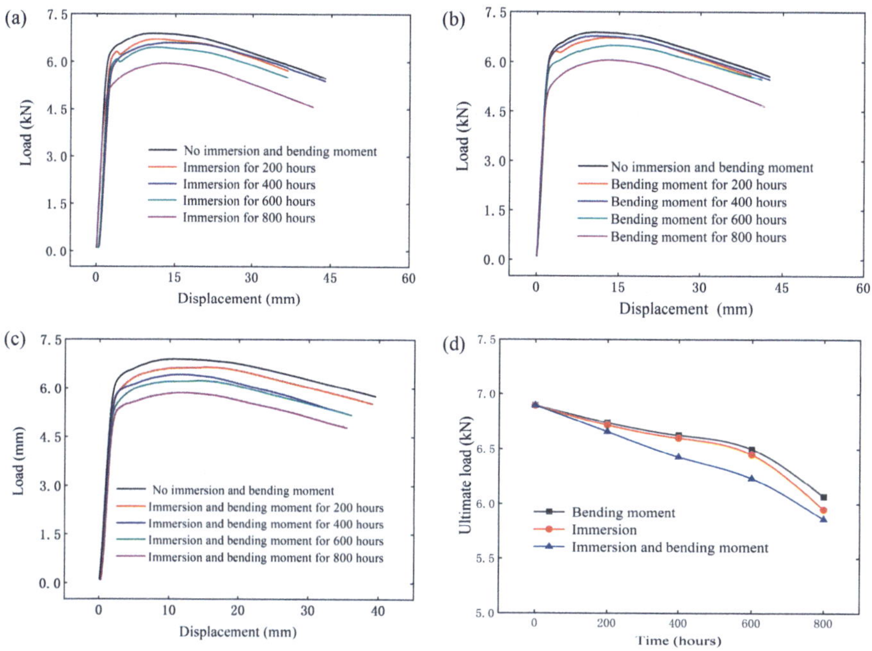

Fig. 5 The load-displacement curves of repaired pipelines. (**a**) Under seawater immersion; (**b**) Under bending moment; (**c**) Under seawater immersion and bending moment; (**d**) The relationship between the ultimate loads and time under different conditions

conditions had almost no change. The ultimate load decreased by 15% after 800 h compared to the repaired pipelines with no immersion or bending moment.

Figure 5d shows the relationships between the ultimate load and the time under different durability conditions. With the increase of the time, the ultimate loads of the CFRP repaired pipelines decreased. It was found that the effect of a long bending moment time on the ultimate loads of the pipelines was less than that of the long seawater immersion time, and the effect of a long time for both the bending moment and the seawater immersion was the largest. The ultimate load decreased by about 15% after 800 h.

The durability design factor refers to the anti-damage ability of materials or structures under different durability conditions. In this research, the durability design factors were the ultimate load ratios between the values after durability experiments and the initial values. About durability factor, Smith [38] proposed that the durability design factor of FRP reinforced concrete structures in a cold environment was 0.85. Stratford [39] proposed that the durability design factor for FRP reinforced metal structures in a cold environment was 0.85. Kabir [40] conducted a 12-month low-temperature experimental study on CFRP strengthened round steel members, for which the durability design factor was 0.90. In our research as shown in Table 4, the durability factor with a bending moment after 800 h was 0.89, and it was 0.88

Table 4 Durability factors of CFRP repaired pipelines under different time and conditions

Condition	Durability time (h)	Durability factor
Under bending moment	0	1
	200	0.98
	400	0.96
	600	0.94
	800	0.89
Under seawater immersion	0	1
	200	0.97
	400	0.96
	600	0.94
	800	0.88
Under bending moment and seawater immersion	0	1
	200	0.97
	400	0.93
	600	0.90
	800	0.85

with seawater immersion, while it was 0.85 for the condition of both immersion and a bending moment.

Why did the ultimate loads of the composite structures attenuate? One reason for this could be the attenuation of the CFRP with a long-term bending moment and seawater immersion. The other reason could be the attenuation of the interface strength between the CFRP and the steel pipelines. These reasons would lead to the earlier debonding of the repaired pipelines. However, the interface strength attenuation was not equal to the adhesive attenuation because the interface was also related to the CFRP and the steel. Therefore, to explain the mechanics of the decrease of the ultimate loads of the CFRP repaired pipelines, the durability of the interface strength between the CFRP and the pipelines and the durability of the CFRP materials was studied, as described below.

3 Durability Study of CFRP Material and Interface Strength of CFRP-Steel Joint

To study the attenuation law of the interface strength between CFRP and steel, a CFRP-steel bonding joint was used. Figure 6a shows the geometry dimensions of the CFRP specimen, and Fig. 6b, c shows the geometry dimensions of the CFRP-steel bonding joint. The adhesive between the CFRP and the steel was Araldite 2015, the same as described above. The strengthened sheet was pasted on the two ends of the specimen to prevent early failure. To obtain good bonding performance, the surface of the steel was polished, and the surfaces were cleaned using acetone. Five groups of CFRP-steel bonding joint experiments and five groups of CFRP experiments were

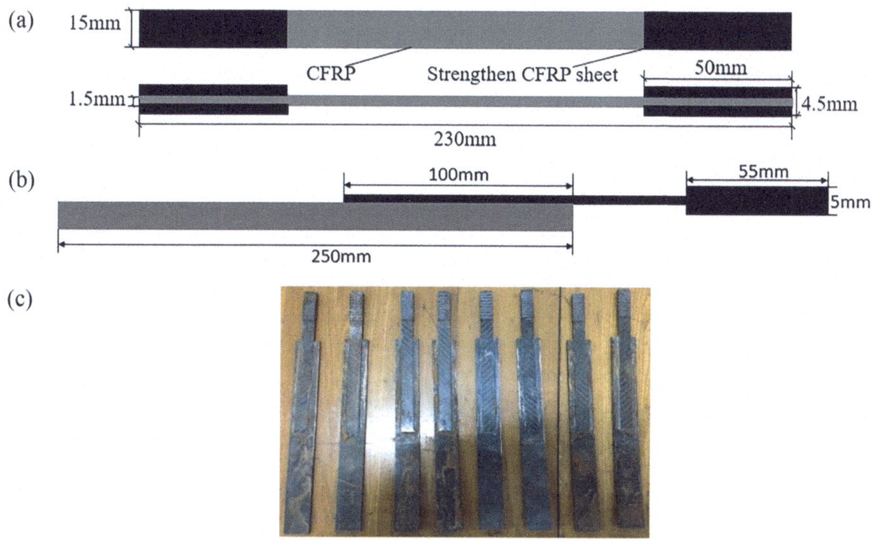

Fig. 6 The geometrical dimensions of CFRP specimen and CFRP-steel bonding joint. (**a**) CFRP specimens; (**b**) CFRP-steel bonding joints; (**c**) CFRP-steel joints from the front view

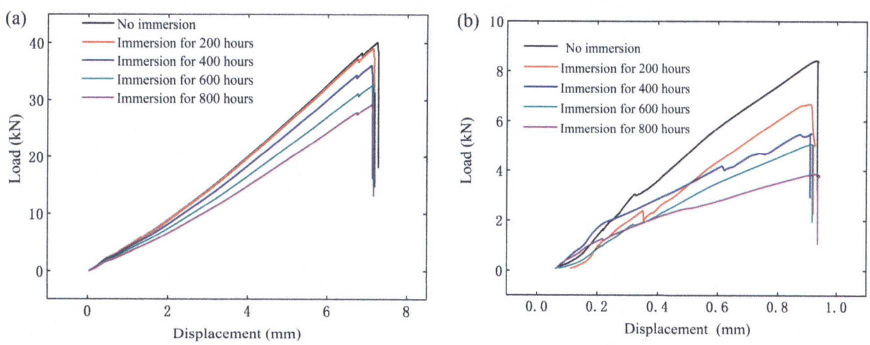

Fig. 7 The load-displacement curves of the CFRP specimen and the CFRP-steel bonding joint for different immersion time. (**a**) The CFRP specimen; (**b**) The CFRP-steel bonding joint

constructed. The CFRP specimen and the CFRP-steel bonding joint were put into a five times seawater concentration at 17 °C for different times (200, 400, 600, and 800 h).

Tension tests were carried out for the CFRP specimens before and after seawater immersion. Test set up and the way of measurement are the same with the test of CFRP repaired pipelines. During the failure of the CFRP specimens, the splitting sound happened. As the load increased, the sound became more and more obvious, and a fiber spatter and a strong burnt smell accompanied the sound.

The load-displacement curves of the CFRP specimens are shown in Fig. 7a. The curves showed almost the linear brittle properties. The ultimate loads were 38.3 kN,

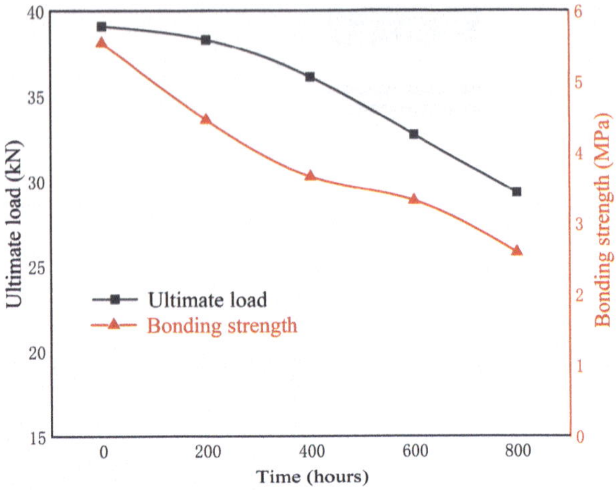

Fig. 8 The changes of ultimate load and bonding strength with immersed time for CFRP specimens and CFRP-steel bonding joints

36.1 kN, 32.7 kN, and 29.3 kN with seawater immersion for 200 h, 400 h, 600 h, and 800 h, respectively, while the ultimate load of CFRP without seawater immersion was 39.1 kN. Figure 14 shows the relationship between the ultimate load and the immersion time. The ultimate load decreased gradually with the increase of the immersion time, and the ultimate load decreased more quickly when the immersion time was longer.

The load-displacement curves of a CFRP-steel joint with different seawater immersion times are shown in Fig. 7b. The ultimate loads were 6.5 kN, 5.4 kN, 4.9 kN, and 3.5 kN after seawater immersion for 200 h, 400 h, 600 h, and 800 h, respectively, while the ultimate load of CFRP-steel joint without seawater immersion was 8.4 kN. The load-displacement curves were all basically linear including some skip before failing. The average bonding strength of a CFRP-steel joint for three experiments is shown in Fig. 8. The bond strength equals to the ultimate loads of CFRP-steel joints divided by the contacting area. It can be found from the figure that the average bonding strength decreased with the increase of the immersion time. The ultimate load of CFRP-steel joints immersed in seawater for 800 h was about 1.6 times larger than the case of 200 h.

Why was the interface strength seriously attenuated? The failure modes of a CFRP-steel joint after different immersion times are shown in Fig. 9. In the short seawater immersion, the water did not come into the interface yet. Part of the adhesive and the CFRP material could be found on the debonded steel plate, as shown in Fig. 9a, which meant that the interfacial failure happened for the steel-adhesive interface, CFRP-adhesive interface, or the (b) fracture of the steel and the CFRP. After longer seawater immersion, the water came into the interface already, as shown in Fig. 9b, resulting in the serious attenuation of the interface strength.

The durability factors refer to the damage ratio of ultimate loads for a CFRP-steel joint and a CFRP specimen, and these factors are shown in Fig. 10. It can be seen

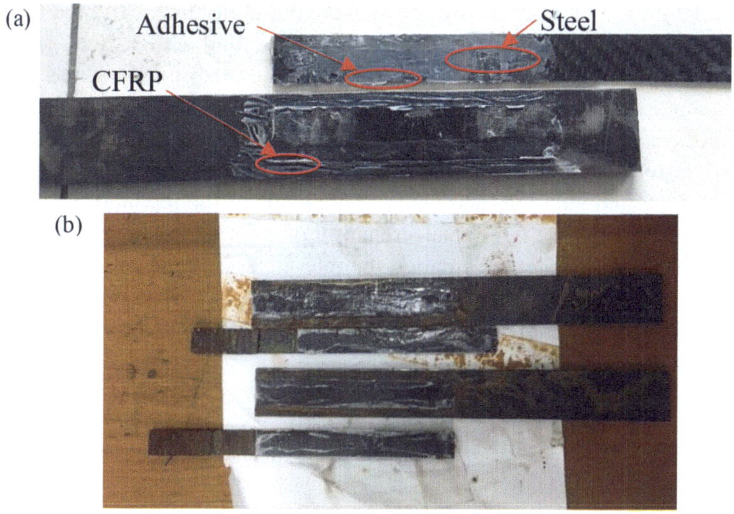

Fig. 9 The failure modes of CFRP-steel joints. (**a**) No immersion; (**b**) Immersion for 800 h

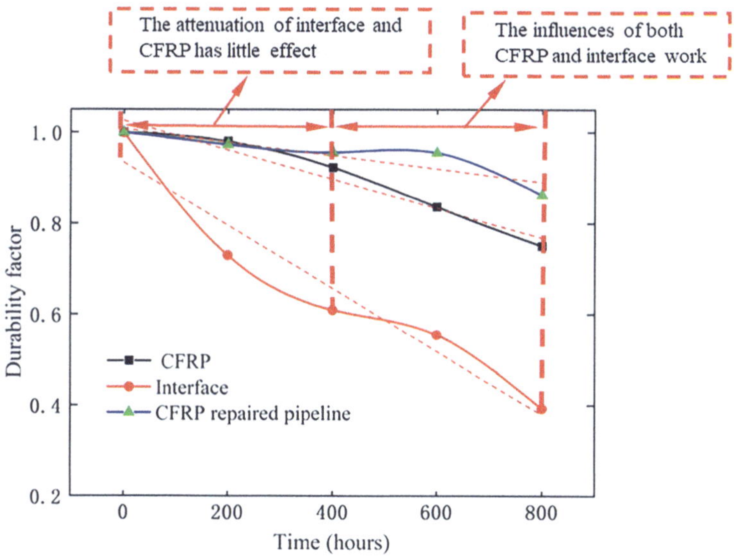

Fig. 10 The relationship between the durability factor and the immersion time for the repaired pipelines, CFRP specimens and CFRP-steel joints

from the figure that the ultimate loads were seriously attenuated in the accelerated experiment in five times seawater concentration. The ultimate load of the CFRP-steel joint was mainly determined by the interface bonding strength, and the durability factor of the CFRP-steel joint after 800 h of immersion was the smallest value, 0.40.

Figure 10 shows the relationship between the durability factor and the immersion time for the repaired pipelines, CFRP specimens, and CFRP-steel joints. The attenuation of the ultimate load of a CFRP repaired steel pipeline was related to the attenuation of the CFRP material properties and the attenuation of the CFRP-steel interface. During 0–400 h, the durability factor of the interface strength decreased faster than the CFRP and the repaired pipelines. But the interfacial strength had enough strength, and the ultimate loads of the repaired pipelines were largely not affected by the interfacial strength. For 400–800 h, the differences of the three decreased speeds become smaller, the interfacial strength might not have had enough strength, and the ultimate loads of the repaired pipelines were affected by both the interfacial strength and the CFRP strength.

4 Numerical Study on the Durability of CFRP Repaired Pipelines

A three-point bending finite element model of the CFRP repaired corroded pipelines was established in ABAQUS 6.14-2, as shown in Fig. 11. The geometry dimensions and the material properties were the same as those of the experiments. In the numerical simulation model of the CFRP repaired pipelines, the failure criterion of the CFRP and the debonding criterion were discussed.

There are several failure criteria for CFRP, including the maximum strain criterion, maximum stress criterion, Tsai-Hill criterion, Tsai-Wu criterion, and Hashin criterion. The maximum strain criterion and the maximum stress criterion don't consider the interaction of components, and the Tsai-Hill criterion considers that the tensile strength and compressive strength of composites have the same value. The Hashin criterion in ABAQUS only can solve the two-dimensional problem. The Tsai-Wu criterion could take into account the interaction of stress components and has good accuracy. In this paper, the Tsai-Wu criterion was used to calculate the failure of the CFRP. When the composite material reached the failure standard, the composite material was damaged. The calculation was carried out through the subroutines in ABAQUS. The cohesive element in ABAQUS was used to simulate the adhesive layer between CFRP and pipe [41–43]. Figure 12 shows the traction-separating displacement curve of cohesive elements. The normal and the two shear

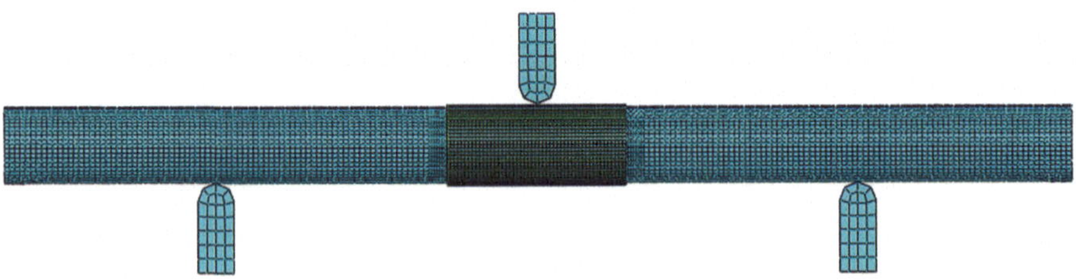

Fig. 11 The FEM model of repaired pipelines with meshes

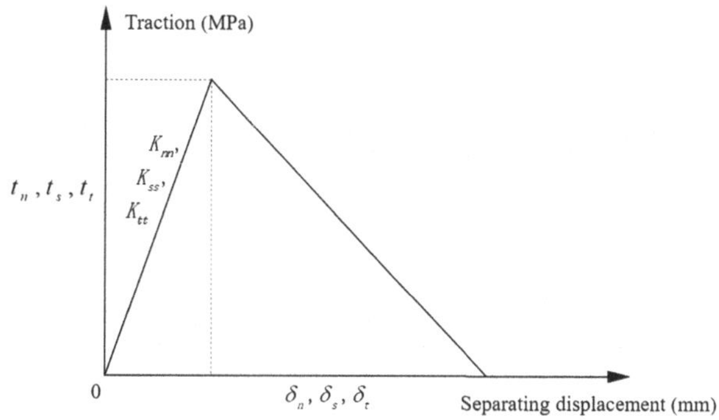

Fig. 12 Traction-separating displacement curve

traction were denoted by t_n, t_s and t_t spectively, while the corresponding separating displacements were denoted by δ_n, δ_s and δ_t respectively, and the thickness of cohesive element T_0 was taken as the original thickness of the adhesive layer.

It was assumed that the cohesive element had linear elasticity before failure [44, 45], and its elastic behavior could be written as

$$\begin{Bmatrix} t_n \\ t_s \\ t_t \end{Bmatrix} = \begin{bmatrix} K_{nn} & 0 & 0 \\ 0 & K_{ss} & 0 \\ 0 & 0 & K_{tt} \end{bmatrix} \begin{Bmatrix} \delta_n \\ \delta_s \\ \delta_t \end{Bmatrix} \tag{1}$$

There were three elastic parameters K_{nn}, K_{ss} and K_{tt} in all three directions, and K_{nn} could be expressed as

$$K_{nn} = \frac{E_a}{T_0} \tag{2}$$

It was assumed that K_{ss} and K_{tt} were the same [46], and they could be expressed as

$$K_{ss} = K_{tt} = 3\left(\frac{G_a}{T_0}\right)^{0.65} \tag{3}$$

E_a is the elastic modulus and G_a is the shear modulus, which were provided by the material manufacturers. There are mainly four failure criteria for the damage initiation for the traction separation rule, which are the maximum nominal stress criterion, maximum nominal strain criterion, quadratic nominal stress criterion, and quadratic nominal strain criterion. The stress criteria are more widely used compared with the strain criteria, because stresses are easier to be obtained compared with strains in experiments, but the maximum nominal stress does not consider the coupling effect of forces in all directions. In this research, the quadratic nominal stress traction damage criterion (QUADS) was adopted, as shown in Eq. (4).

$$\left\{\frac{\langle t_n \rangle}{t_n^0}\right\}^2 + \left\{\frac{t_s}{t_s^0}\right\}^2 + \left\{\frac{t_t}{t_t^0}\right\}^2 = 1 \quad (4)$$

where t_n^0, t_s^0, and t_t^0 represent the material strength in three directions, namely, the bonding peel strength and the bonding shear strengths in two directions. The two shear strengths are typically the same [47]. The failure parameters of the cohesive elements are shown in Table 5. t_n^0, t_s^0, and t_t^0 were obtained from the material producer, and K_{nn}, K_{ss}, K_{tt} were calculated by Eqs. (2) and (3).

The mesh type of the steel pipeline was C3D8R, the adhesive mesh was COH3D8, and the CFRP mesh was S4R. The grid sizes of the pipeline, CFRP, and adhesive were 1 mm, 1.5 mm, and 1 mm, respectively. The maximum increment step was 1000, and the initial increment step was 0.01, with the minimum analysis step 1e-15. The pipeline layer, adhesive layer, and CFRP surface were bounded by the type of tie, the pipeline and the CFRP were set as the main surfaces, and the adhesive was set as the secondary surface in the model.

The failure modes of CFRP repaired pipelines mainly includes three kinds: the bending failure of steel pipeline, the failure of CFRP and the debonding failure. The failure of the steel pipeline was based on the plastic failure criterion. The Tsai-Wu criterion was used as the failure criterion of CFRP, and it was implemented using the subroutine (USDFLD) of ABAQUS. The adhesive layer was simulated using cohesive elements, and the debonding failure was based on the quadratic nominal stress criterion (QUADS).

Figure 13a is the damage nephogram of the adhesive layer, and Fig. 13b shows the damage nephogram of CFRP in the failure of repaired pipelines. The failure of pipelines, CFRP and adhesive layer interacted with each other, when the loads at the

Table 5 Material parameters of cohesive elements

Material parameters	Numerical value
t_n^0 (MPa)	23
t_s^0 (MPa)	23
t_t^0 (MPa)	23
K_{nn} (N/mm^3)	2.8×10^3
K_{ss} (N/mm^3)	1.4×10^3
K_{tt} (N/mm^3)	1.4×10^3

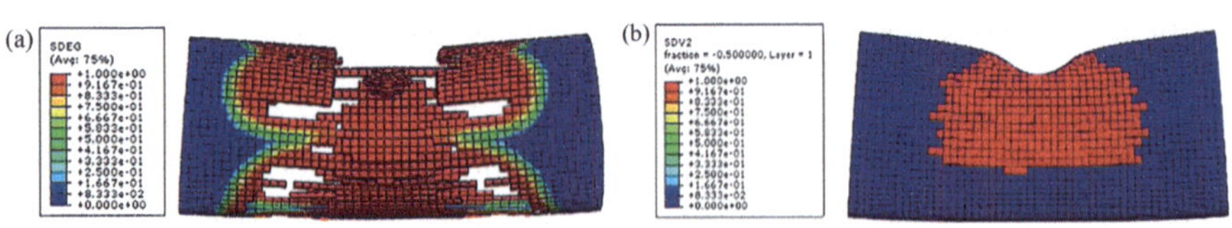

Fig. 13 The damage nephograms. (**a**) The damage nephogram of adhesive layer; (**b**) The damage nephogram of CFRP

corroded pipelines were shared by CFRP and adhesive, the failure of pipelines would be proposed. But if the values of t_n^0, t_s^0 and t_t^0 in Table 5 were small, the failure of adhesive layer would happen partially leading to the debonding, so the loads subjected to CFRP would be decreased, and the failure of pipelines would be accelerated.

These failure modes are shown in Fig. 14 for comparison with the experimental results. It can be seen from Fig. 14a, b that the failure modes obtained with the experiments and with FEM were almost the same for both the corroded pipelines and the repaired pipelines. The stresses near the buckling area were significantly decreased by CFRP repairing. It can be seen from Fig. 14c that under the bending load, buckling happened and debonding occurred at the end of the interface between the CFRP and the pipeline in the experiments, which led to the slipping of the CFRP. The same phenomenon could be found in the FEM simulation results, as shown in Fig. 14c.

The results for the FEM were compared with the experimental results in this research. Figure 15a shows the load-displacement curves of the CFRP repaired pipelines that were obtained from the experiments and the FEM. It could be seen that the load-displacement curves obtained with FEM and the experiments were basically consistent in the elastic stage for both the corroded pipelines and the repaired pipelines. The ultimate loads obtained with FEM were slightly higher than those obtained with the experiments. Figure 15b shows the ultimate loads of the corroded pipelines and the repaired pipelines obtained with the experiments and

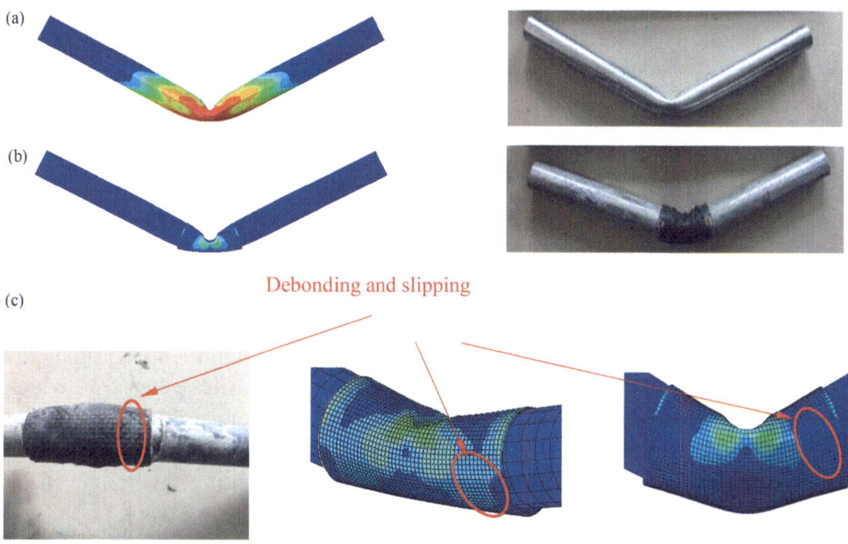

Fig. 14 Comparison of experimental and finite element pipeline failure modes. (**a**) The corroded pipelines under a bending moment; (**b**) The CFRP repaired pipelines under a bending moment; (**c**) Debonding of CFRP repaired pipelines

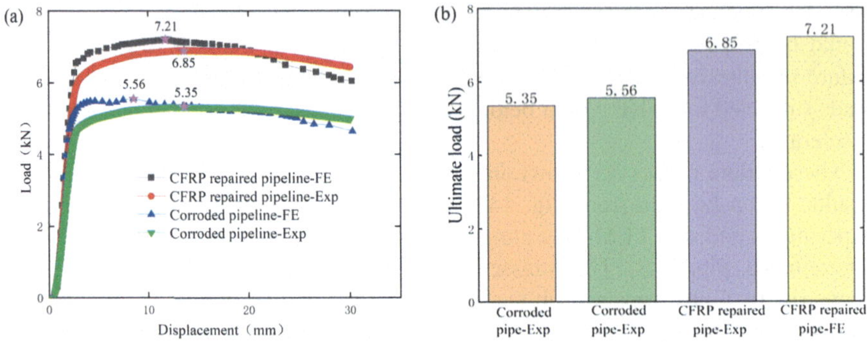

Fig. 15 Comparison of experimental and FEM resutls. (**a**) The load-displacement curves; (**b**) The ultimate loads

with FEM. The ultimate bending loads of the corroded pipeline obtained with the experiments and with the FEM were 5.35 kN and 5.56 kN, respectively, with an error of 4%. The ultimate bending loads of the repaired pipeline obtained with the experiments and the FEM were 6.90 kN and 7.21 kN, respectively, with an error of 4.5%. The ultimate loads of the repaired pipelines increased by about 30% compared with those of the corroded pipelines, which illustrated that the CFRP has a good repairing effect. Based on the above analysis, FEM used in this research could be used to obtain almost the same results as the experiments.

In the FEM model of the repaired pipelines for different durability conditions, the ultimate strength of the CFRP, and the interface strength were obtained by multiplying the durability factors (as shown in Fig. 15) and their original strength. Using the damage FEM model for the above, the load-displacement curves of the repaired pipelines under different durability conditions were obtained numerically, and these curves were compared with the experimental results shown in Fig. 16. The load-displacement curves for the experiments and the FEM were almost the same, except for the fact that the ultimate loads obtained with FEM were slightly higher than the experimental results. When the time was 800 h, the ultimate load was reduced by 12.7% compared with the repaired pipeline without seawater immersion.

The ultimate loads of the CFRP repaired pipelines obtained with the experiments and the FEM for different immersion times (200, 400, 600, and 800 h) are shown in Fig. 17. The result for the finite element method was slightly higher than the experimental result, and the largest error between them was less than 6%, which proved that the FEM model used in this research could simulate the damage and failure of CFRP repaired pipelines. The load-displacement curves of repaired structures could be obtained for different conditions and times using this model if the durability factors of the materials and the interfaces were known.

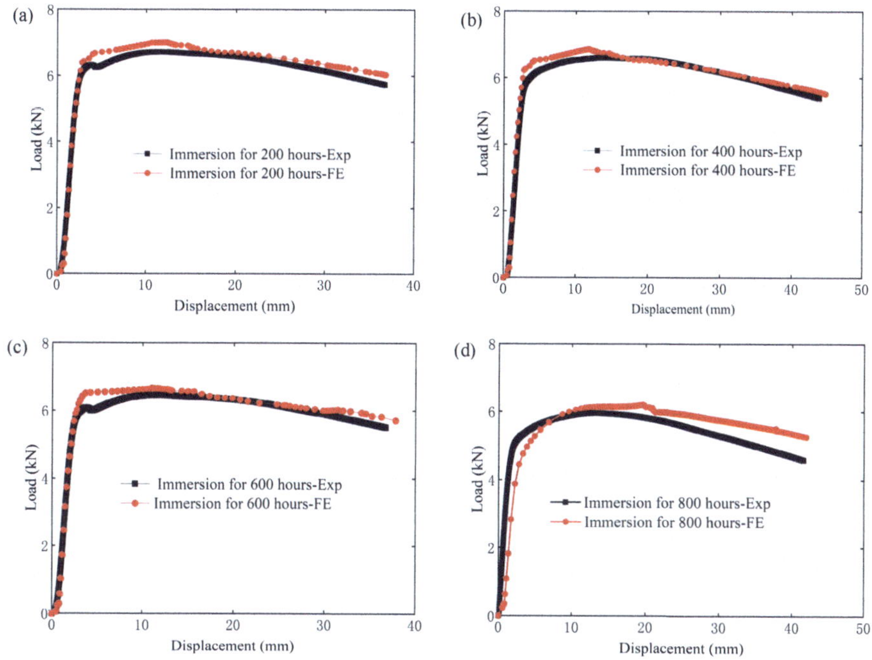

Fig. 16 The experiment and FEM load-displacement curves for different time. (**a**) 200 h; (**b**) 400 h; (**c**) 600 h; (**d**) 800 h

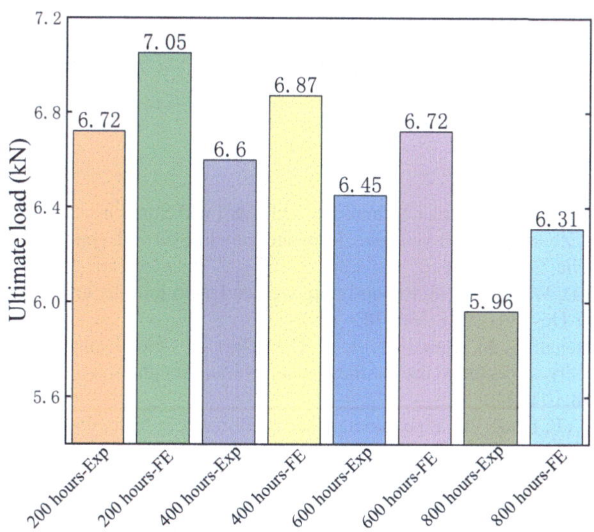

Fig. 17 Experiment and finite element ultimate loads under different immersion time

5 Conclusions

In this research, the attenuation law of CFRP repaired corroded pipelines was studied. The durability experiments of CFRP repaired corroded pipelines, CFRP material, and CFRP-steel interfacial strength with a long-term bending moment and seawater immersion were constructed. The ultimate loads of the repaired pipelines decreased with the increase of the time of the bending moment and the seawater immersion, and the effect of the bending moment on the degradation of the ultimate loads was less than that of the seawater immersion. The attenuation of the ultimate load of the CFRP repaired corroded pipeline could be broken into the attenuation of the CFRP material properties and the CFRP-steel interfacial strength. The attenuation of the CFRP-steel interface was more serious than the attenuation of the CFRP, but the ultimate loads of the repaired pipelines were not affected by the interfacial strength largely after the short immersion time. It could be explained that the interface is not directly contacted with seawater, as the effect of immersion in seawater was more than longtime static loads on the durability of repaired pipelines.

The damage model of the CFRP repaired corroded pipelines was constructed using FEM, and the load-displacement curves of the repaired corroded pipelines were obtained using the durability factors of the CFRP and the interfacial strength obtained from the experiments. The ultimate loads obtained from the FEM were slightly higher than the experimental results, the error between them was less than 6%. The failure modes were almost the same as those of the experiments, which proved the accuracy of the FEM model. This paper proposes a numerical method for obtaining the mechanical properties of composite structures with different durability conditions if the durability laws of materials and interfaces are known.

References

1. Wang HQ. Corrosion pipe repair technology. Oil Gas Field Surf Eng. 2014;33(8):109.
2. Sun C, Mao D, Zhao T, et al. Investigate deepwater pipeline oil spill emergency repair methods. Aquatic Procedia. 2015;3:191–6.
3. Teng J, Wang D, Wang Z, et al. Repair of arc welded DH36 joint by underwater friction stitch welding. Mater Des. 2017;118:266–78.
4. Sahraoui Y, Benamira M, Nahal M, et al. The effect of welded joint repair on a corroded pipeline reliability subjected to the hardness spatial variability and soil aggressiveness. Eng Fail Anal. 2020;118:104854.
5. Lu MX, Chen YF, Dong SH. Pipe maintenance reinforcement technology and its development trend. City Gas. 2006;24(2):129–32.
6. Michael BJ. Mechanical damage in pipelines. Final Rep. 2009:13–5.
7. Emerson MR, Hall DM, Gilbertz SJ. Pipeline pipedreams: oil spills, pipeline accidents, and the local truths embedding fossil fuels in the Yellowstone River Valley, United States. Energy Res Soc Sci. 2021;72:101859.
8. Chen J, Di Z, Shi J, et al. Marine oil spill pollution causes and governance: a case study of Sanchi tanker collision and explosion. J Clean Prod. 2020;273:122978.

9. Rekadwad BN, Khobragade CN. A case study on effects of oil spills and tar-ball pollution on beaches of Goa (India). Mar Pollut Bull. 2015;100(1):567–70.
10. Peck JA, Li G, Pang SS, et al. Light intensity effect on UV cured FRP coupled composite pipe joints. Compos Struct. 2004;64(3–4):539–46.
11. Ramirez FA, Carlsson LA, Acha BA. Evaluation of water degradation of vinylester and epoxy matrix composites by single fiber and composite tests. J Mater Sci. 2008;43:5230–42.
12. Abanilla MA, Li Y, Karbhari VM. Durability characterization of wet layup graphite/epoxy composites used in external strengthening. Compos Part B. 2005;37(2–3):200–12.
13. Nguyen TC, Bai Y, Zhao XL, et al. Mechanical characterization of steel/CFRP double strap joints at elevated temperatures. Compos Struct. 2011;93(6):1604–12.
14. Al-Mosawe A, Al-Mahaidi R, Zhao XL. Bond behaviour between CFRP laminates and steel members under different loading rates. Compos Struct. 2016;148:236–51.
15. Batuwitage C, Fawzia S, Thambiratnam D, et al. Durability of CFRP strengthened steel plate double-strap joints in accelerated corrosion environments. Compos Struct. 2017;160:1287–98.
16. Li F, Du Y, Sun X, et al. Sensing performance assessment of twisted CFRP with embedded fiber Bragg grating sensors subjected to monotonic and fatigue loading. Sensors Actuators A Phys. 2018;271:153–61.
17. Zhang Y, Zhang K, Zhao H, et al. Stress analysis of adhesive in a cracked steel plate repaired with CFRP. J Constr Steel Res. 2018;145:210–7.
18. Sun G, Yu H, Wang Z, et al. Energy absorption mechanics and design optimization of CFRP/aluminium hybrid structures for transverse loading. Int J Mech Sci. 2019;150:767–83.
19. Galvez P, Abenojar J, Martinez MA. Durability of steel-CFRP structural adhesive joints with polyurethane adhesives. Compos Part B. 2019;165:1–9.
20. Pang Y, Wu G, Wang H, et al. Experimental study on the bond behavior of CFRP-steel interfaces under quasi-static cyclic loading. Thin-Walled Struct. 2019;140:426–37.
21. Fawad M, Kalman K, Khushnood RA, et al. Retrofitting of damaged reinforced concrete bridge structure. Procedia Struct Integrity. 2019;18:189–97.
22. Wang X, Petrů M. Mode I fracture evaluation of CFRP-to-concrete interfaces subject to aggressive environments agents: freeze-thaw cycles, acid and alkaline solution. Compos Part B. 2019;168:581–8.
23. Tie Y, Hou Y, Li C, et al. Optimization for maximizing the impact-resistance of patch repaired CFRP laminates using a surrogate-based model. Int J Mech Sci. 2020;172:105407.
24. Wu C, Yu YZ, Tam L, et al. Effect of glass fiber sheet in adhesive on the bond and galvanic corrosion behaviours of CFRP-steel bonded system. Compos Struct. 2021;259:113218.
25. Liu ZQ, Luo B, Wang Q, et al. Experimental and numerical investigation of the anti-debonding performance for novel CFRP-steel tube composite member under tension. J Build Eng. 2021;35:102004.
26. He J, Xian G, Zhang YX. Numerical modelling of bond behaviour between steel and CFRP laminates with a ductile adhesive. Int J Adhes Adhes. 2021;104:102753.
27. Ke L, Li C, He J, et al. Effects of elevated temperatures on mechanical behavior of epoxy adhesives and CFRP-steel hybrid joints. Compos Struct. 2020;235:111789.
28. Raza A, Rafique U, Masood B, et al. Performance evaluation of hybrid fiber reinforced low strength concrete cylinders confined with CFRP wraps. Structure. 2021;31:182–9.
29. Kafodya I, Xian G, Li H. Durability study of pultruded CFRP plates immersed in water and seawater under sustained bending: water uptake and effects on the mechanical properties. Compos Part B. 2015;70:138–48.
30. Borrie D, Liu HB, Zhao XL, et al. Bond durability of fatigued CFRP-steel double-lap joints pre-exposed to marine environment. Compos Struct. 2015;131:799–809.
31. Correia L, Sena-Cruz J, Michels J, et al. Durability of RC slabs strengthened with prestressed CFRP laminate strips under different environmental and loading conditions. Compos Part B. 2017;125:71–88.
32. Yang Y, Silva MAG, Biscaia H, et al. Bond durability of CFRP laminates-to-steel joints subjected to freeze-thaw. Compos Struct. 2019;212:243–58.

33. Zhang X, Wang Y, Wan B, et al. Effect of specimen thicknesses on water absorption and flexural strength of CFRP laminates subjected to water or alkaline solution immersion. Constr Build Mater. 2019;208:314–25.
34. Miyano Y, Nakada M. Accelerated testing methodology for durability of CFRP. Compos Part B. 2020;191:107977.
35. Lu Z, Li J, Xie J, et al. Durability of flexurally strengthened RC beams with prestressed CFRP sheet under wet–dry cycling in a chloride-containing environment. Compos Struct. 2021;255: 112869.
36. Mu W, Na J, Tan W, et al. Durability of adhesively bonded CFRP-aluminum alloy joints subjected to coupled temperature and alternating load. Int J Adhes Adhes. 2020;99:102583.
37. American Society of Mechanical Engineers. Boiler, Pressure Vessel Committee. 2010 ASME boiler and pressure vessel code. American Society of Mechanical Engineers; 2011.
38. Smith ST, Kaul R, Ravindrarajah R, et al. Durability considerations for FRP-strengthened RC structures in the Australian environment. In: Proceedings of Australian structural engineering conference. 2005. p. 952–961.
39. Stratford T, Cadei J, Hollaway L. CIRIA C595–STRENGTHENING METALLIC STRUCTURES USING EXTERNALLY-BONDED FRP. In: Advanced polymer composites for structural applications in construction. Woodhead Publishing; 2004. p. 693–700.
40. Kabir MH, Fawzia S, Chan THT. Durability of CFRP strengthened circular hollow steel members under cold weather: experimental and numerical investigation. Constr Build Mater. 2016;123:372–83.
41. Bertolini J, Castanié B, Barrau JJ, et al. An experimental and numerical study on omega stringer debonding. Compos Struct. 2008;86(1–3):233–42.
42. Zhou Y, Chen X, Fan Z, et al. Bond behaviors of FRP-to-concrete interface under the control of a novel end-anchorage system. Compos Struct. 2017;168:130–42.
43. Xu Y, Gao Y, Wu C, et al. On design of carbon fiber reinforced plastic (CFRP) laminated structure with different failure criteria. Int J Mech Sci. 2021;196:106251.
44. Fernando ND. Bond behaviour and debonding failures in CFRP-strengthened steel members. 2010.
45. De Lorenzis L, Zavarise G. Cohesive zone modeling of interfacial stresses in plated beams. Int J Solids Struct. 2009;46(24):4181–91.
46. Teng JG, Fernando D, Yu T. Finite element modelling of debonding failures in steel beams flexurally strengthened with CFRP laminates. Eng Struct. 2015;86:213–24.
47. Koloor SSR, Khosravani MR, Hamzah RIR, et al. FE model-based construction and progressive damage processes of FRP composite laminates with different manufacturing processes. Int J Mech Sci. 2018;141:223–35.

The Research of Strength Degradation Characteristics of CFRP-Repaired Pipes Under Fatigue Loads Based on Experimental and Numerical Analysis

1 Introduction

Marine pipes are critical structures in the development of oil and gas resources and known as the "lifeline" of offshore oil and gas engineering [1]. However, marine pipes are in complex environments and are subjected to the effects of current erosion, seabed movement, corrosion, third-party activity, structural defect and operational error, all of which can lead to pipe failures [2, 3]. At present, the main repair methods for subsea damaged pipes include fixture repair, clamp repair, welding repair and CFRP repair (as shown in Fig. 1). Compared with other repair methods, such as welding, bolted connection, riveting, the CFRP has the advantages of light weight, high strength, high modulus, excellent fatigue resistance, simple operation and no welding heat input during repair [4]. Moreover, CFRP are resistant to corrosion by acid, alkali, and salt media, which makes them more suitable for the repair of marine structures. The unevenness of the seabed and the scouring of the current on the seabed can cause pipelines to be suspended [2]. When the frequency of vortex shedding is close to the natural frequency of the spanning pipeline, vortex induced resonance will occur, which can cause fatigue damage to the pipe [3]. The interface layer is the weakest part of the CFRP-repaired structure. The strength of the CFRP-steel interface and the steel will decay under cyclic loads [5, 6], which will lead to a decrease in the bearing capacity of the structure repaired by CFRP and make the structure more prone to failure. Therefore, it is of great significance to study the strength degradation law of CFRP-repaired pipes under fatigue loads.

In the durability studies of CFRP-strengthened structures, Kafodya et al. [7] investigated the long-term performance of pultruded unidirectional CFRP plates immersed in water and seawater at room temperature under sustained bending strains of 30% and 50% of ultimate strain. Kabir et al. [8] examined the durability of CFRP-strengthened tubular steel structures exposed to cold weather for durations of 3 and 6 months, assessing their structural responses in terms of failure load, stiffness, and failure modes under four-point bending. Correia et al. [9] analyzed the durability and

Y. Zhang, *Interfacial Mechanics of Offshore Engineering Equipment*,
https://doi.org/10.1007/978-981-95-3200-1_12

Fig. 1 CFRP repair method for damaged pipes

long-term behavior of reinforced concrete structures strengthened with CFRP laminate strips under chloride exposure and wet-dry cycling at a water temperature of 20 °C. Yang et al. [10] evaluated the bond durability of CFRP laminates-to-steel joints subjected to freeze-thaw cycles for 10,000 h, confirming that pure thermal cycle aging had minimal impact on joint degradation under the imposed conditions. Miyano et al. [11] provided a comprehensive review of accelerated testing methodologies for assessing CFRP durability. Mu et al. [12] experimentally investigated the durability of adhesively bonded CFRP-aluminum alloy joints under coupled temperature and alternating load conditions, analyzing failure modes. Zhang et al. [13] studied the bond durability of CFRP reinforcements bonded to concrete using epoxy and cement adhesives under water immersion for durations ranging from 0 to 6 months, revealing an exponential decline in interfacial fracture energy with increased immersion time. Zhang et al. [14] explored the residual strength of CFRP-repaired pipes subjected to bending moments and seawater immersion, incorporating CFRP degradation and interface deterioration into their numerical model. Shao et al. [15] examined the durability lifespan, stiffness degradation, and horizontal bearing capacity of CFRP composite piles in ocean corrosion environments. Lee et al. [16] assessed the durability of CFRP-strengthened wood members under seawater dry-wet cycling and freeze-thaw cycling conditions, evaluating their performance through SEM analysis, compressive shear strength, pull-off strength, and flexural strength tests.

According to the literature review, significant progress has been made in investigating the durability of CFRP-repaired structures under marine corrosive environments, dry-wet cycling, cold weather, freeze-thaw cycling, and static loading. However, there remains a notable research gap regarding the strength degradation characteristics of CFRP-repaired structures subjected to fatigue loading.

Currently, studies on the strength degradation and fatigue life of CFRP-repaired structures lack a comprehensive approach to simulate fatigue damage in the interface layer [17, 18]. The interface layer is the weakest component in CFRP-repaired structures due to its comparatively lower strength within the system and the substantial deformations it experiences when the two materials deform non-uniformly [19, 20]. Existing research on interface strength, fatigue life, and crack propagation predominantly employs the cohesive zone model (CZM), initially proposed by Dugdale [21] in 1960. The CZM represents the traction as a function of crack separation displacement, effectively mitigating the singularity at the crack tip [22]. Over the years, the CZM theory under static loading has matured, with significant advancements in related numerical simulation techniques. Although the application of CZM under cyclic loading is still developing, it has shown potential in several domains, including fatigue damage in adhesively bonded joints [23], fatigue life of asphalt concrete [24], fatigue damage in solder joints of electronic chip packages [25], fatigue crack growth in multi-layered structures [26], and intergranular fatigue crack propagation [27]. This study employs a CZM-based numerical approach to simulate fatigue damage accumulation, implemented through customized subroutines.

In this paper, the strength degradation characteristics of CFRP-repaired pipes under cyclic loads are analyzed using both experimental and numerical modeling methods. This work aims to provide new insights into the fatigue behavior of CFRP-repaired structures.

2 Experiment

In this section, experiments on ultimate bending strength and fatigue performance are conducted on defective pipes and CFRP-repaired pipes. Additionally, the residual bending strengths of CFRP-repaired pipes after varying fatigue cycles are tested to investigate the characteristics of strength degradation.

2.1 Experiment Model

The experimental specimen selected for this study is an API 5L X60 series steel pipe (shown in Fig. 2) [28]. The specimen has a length of 300.00 mm, a diameter of 21.30 mm, and a wall thickness of 1.90 mm. The material properties of the X60 steel pipe are as follows: a modulus of elasticity of 210.00GPa, a Poisson's ratio of 0.3, a

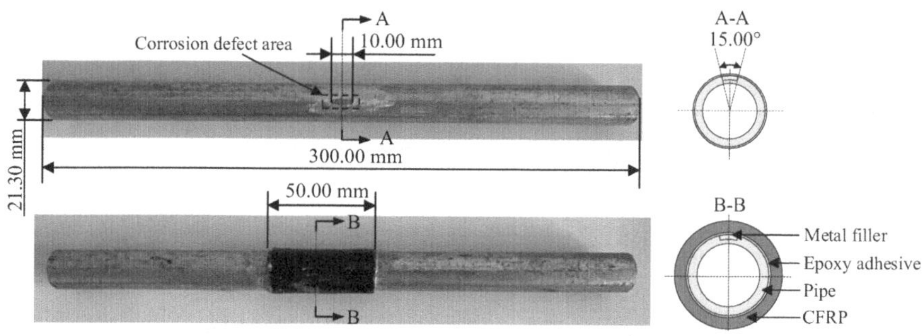

Fig. 2 Schematic diagram of experimental specimen

density of 7.80 g/cm^3, and a yield strength of 483.00 MPa. Stress-plastic strain data for the pipe, obtained through tensile testing, are presented in Table 1.

To simulate corrosion damage, a rectangular defect with a length of 10.00 mm, a depth of 1.00 mm, and an angle of 15.00° is introduced at the pipe's center. Before CFRP repair, the defect is filled with PH2150 metal filler and cured for 72 h. Following the curing process, the defect is repaired using Araldite® 2015 epoxy adhesive and T1200 3K-200g 0/90° CFRP prepreg. Six layers of CFRP prepreg are applied, with a winding width of 50.00 mm. The repaired specimen is then cured in a dry, ventilated environment for 72 h. The number of CFRP layers and the repair length are determined following the guidelines of ASME PCC-2 [29]. The material properties of the cured metal filler, provided by the manufacturer (Dongguan Pinheng Adhesive Co., Ltd.), are detailed in Table 2. The properties of the epoxy adhesive are obtained through double cantilever beam (DCB) tests and end-notched flexure (ENF) tests and are summarized in Table 3. The CFRP used in the repair is an orthotropic material and its parameters are shown in Table 4, where E_1, E_2, and E_3 represent the elastic moduli in the longitudinal, transverse, and through-thickness directions, respectively; v_{12}, v_{13}, and v_{23} denote the Poisson's ratios; and G_{12}, G_{13}, and G_{23} indicate the shear moduli.

Table 1 Stress-plastic strain data of API 5L X60 steel pipe

Yield stress (MPa)	Plastic strain (%)	Yield stress (MPa)	Plastic strain (%)	Yield stress (MPa)	Plastic strain (%)
483.00	0.00	541.94	2.06	570.97	5.59
488.71	0.55	550.00	2.49	574.19	6.51
495.16	0.62	554.84	2.81	588.71	7.45
501.61	0.75	559.68	3.13	590.32	7.98
509.68	0.92	564.52	3.58	593.55	9.34
516.13	1.09	570.97	4.22	595.16	10.61
525.81	1.34	574.19	4.60	596.00	14.21
535.48	1.76	577.42	4.92	597.00	17.32

Table 2 Material parameters of metal filler

Parameter	Value	Parameter	Value
Density (g/cm$^{3)}$	1.89	Modulus (GPa)	5.50
Compressive strength (MPa)	96.50	Poisso's ratio	0.43
Tensile strength (MPa)	37.50	Bending strength (MPa)	57.80
Shear strength (MPa)	24.60	Working temperature (°C)	−60.00~206.00

Table 3 Material parameters of epoxy adhesive

Parameter	Value	Parameter	Value	Parameter	Value
K_n (MPa)	2.80×10^3	K_s (MPa)	1.40×10^3	K_t (MPa)	1.40×10^3
t_n (MPa)	23.20	t_s (MPa)	23.60	t_t (MPa)	23.60
G_n^c (mJ/mm^2)	0.23	G_s^c (mJ/mm^2)	0.24	G_t^c (mJ/mm^2)	0.24

Table 4 Material parameters of CFRP

Parameter	Value	Parameter	Value
Density (g/cm^3)	1.78	Elastic modulus E_1 (GPa)	114.00
Elastic modulus E_2 (GPa)	8.61	Elastic modulus E_3 (GPa)	8.61
Shear modulus G_{12} (GPa)	4.16	Poisso's ratio v_{12}	0.30
Shear modulus G_{13} (GPa)	4.16	Poisso's ratio v_{13}	0.30
Shear modulus G_{23} (GPa)	3.00	Poisso's ratio v_{23}	0.45

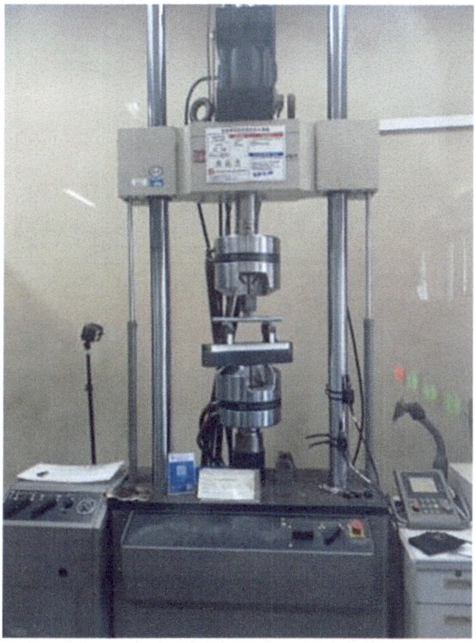

Fig. 3 MTS hydraulic servo fatigue tester

Three-point bending experiments are conducted to evaluate the ultimate bending strength, apply bending fatigue loads, and assess residual strength. These tests are performed using an MTS 810-25 hydraulic servo fatigue tester (shown in Fig. 3). In the experimental setup, the pipes are positioned on a three-point bending fixture with the corrosion defect oriented on the tensioned side. The testing followed the guidelines of ISO 7438 [30], with the span between the support rollers set to 180.00 mm. The diameters of both the loading and support rollers are 20.00 mm. The loading speed is maintained at 2.00 mm/min throughout the tests.

2.2 *Results and Discussion*

Experiments are conducted to evaluate the ultimate bending strength of both defective and CFRP-repaired pipes. The load-displacement curves are shown in Fig. 4, indicating that the defective pipe's ultimate bending strength is 7.40 kN, whereas the

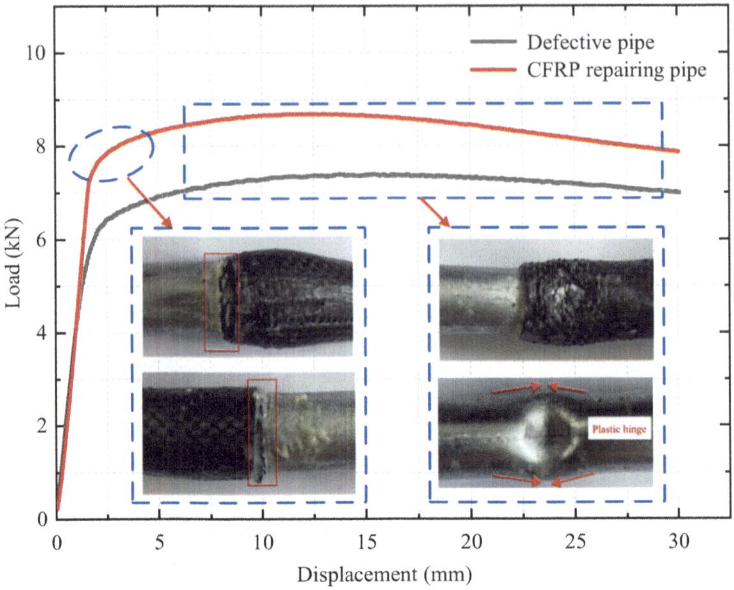

Fig. 4 Load-displacement curves of ultimate bending strength experiment

Table 5 Ultimate bending strength and fatigue life

Pipe type	Defective pipe	CFRP-repaired pipe	Increase ratio
Ultimate strength (kN)	7.40	8.71	17.70%
Fatigue life	157,824	408,040	158.54%

CFRP-repaired pipe achieves an ultimate bending strength of 8.71 kN. This corresponds to a 17.70% improvement in bending strength after repair. The failure mode of the CFRP-repaired pipe is characterized by CFRP debonding and pipe buckling.

Fatigue experiments are also performed on defective and CFRP-repaired pipes, with results presented in Table 5. In these experiments, the maximum load is set at 7.30 kN, with a stress ratio of 0.5, a loading frequency of 10 Hz, and a sinusoidal waveform. The defective pipe exhibited a fatigue life of 157,824 cycles, while the CFRP-repaired pipe achieved a fatigue life of 408,040 cycles—1.58 times that of the defective pipe. These results demonstrate that the CFRP repair method is more effective at enhancing the fatigue life of the repaired structure than at improving its ultimate bending strength, aligning with findings from previous studies [19, 31].

Residual bending strength experiments are conducted on CFRP-repaired pipes to assess the impact of cyclic loading. The ultimate bending strength of CFRP-repaired pipe is measured after different numbers of loading cycles (50, 100, 500, 1000, 5000, and 10,000) using the three-point bending test. The load-displacement curve is presented in Fig. 5, while the residual strength and the degradation rate are illustrated in Table 6 and Fig. 6. The results indicate that the residual strength decreases exponentially with the increase in the number of cyclic loads (coefficient of

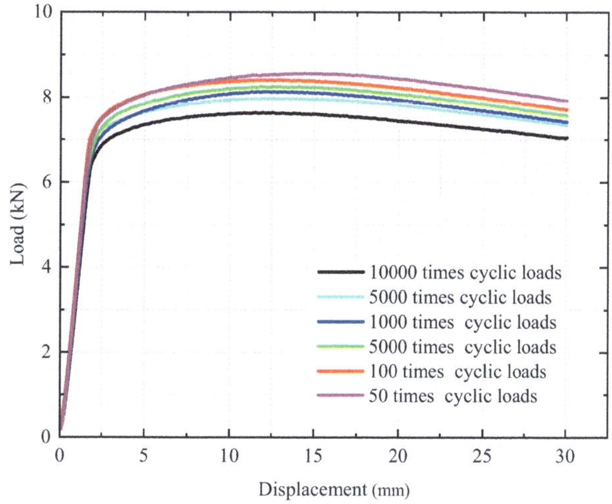

Fig. 5 Load-displacement curves after applying different number of cyclic loads

Table 6 Residual strength and degradation rate obtained from experiments

Number of cyclic loads	0	50	100	500	1000	5000	10,000
Residual strength (kN)	8.71	8.58	8.43	8.26	8.15	7.98	7.64
Degradation rate (%)	\	1.49	3.21	5.17	6.43	8.38	12.28
Degradation rate caused by each cyclic load (‰)	\	2.98	3.44	0.49	0.25	0.05	0.08

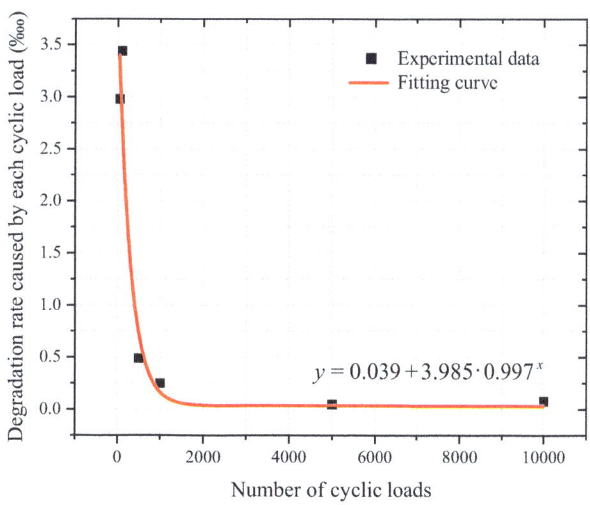

Fig. 6 The relationship between the degradation rate and the number of cyclic loads

determination $R^2 = 0.95$). When the number of cyclic loads is less than 100, the residual strength declines rapidly. The reason for the above phenomenon is that the displacement at the loading point is large in the early stage of applying cyclic loads, and the pipe has a large deformation, which has a greater damage to the CFRP-repaired pipe.

3 Numerical Analysis of Ultimate Strength and Residual Strength

In this section, a numerical model for analyzing the ultimate strength and residual strength of CFRP-repaired pipes under fatigue loading is developed using ABAQUS®. VUSDFLD and VUMAT Subroutines are implemented to simulate the damage and failure of the interface, CFRP, and pipe. The model is employed to investigate the failure modes, ultimate bending strength, and residual strength of the repaired pipes. Numerical analysis results are compared with experimental data to validate the model's accuracy.

3.1 Finite Element Model

The numerical model for the three-point bending test of the CFRP-repaired pipe is illustrated in Fig. 7. The model dimensions, including the CFRP-repaired pipe specimen, the support rollers, and the indenter, as well as the material properties of the pipe, CFRP, metal filler, and epoxy adhesive, are consistent with the experimental setup described in Sect. 2.1. The number of CFRP layers, their layup angles, and the span of the support rollers in the numerical model are also aligned with the experimental configuration.

In the mesh setup, C3D8R elements are used to model the pipe, CFRP, and metal filler, while the COH3D8 cohesive elements are utilized for the epoxy adhesive. In the interaction settings, tie constraints are applied between the pipe and epoxy adhesive surfaces, and between the epoxy adhesive and CFRP surfaces. Surface-to-surface contact interactions are defined between the support rollers and the pipe surface, as well as between the indenter and the CFRP surface. The indenter and support rollers are modeled as rigid bodies, with reference points RP1, RP2, and RP3 representing their control points. Boundary conditions are defined as follows: RP2 and RP3 are fully constrained in all degrees of freedom, while the pipe ends are

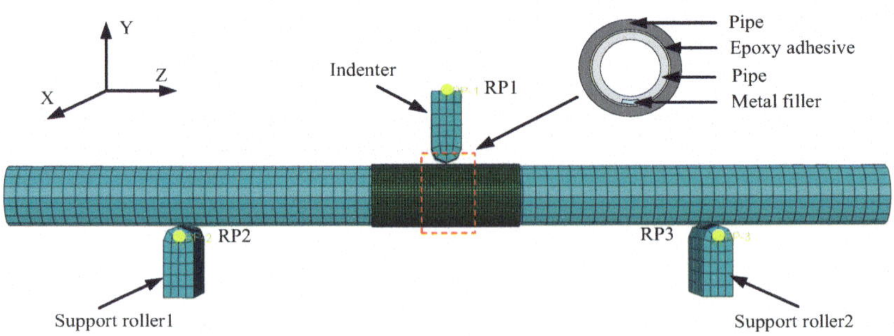

Fig. 7 Finite element model

restricted in the X-direction (refer to Fig. 7 for the coordinate system). From the experimental load-displacement curve, the CFRP-repaired pipe is observed to fail when the displacement reached 30.00 mm. Consequently, in the numerical model, a displacement of −30.00 mm in the Y-direction is applied to RP1 to calculate the ultimate bending strength of the CFRP-repaired pipe. To simulate the residual strength after applying cyclic loads, cyclic loading is first applied to RP1, followed by a displacement of −30.00 mm to assess the ultimate strength. The amplitude, stress ratio, and loading frequency of the cyclic loads in the numerical model are consistent with those used in the experimental setup. This approach ensures an accurate comparison between experimental and numerical results.

3.2 Damage and Failure Criteria

3.2.1 Steel Pipe

The calculation equation for the residual strength of the steel pipe $f_y(N)$ after N cycles of loading is written as [35]:

$$f_y(N) = \begin{cases} f_y, & N = 0 \\ (\sigma_{\min} + \Delta\sigma)\left[1 - \dfrac{\lg N}{7.253 - 0.0056\Delta\sigma}\left(1 - \dfrac{\Delta\sigma + \sigma_{\min}}{f_y}\right)\right], & N > 0 \end{cases} \tag{1}$$

where $\sigma_{\min}$ is the minimum stress, $\Delta\sigma$ is the stress amplitude, f_y is the initial strength of the steel pipe. Weakening of steel pipe strength under cyclic load is achieved by writing the user subroutine VUSDFLD.

3.2.2 CFRP

The Tsai-Wu criterion (as shown in Eqs. (2) and (3)) that considers the difference between tensile strength and compressive strength is chosen as the failure criterion for CFRP, which can more realistically reflect the damage of the composite material [32]. In Eqs. (2) and (3), F_{ij} $(i, j = 1,\ldots, 6)$ are the strength coefficients, and the failure parameters of CFRP are shown in Table 7. The user subroutine VUMAT is written to define the orthogonal anisotropy constitutive relationship and failure criteria of CFRP.

$$\begin{aligned} F_1\sigma_1 + F_2\sigma_2 + F_3\sigma_3 + F_{11}\sigma_1^2 + F_{22}\sigma_2^2 + F_{33}\sigma_3^2 + 2F_{12}\sigma_1\sigma_2 + 2F_{23}\sigma_2\sigma_3 \\ + 2F_{31}\sigma_3\sigma_1 + F_{44}\sigma_4^2 + F_{55}\sigma_5^2 + F_{66}\sigma_6^2 \leq 1 \end{aligned} \tag{2}$$

Table 7 CFRP failure parameters

Parameter	Value	Parameter	Value	Parameter	Value
X_T (MPa)	2688.00	X_C (MPa)	1458.00	S_{12} (MPa)	136.00
Y_T (MPa)	69.50	Y_C (MPa)	236.00	S_{31} (MPa)	136.00
Z_T (MPa)	55.50	Z_C (MPa)	175.00	S_{23} (MPa)	95.60

$$\begin{cases} F_1 = \frac{1}{X_C} - \frac{1}{X_T}; F_2 = \frac{1}{Y_C} - \frac{1}{Y_T}; F_3 = \frac{1}{Z_C} - \frac{1}{Z_T}; \\ F_{11} = \frac{1}{X_C X_T}; F_{22} = \frac{1}{Y_C Y_T}; F_{33} = \frac{1}{Z_C Z_T}; F_{44} = \frac{1}{S_{23}^2}; F_{55} = \frac{1}{S_{31}^2}; F_{66} = \frac{1}{S_{12}^2}; \\ F_{12} = -\frac{1}{2\sqrt{X_T X_C Y_T Y_C}}; F_{23} = -\frac{1}{2\sqrt{Y_T Y_C Z_T Z_C}}; F_{31} = -\frac{1}{2\sqrt{Z_T Z_C X_T X_C}} \end{cases} \tag{3}$$

3.2.3 Interface

The CZM incorporating fatigue damage is employed to simulate the mechanical behavior of the interface layer between CFRP and steel pipes under cyclic loads. Within this framework, both the fracture toughness and stiffness of the interface layer are progressively degraded as a function of the accumulated damage. This damage evolution is implemented through a custom user-defined field subroutine (VUSDFLD). The fatigue cohesive zone model written by VUSDFLD can dynamically adjusts the material properties of the interface layer during the simulation, allowing the model to capture the degradation of mechanical performance under cyclic loading conditions accurately.

The relationship between traction stress and separation displacement considering damage is shown in Eq. (4), where D is the damage parameter, δ_n is the normal separation displacement, δ_s and δ_t are the tangential separation displacement; t_n is the normal traction stress, t_s and t_t are the tangential traction stress; K_n is the normal stiffness, K_s and K_t are the tangential stiffness.

$$\begin{Bmatrix} t_n \\ t_s \\ t_t \end{Bmatrix} = (1 - D) \begin{bmatrix} K_n & & \\ & K_s & \\ & & K_t \end{bmatrix} \begin{Bmatrix} \delta_n \\ \delta_s \\ \delta_t \end{Bmatrix} \tag{4}$$

The secondary nominal stress criterion (as shown in Eq. (2)) is chosen as the damage initiation criteria of the interface layer and no damage occurs when the interface layer is pressurized. In Eq. (5), t_n^0, t_s^0, and t_t^0 are the damage initiation traction in the three directions, respectively.

$$\left\{\frac{\langle t_n\rangle}{t_n^0}\right\}^2+\left\{\frac{t_s}{t_s^0}\right\}^2+\left\{\frac{t_t}{t_t^0}\right\}^2\leq 1 \tag{5}$$

The B-K criterion (as shown in Eq. (6)) is chosen as the failure criterion of the interface layer. In Eq. (6), G_T is the total energy release rate, G_T^C is the mixed-mode critical energy release rate, G_s is the tangential energy release rate, η is the coefficient of the B-K criterion, G_n^C and G_s^C are normal and tangential fracture toughness, respectively. The above failure parameters that provided by the manufacture of epoxy adhesive are shown in Table 6.

$$G_n^C+\left(G_s^C-G_n^C\right)\left(\frac{G_s}{G_T}\right)^\eta\leq G_T^C \tag{6}$$

The calculation equation for the damage of the interface layer due to monotonic loading is written as:

$$D_m{}'=\frac{\delta_m^f\left(\Delta\overline{u}'-\delta_m^0\right)}{\Delta\overline{u}'\left(\delta_m^f-\delta_m^0\right)}\text{ if }\Delta\overline{u}'\geq\delta_m^0 \tag{7}$$

where $\Delta\overline{u}'$ is the separation displacement, δ_m^0 is the damage initiation displacement, δ_m^f is the damage failure displacement.

The calculation of the damage of the interface layer under cyclic loads should satisfy the continuous damage evolution law [33], and the calculation equation for the damage of the interface layer under cyclic loading is written as:

$$D_c{}'=\frac{|\Delta\overline{u}'|}{a\delta_0}\left(\frac{\overline{T}}{\sigma_{\max}}-\frac{\sigma_f}{\sigma_{\max,0}}\right)H(\Delta u-\delta_0)\text{ and }D_c{}'\geq 0 \tag{8}$$

where $\overline{T}$ is the current traction, $\sigma_{\max}$ is the current cohesive strength, σ_f is the cohesive zone endurance limit, $\sigma_{\max,\,0}$ is the initial cohesive strength, $\Delta\overline{u}$ is the accumulated separation displacement ($\Delta\overline{u}=\int\Delta\overline{u}'\mathrm{d}t$), δ_0 is the characteristic distance and H designating the Heaviside function.

The greater of the $D_c{}'$ and $D_m{}'$ is stored as the damage $D_i=\max(D_c{}',D_m{}')$ caused by one load [34]. The calculation equation for the total damage of the interface layer D caused by N cycles of loading is shown in Eq. (9), and the interface layer will fail when the total damage value is greater than or equal to 1.

$$D=\sum\nolimits_{i=1}^{N}D_i \tag{9}$$

3.3 Results and Discussion

3.3.1 Analysis of Ultimate Bending Strength

The comparison of the load-displacement curves obtained from the experiment and the numerical model is presented in Fig. 8. The results reveal that the experimental and numerical trends are consistent, particularly in the elastic stage where the curves align closely. As the displacement increases, the numerical results slightly exceed the experimental ones. For the defective pipe, the ultimate bending strength values derived from the experiment and the numerical model are 7.40 kN and 7.57 kN, respectively, yielding an error of 2.30%. For the CFRP-repaired pipe, the ultimate bending strength values are 8.71 kN and 9.07 kN, respectively, with an error of 4.13%. These small discrepancies validate the accuracy and reliability of the numerical model.

The deformation of the defective pipe, as observed from the experiment and the numerical model, is shown in Fig. 9. Both results indicate that the pipe undergoes buckling failure, and the pipe has an obvious depression at the contact position between the indenter and the pipe surface.

The comparison of the deformation of the CFRP-repaired pipe between experimental and numerical results is shown in Fig. 10. The numerical model successfully simulates the damage failure of the CFRP and the interface. Figure 10b highlights the collapse failure at the contact region between the CFRP and the indenter. Figure 10c further shows evidence of sliding and debonding failure between the CFRP and the pipe, attributed to the degradation of the interface layer between the two materials. These observations demonstrate the capability of the numerical model to accurately replicate the mechanical behavior and failure modes of CFRP-repaired pipes under bending loads.

Fig. 8 Load-displacement curves under static load

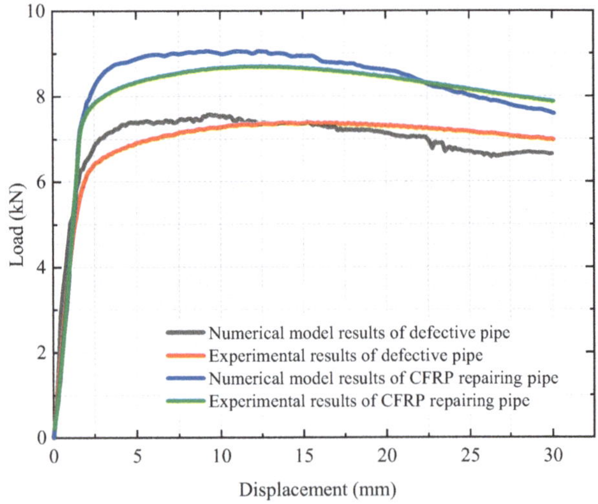

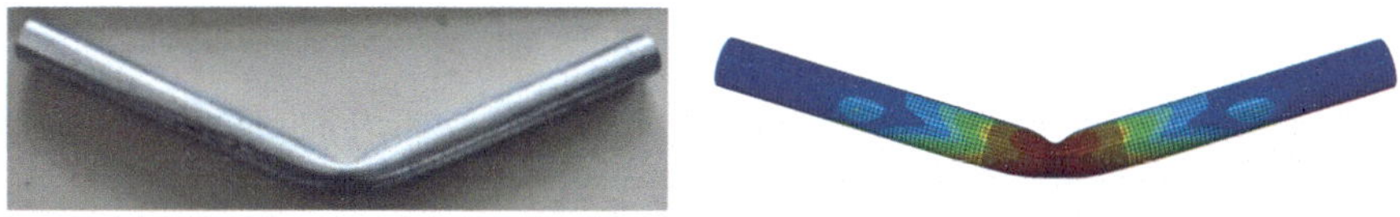

(a) Overall deformation diagram

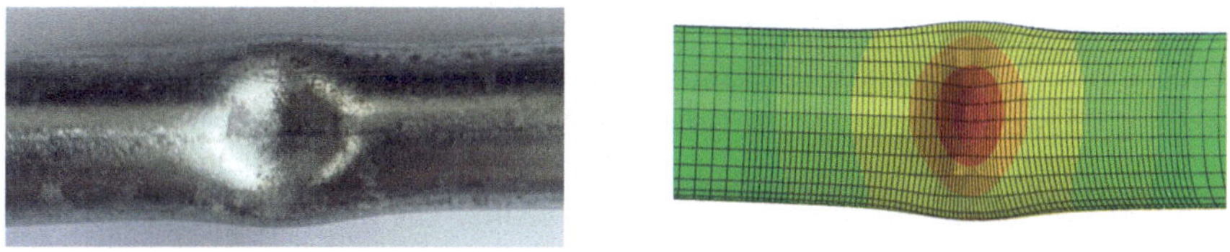

(b) Local enlarged diagram

Fig. 9 Deformation diagram of defective pipe. (**a**) Overall deformation diagram. (**b**) Local enlarged diagram

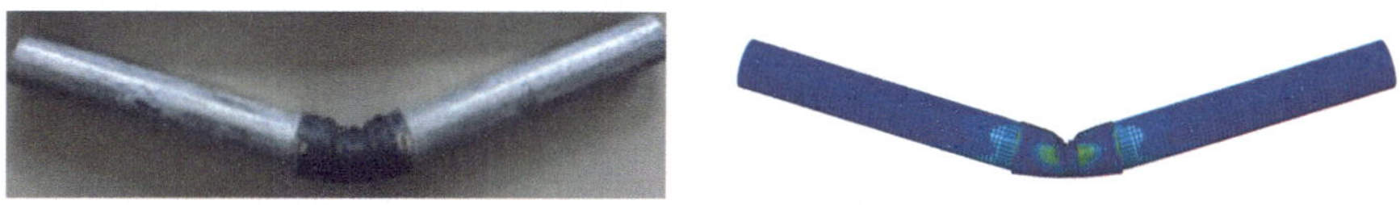

(a) Overall deformation diagram

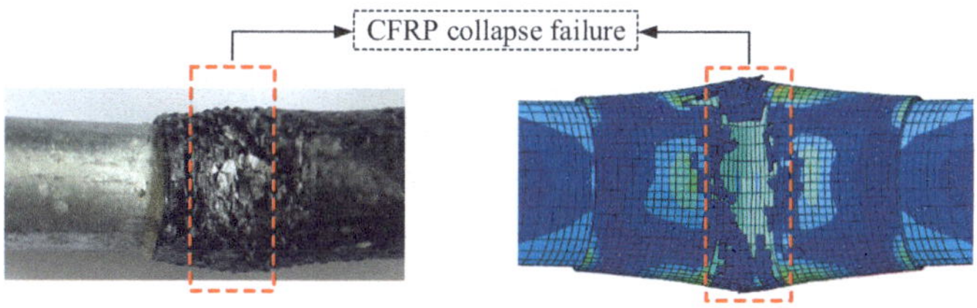

(b) Collapse failure

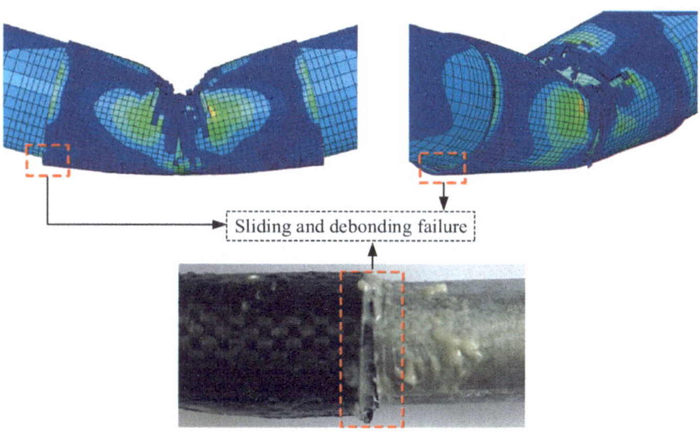

(c) Sliding and debonding failure

Fig. 10 Deformation diagram of pipe repaired by CFRP. (**a**) Overall deformation diagram. (**b**) Collapse failure. (**c**) Sliding and debonding failure

3.3.2 Analysis of Residual Strength

The residual strength of the CFRP-repaired pipe under cyclic loading conditions is evaluated by applying bending displacements after subjecting the pipe to 0, 50, 100, 500, 1000, 5000, and 10,000 fatigue cycles.

The comparison of residual strength obtained from experiments and numerical simulations is illustrated in Fig. 11 and Table 8. A power function is used to fit the relationship between residual strength and the number of fatigue cycles. The coefficients of determination R^2 for the experimental and numerical fitting formulas are 0.96 and 0.97, respectively. This high degree of correlation demonstrates the consistency between experimental and numerical results. The maximum error between experimental and numerical residual strength values is 4.13%, confirming the accuracy of the numerical model.

The results show a rapid decrease in residual strength for the first 100 fatigue cycles, followed by a gradual decline as the number of cycles increases. The reason for the above phenomenon is that the bending displacement is relatively large, and the increase rate of bending displacement is fast at the loading point in the early stage of applying cyclic loads. As the number of fatigue loads increases, the increase rate

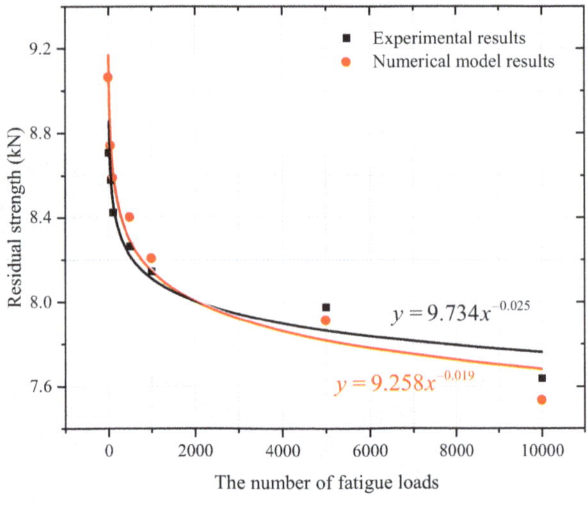

Fig. 11 Residual strength for different number of fatigue loads

Table 8 Residual strength and degradation rate obtained from numerical models

The number of fatigue loads	0	50	100	500	1000	5000	10,000
Residual strength (kN)	9.07	8.75	8.60	8.41	8.21	7.92	7.54
Degradation rate (%)	\	3.53	5.18	7.28	9.48	12.68	16.87
Degradation rate caused by each cyclic load (‰)	\	7.06	3.30	0.53	0.44	0.08	0.08
Error between experiment and numerical model (%)	4.13	1.98	2.02	1.82	0.74	−0.75	−1.31

of bending displacement at the loading point decreases, and the decline rate of the residual strength gradually decreases.

The interface layer, being the weakest component of the CFRP-repaired structure, undergoes damage under fatigue loading. This leads to a reduction in interfacial strength, adversely affecting the overall structural strength. Experimental observation of interface damage is challenging due to its location between the CFRP and the steel pipe. However, the numerical model captures the damage evolution, as shown in Fig. 12. The degradation of interface fracture energy and stiffness under fatigue loads are achieved using user subroutines, where the solution dependent variables 1 (SDV1) represents the total damage value of the interface. From Fig. 12, it can be seen that the interface layer begins to fail at the contact point between the indenter and the pipe, and the failure area gradually expands around with the continuous application of fatigue loads.

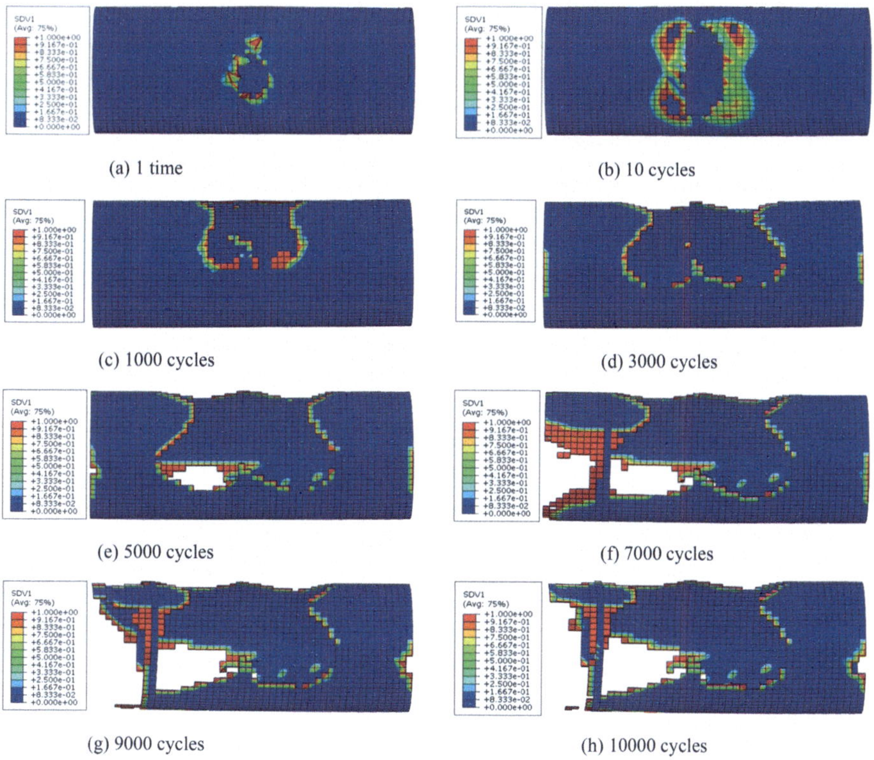

Fig. 12 Damage evolution of the interface layer under different number of cyclic loads. (**a**) 1 time; (**b**) 10 cycles; (**c**) 1000 cycles; (**d**) 3000 cycles; (**e**) 5000 cycles; (**f**) 7000 cycles; (**g**) 9000 cycles; (**h**) 10,000 cycles

A numerical model considering the fatigue damage of the interface between the CFRP and the pipe is established, and the degradation law of residual strength under different number of cyclic loads is studied through the numerical mode. The CZM is used to define the constitutive relation, damage and failure of the interface, which can accurately describe the mechanical behavior of the interface. The Explicit solver is used to solve the numerical model, which effectively handles the complex fatigue damage evolution but requires significant computational resources. This study provides a comprehensive analysis of the residual strength degradation in CFRP-repaired structures under cyclic loading, offering valuable insights into their durability. Future research should aim to develop simpler, more efficient numerical models that maintain high accuracy while reducing computational demand.

4 Conclusions

The degradation characteristics of CFRP-repaired pipe under fatigue loads are investigated through experiments and numerical models, providing a valuable reference for durability analysis under cyclic loading conditions. The main work and related conclusions are as follows:

1. The experiments of ultimate bearing load, fatigue life and strength degradation are conducted. Compared to the defective pipe, the ultimate bending strength and fatigue life of CFRP-repaired pipe increased by 17.70% and 158.54% respectively. The strength degradation rate exhibited a gradual decrease with an increasing number of fatigue load cycles.
2. An interfacial fatigue damage model is developed, and a numerical model for analyzing the residual strength of CFRP-repaired steel pipes under cyclic loading is established by considering the damage to the interface between CFRP and steel pipes, as well as the damage to the steel pipe and CFRP.
3. The correctness of the numerical model is verified through experiments. The errors of the ultimate bending strengths between experiment and numerical model for the defective pipe and CFRP-repaired pipe are 2.30% and 4.13% respectively. The maximum error in residual strengths between experiments and numerical simulations for CFRP-repaired pipes is 4.13%.
4. Both experimental and numerical results confirmed that the residual strength of CFRP-repaired pipes follows a power-law decline with the number of fatigue load cycles. The degradation rate of residual strength decreases nonlinearly as the number of fatigue cycles increases.

This study focuses on the bending fatigue loading of CFRP-repaired pipes. Future research could extend to investigate the strength degradation characteristics under tensile loads, torsional loads, and complex multi-axial stress states to provide a more comprehensive understanding of CFRP-repaired structures' durability.

References

1. Xin J, Zhang Y, Zhong C, Cheng Z, Jia Z. Analysis of ultimate pressure of corroded subsea pipe repaired with composite considering temperature and material degradation. J Mar Sci Technol. 2020;25:285–97.
2. Zhang Y, Cheng N, Cheng Z, Luo Q. The debonding failure analysis of corroded pipes repaired with CFRP under tension and the bending moment. Int J Adhes Adhes. 2020;103:102714.
3. Yan YF. Axial load on the submarine pipeline suspended law of vortex induced vibration. Appl Mech Mater. 2011;94:1511–4.
4. Huang C, Chen T, Xia Z, Jiang L. Numerical study of surface fatigue crack growth in steel plates repaired with CFRP. Eng Struct. 2022;268:114743.
5. Borrie D, Liu HB, Zhao XL, Raman RS, Bai Y. Bond durability of fatigued CFRP-steel double-lap joints pre-exposed to marine environment. Compos Struct. 2015;131:799–809.
6. Wu C, Zhao XL, Al-Mahaidi R, Emdad MR, Duan W. Fatigue tests of cracked steel plates strengthened with UHM CFRP plates. Adv Struct Eng. 2012;15(10):1801–15.
7. Kafodya I, Xian G, Li H. Durability study of pultruded CFRP plates immersed in water and seawater under sustained bending: water uptake and effects on the mechanical properties. Compos Part B. 2015;70:138–48.
8. Kabir MH, Fawzia S, Chan THT, Gamage JCPH. Comparative durability study of CFRP strengthened tubular steel members under cold weather. Mater Struct. 2016;49:1761–74.
9. Correia L, Sena-Cruz J, Michels J, França P, Pereira E, Escusa G. Durability of RC slabs strengthened with prestressed CFRP laminate strips under different environmental and loading conditions. Compos Part B. 2017;125:71–88.
10. Yang Y, Silva MA, Biscaia H, Chastre C. Bond durability of CFRP laminates-to-steel joints subjected to freeze-thaw. Compos Struct. 2019;212:243–58.
11. Miyano Y, Nakada M. Accelerated testing methodology for durability of CFRP. Compos Part B. 2020;191:107977.
12. Mu W, Na J, Tan W, Wang G, Shen H, Li X. Durability of adhesively bonded CFRP-aluminum alloy joints subjected to coupled temperature and alternating load. Int J Adhes Adhes. 2020;99: 102583.
13. Zhang F, Dai JG, Wang Z, Wang M, Leng Y, Xu Q. Bond durability of epoxy and cement-bonded CFRP reinforcement to concrete interfaces subject to water immersion. Mater Struct. 2021;54:1–12.
14. Zhang Y, Liu Z, Xin J, Wang Y, Zhang C, Zhang Y. The degradation mechanism of CFRP repaired corroded marine pipelines based on experiments and FEM. Thin-Walled Struct. 2021;169:108469.
15. Shao W, Sun Q, Xu X, Yue W, Shi D. Durability life prediction and horizontal bearing characteristics of CFRP composite piles in marine environments. Constr Build Mater. 2023;367:130116.
16. Lee MG, Huang Y, Kan YC, Wang YC, Chen YS, Kao SC. Experimental study on durability of CFRP-strengthened wood members. J Mater Res Technol. 2023;24:3704–16.
17. Mu W, Na J, Qin G, Tan W, Shen H. Effect of alternating load on residual strength of adhesively bonded CFRP composite-aluminum alloy joints. Acta Mater Compos Sin. 2019;36(5):1124–31.
18. Doroudi Y, Fernando D, Hosseini A, Ghafoori E. Behavior of cracked steel plates strengthened with adhesively bonded CFRP laminates under fatigue loading: experimental and analytical study. Compos Struct. 2021;266:113816.
19. Jones SC, Civjan SA. Application of fiber reinforced polymer overlays to extend steel fatigue life. J Compos Constr. 2003;7(4):331–8.
20. Chen B, Zhang Y, Wang Q, Zhang H, Wang Y. The signal analysis of lamb wave in steel plates strengthened by CFRP with interface debonding. Ocean Eng. 2023;283:115061.
21. Dugdale DS. Yielding of steel sheets containing slits. J Mech Phys Solids. 1960;8(2):100–4.
22. Roe KL, Siegmund T. An irreversible cohesive zone model for interface fatigue crack growth simulation. Eng Fract Mech. 2003;70(2):209–32.

23. Khoramishad H, Crocombe AD, Katnam KB, Ashcroft IA. Predicting fatigue damage in adhesively bonded joints using a cohesive zone model. Int J Fatigue. 2010;32(7):1146–58.
24. Jin G, Huang X. Predicting fatigue damage of asphalt concrete using a cohesive zone model. Int J Pavement Res Technol. 2014;7(5):361.
25. Benabou L, Sun Z, Dahoo PR. A thermo-mechanical cohesive zone model for solder joint lifetime prediction. Int J Fatigue. 2013;49:18–30.
26. Wang B, Siegmund T. A numerical analysis of constraint effects in fatigue crack growth by use of an irreversible cohesive zone model. Int J Fract. 2005;132(2):175–96.
27. Le VN, Benabou L, Tao QB, Etgens V. Modeling of intergranular thermal fatigue cracking of a lead-free solder joint in a power electronic module. Int J Solids Struct. 2017;106:1–12.
28. Ruggieri C, Dotta F. Numerical modeling of ductile crack extension in high pressure pipelines with longitudinal flaws. Eng Struct. 2011;33(5):1423–38.
29. ASME PCC-2-2015. Repair of pressure equipment and piping. American Society of Mechanical Engineers.
30. ISO 7438. Metallic materials-bend test. International Organization for Standardization, Switzerland; 2010.
31. Al-Saidy AH. Structural behavior of composite steel beams strengthened/repaired with carbon fiber reinforced polymer plates. Iowa State University; 2001.
32. Chen X, Sun X, Chen P, Wang B, Gu J, Wang W, Chai Y, Zhao Y. Rationalized improvement of Tsai–Wu failure criterion considering different failure modes of composite materials. Compos Struct. 2021;256:113120.
33. Lemaitre J. A course on damage mechanics. Berlin, Heidelberg, New York: Springer; 1996.
34. Siegmund T. A numerical study of transient fatigue crack growth by use of an irreversible cohesive zone model. Int J Fatigue. 2004;26(9):929–39.
35. Zhu J, Zhu X. Study on simplified method for the analysis of fatigue failure process of RC bridges. Engineering mechanics. 2012;29(5):107–14.

Research on the Damage Mechanisms of Interface in Offshore Floating LCO_2 Cryogenic Hose under LCO_2 Corrosion

1 Introduction

In recent years, with the increasingly severe issues of global climate change and carbon emissions, Carbon Capture, Utilization and Storage (CCUS) technology has become an important technical means to address climate change and achieve carbon neutrality goals [1, 2]. CCUS technology achieves the dual goals of greenhouse gas reduction and resource utilization by capturing carbon dioxide from industrial emission sources, converting and utilizing it or sequestering it in geological structures [3, 4]. The safe transportation of liquefied carbon dioxide (LCO_2) is one of the key links in the entire CCUS chain. In response to the LCO_2 transport problem between transport ships and receiving equipment in offshore CO_2 storage operations, the offshore floating LCO_2 cryogenic hose that can achieve floating, insulation, and corrosion resistance has become a concerning hot point in global CCUS field, and its long-term performance and reliability in harsh deep-sea environments have become the core issue in ensuring the safe operation of transportation systems (Fig. 1).

The offshore floating LCO_2 cryogenic hose is similar to the floating hose for offshore oil and gas transportation in Fig. 2. Serving in complex marine environments, the offshore floating LCO_2 cryogenic hose is typically subjected to a combination of internal pressure, tension, torsion, and bending loads. Analyzing the mechanical properties of hoses under various load combinations is of great significance for ensuring the safe operation of transportation systems, and relevant research about similar hoses can provide reference for the mechanical performance analysis and ultimate bearing capacity evaluation. For floating hose used for offshore oil and gas transportation, Gao et al. [5] and Tonatto et al. [6] used the Rebar element and Embedded element embedding techniques in finite element software to establish a simulation analysis model of a floating hose, analyzed the ultimate bearing capacity of the hose, and discussed the key parameters of the reinforcement components on the ultimate internal pressure bearing characteristics. Gao et al. [7] studied the distribution characteristics of stress and radial displacement of floating

Y. Zhang, *Interfacial Mechanics of Offshore Engineering Equipment*,
https://doi.org/10.1007/978-981-95-3200-1_13

Fig. 1 Marine storage of CO_2

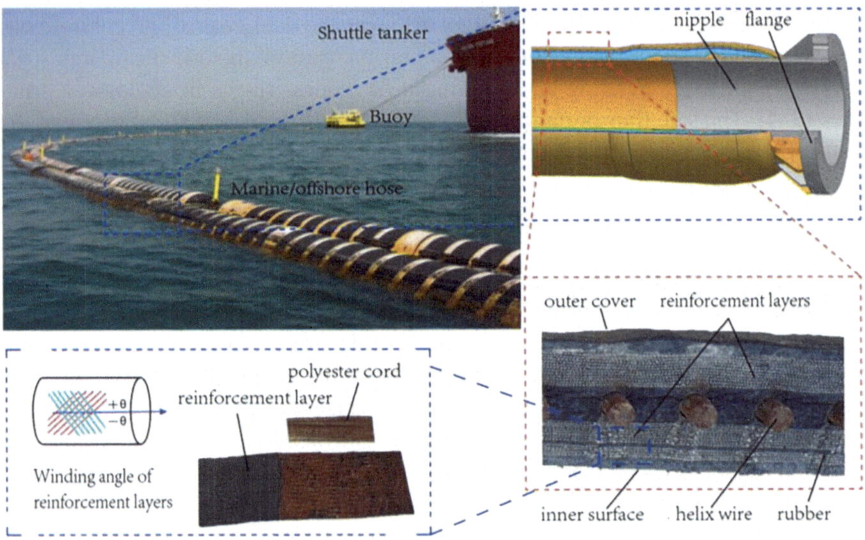

Fig. 2 The floating hose for offshore oil and gas transportation [16]

hoses under internal pressure, and parametrically analyzed the effects of pipeline structural parameters and material mechanical properties on axial, circumferential, and shear stress and radial displacement of pipelines. Amaechi et al. [8, 9] studied the bending characteristics of a floating hose in a single point mooring system operation mode under loads such as waves and currents using hydrodynamic analysis methods. For reinforced thermoplastic composite pipe, Xia et al. [10, 11]

proposed an accurate method for analyzing the stress-strain and deformation properties of thick walled fiber-reinforced sandwich tubes under pure bending conditions, and provided precise elastic solutions for the stress and deformation of multi-layer fiber-reinforced tubes under internal pressure based on three-dimensional anisotropic elastic theory. Bai et al. [12–14] studied the mechanical response of RTP under different load conditions, including hose external pressure, axial tensile load, bending load, and combination of external pressure and tensile load. Xu et al. [15] combined full-scale experiments, theoretical models, and finite element methods to study the mechanical behavior of glass fiber reinforced hoses under tension.

The floating LCO_2 cryogenic hose is redesigned in which the inner and outer pipes are made of high-density polyethylene, reinforced cord, and reinforced steel wire thermoplastic processing [17]. During the long-term service of the hose, its interior is filled with LCO_2 for transportation operations, and its working environment is located in the target sea area for CO_2 sequestration, usually deep offshore. In the long-term use process, CO_2 encountering water forms an acidic environment, which leads to the degradation of the mechanical properties of the reinforced steel wire and the adhesive interface between the reinforced steel wire and HDPE of the hose. This will inevitably directly affect the mechanical properties of the hose. Considering the effects of low temperatures and corrosion, many scholars have conducted research on the corrosion of metals in CO_2 and water environments, pointing out that over the long term, CO_2 corrosion has a significant impact on the mechanical properties of metal structures [18–20]. However, the research mainly focuses on the effect of CO_2 corrosion on the mechanical properties of metal structures, and few studies have been found on the influence of CO_2 corrosion on the strength of metal non-metal bonding interfaces. Moreover, the present analysis of interface strength is mainly aimed at the composite structure formed by vulcanization of cord rubber or CFRP, and the mechanical properties of different structures with interfaces are analyzed and evaluated through theoretical analysis, experimental verification, and simulation analysis [21–23].

Based on previous research, studying the effects of LCO_2 corrosion on the mechanical properties of metal structures and the strength of metal-nonmetal bonding interfaces is necessary for evaluating the mechanical properties of hoses, which can provide a more accurate evaluation of the overall mechanical behaviors and damage mechanisms of the pipeline. In this paper, experiments of LCO_2 corrosion and mechanical property testing on the specimen for metal-nonmetal bonding interfaces of floating LCO_2 cryogenic hose are conducted, and comparative analysis of changes in mechanical indicators before and after corrosion is made, to explore the damage mechanism under combined load condition and provide reference for the safety assessment of hose assemblies after long-term service.

2 Interface in Offshore Floating LCO_2 Cryogenic Hose

2.1 Structure and Material of Floating LCO_2 Cryogenic Hose

The floating reinforced thermoplastic hose in this paper is designed to minimize the corrosive effect of CO_2 and improve its load-bearing capacity. The cross-sectional structure of hose mainly consists of three layers: an inner tube, a floating insulation layer, and an outer protective layer (outer tube). More specifically, the inner tube is composed of high-density polyethylene, reinforced steel wire, and reinforced aramid cord, forming a fully bonded composite hose structure through hot melting; The reinforcement layer is distributed in the inner tube at a specific angle (such as winding angle); The floating insulation layer is made of high-density polyethylene with closed cell foam to ensure the buoyancy and insulation of the hose; The outer tube is composed of high-density polyethylene and reinforced aramid cord, which are bonded together in the same way as the inner tube, in order to protect the inner layer and provide a certain load-bearing capacity, as shown in Fig. 3.

Based on the basic structure of the floating LCO_2 cryogenic hose, the mechanical performance parameters of the main materials of the hose (HDPE and reinforced steel wire) are determined through experimental methods, as shown in Tables 1, 2 and 3. The main mechanical performance parameters of the reinforced cord and foamed high-density polyethylene are shown in Tables 4 and 5. It should be noted that the mechanical performance parameters at room temperature in Tables 2 and 3 were determined through experiments. The minimum temperature that the hose can withstand during operation is −27 °C, which has a relatively small impact on the mechanical properties of the material, especially for the reinforced cord, the influence of low temperature on its mechanical performance parameters has not been considered. The mechanical performance parameters of HDPE and reinforced steel wire at low temperatures are also given in the tables.

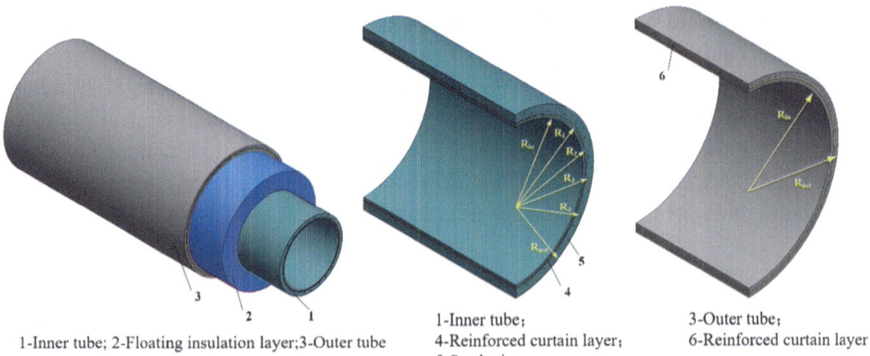

Fig. 3 The floating hose for offshore oil and gas transportation

Table 1 Parameters of offshore LCO_2 floating hose

Parameter	Value
Inner diameter of inner layer ($2R_{in}$/mm)	150
The number of aramid cord layers in the inner layer	2
Laying angle of cord in inner layer (α_i/°)	±55
Radius of reinforced steel wire (mm)	0.5
The number of layers of reinforced steel wire	2
Helix angle of steel wire (α_i/°)	±55
Outer diameter of inner layer ($2R_{out}$/mm)	174
Thickness of floating insulation layer (mm)	41
Inner diameter of outer layer ($2R_{in}$/mm)	256
The number of aramid cord layers in outer layer	4
Laying angle of cord in outer layer (α_i/°)	±55
Diameter of cord (mm)	0.5
Outer diameter of outer layer ($2R_{out}$/mm)	280

Table 2 Mechanical parameters of HDPE

Temperature (°C)	Density ($kg{\cdot}m^{-3}$)	Young's modulus (MPa)	Tensile strength (MPa)	Poisson's ratio
30	950	1100	28	0.4
−27		2114	35	

Table 3 Mechanical parameters of reinforced steel wire

Temperature (°C)	Density ($kg\ m^{-3}$)	Young's modulus (GPa)	Yield strength (MPa)	Tensile strength (MPa)	Poisson's ratio
30	7850	150	1975	2091	0.3
−27		159	2055	2191	

Table 4 Mechanical parameters of aramid cord

Density ($kg\ m^{-3}$)	Young's modulus (GPa)	Shear modulus (MPa)	Tensile strength (MPa)	Poisson's ratio
1400	$E_1 = 87$ $E_2 = 5.8$	$G_{12} = G_{13} = 24$ $G_{23} = 20$	$\sigma_{b1} = 2400$ $\sigma_{b2} = 468$ $\tau = 120$	$\mu_{12} = 0.22$

Table 5 Mechanical parameters of dense cellular foamed HDPE

Density ($kg\ m^{-3}$)	Young's modulus (MPa)	Tensile strength (MPa)	Poisson's ratio
3.2	10	25	0.225

2.2 Experiments of LCO_2 Corrosion and Mechanical Property Testing

To investigate the mechanical properties of the reinforcing steel wire and interface of steel wire-HDPE in floating LCO_2 cryogenic hose after long-term CO_2 corrosion, this paper designs experimental specimens for steel wire and steel wire-HDPE matrix based on the existing steel wire-reinforced thermoplastic composite pipe model. The main structures are shown in Fig. 4. The specimen for reinforced steel wire is made according to the standard GB/T 228.1-2021 for tensile testing of metal materials [24], and the specimen for interface is designed according to the standard GB/T 32439-2015 [25].

The LCO_2 corrosion experimental equipment mainly includes LCO_2 reaction kettle, constant temperature refrigeration cycle bath, constant temperature jacket, CO_2 gas cylinder and pressure divider, temperature detection device, vacuum pump, and pressure sensor inside the reaction kettle. The LCO_2 reactor uses a 500 mL volume stainless steel reactor produced by Century Senlong Company, equipped with internal pressure monitoring interface, temperature monitoring interface, and automatic stirring motor. In addition, the reactor also has a sapphire window for real-time monitoring of the status of samples and liquids inside the reactor. The CO_2 cylinder and pressure divider are provided by the CO_2 gas manufacturer with a purity of 99.9%. The cylinder is equipped with a gas pressure divider and filled with 7 MPa CO_2 gas. The overall schematic diagram is shown as follows in Fig. 5.

The Experiment of LCO_2 corrosion on specimens is conducted with temperature inside the reaction vessel of −15.5 °C and the CO_2 pressure of 2.5 MPa, forming a corrosive solution of LCO_2 and water. The controlled variable for this experiment is corrosion time. After the corrosion of LCO_2 for 60 and 120 h, three sets of specimens for the steel wire and steel wire-HDPE matrix are taken out of the high-pressure reaction vessel, and it is obvious that there are many rust stains on the surface of specimens.

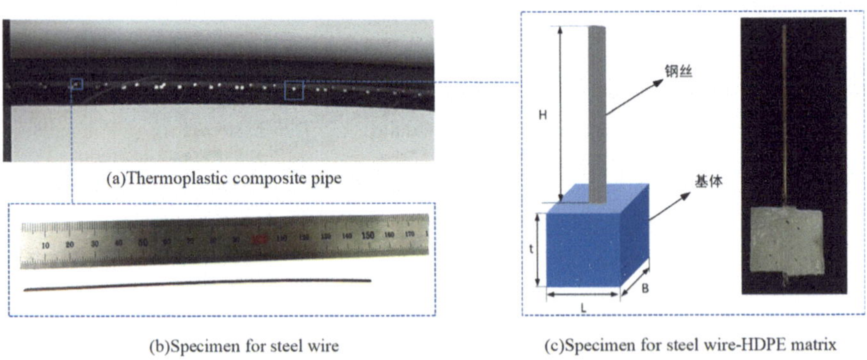

Fig. 4 Structures of specimens for steel wire and steel wire-HDPE matrix. (**a**) Thermoplastic composite pipe. (**b**) Specimen for steel wire. (**c**) Specimen for steel wire-HDPE matrix

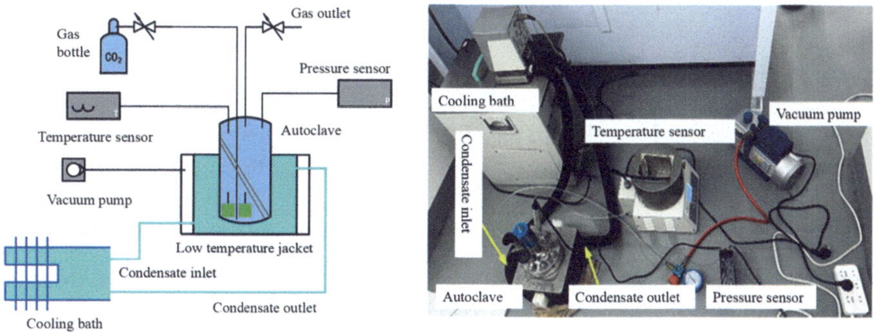

(a)Design of LCO_2 corrosion experiment

(b)Layout of LCO_2 corrosion experiment

Fig. 5 Overall schematic diagram of LCO_2 corrosion experiment. (**a**) Design of LCO_2 corrosion experiment. (**b**) Layout of LCO_2 corrosion experiment

Fig. 6 Tensile testing on steel wire specimens

The tensile testing machine as shown in Fig. 6 is used to obtain the mechanical properties of specimens corroded by LCO_2. During the loading process of the pull-out test, displacement control is used, with a loading speed of 2 mm/min until specimens' break, to obtain the load displacement curve of under tensile load, and the maximum tensile force F is recorded at room temperature. As shown in Fig. 7, it can be clearly observed that corrosion products adhere to the specimens of both steel wire and HDPE matrix after corrosion.

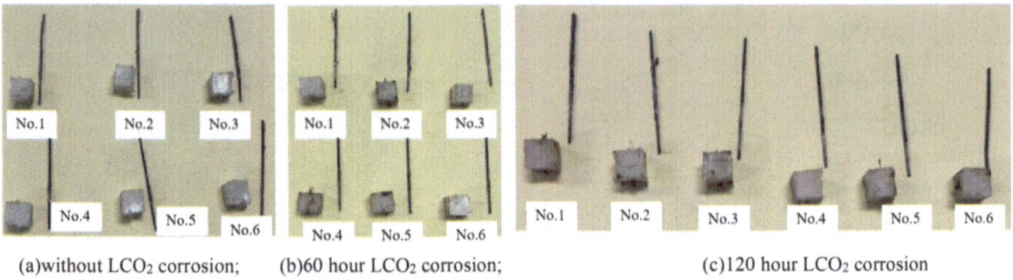

Fig. 7 Corrode specimens for steel wire-HDPE matrix. (**a**) Without LCO_2 corrosion; (**b**) 60 h LCO_2 corrosion; (**c**) 120 h LCO_2 corrosion

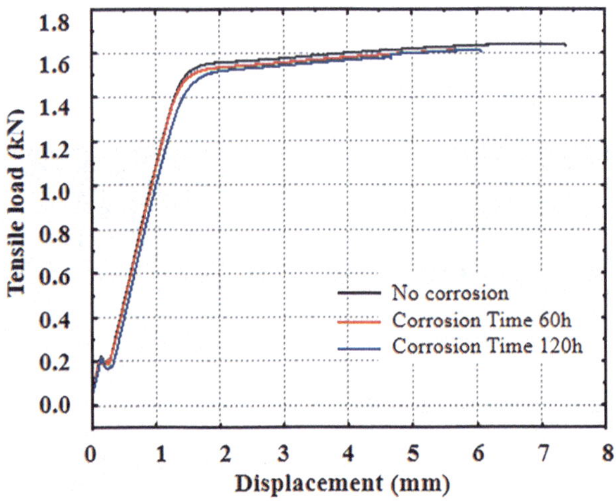

Fig. 8 Tensile test results of steel wire after LCO_2 corrosion

Table 6 Comparison of mechanical parameters of reinforced steel wire after corrosion

Parameter	Value		
Corrosion time (h)	0	60	120
Density (kg m^{-3})	7850	7850	7850
Young's modulus (GPa)	150	163	160
Yield strength (MPa)	1975	1946	1935
Tensile strength (MPa)	2091	2044	2031

2.3 *Experimental Results*

For the reinforced steel wire subjected to LCO_2 corrosion test, the result for load displacement curve of no corrosion, corrosion time 60 h and corrosion time 120 h is shown in Fig. 8, and the corresponding values are shown in Table 6. It can be observed that LCO_2 corrosion reduces both the yield strength and tensile strength of the reinforced steel wire, and the degree of reduction increases with the longer the corrosion time. When the corrosion time is 60 h, the yield strength of the steel wire

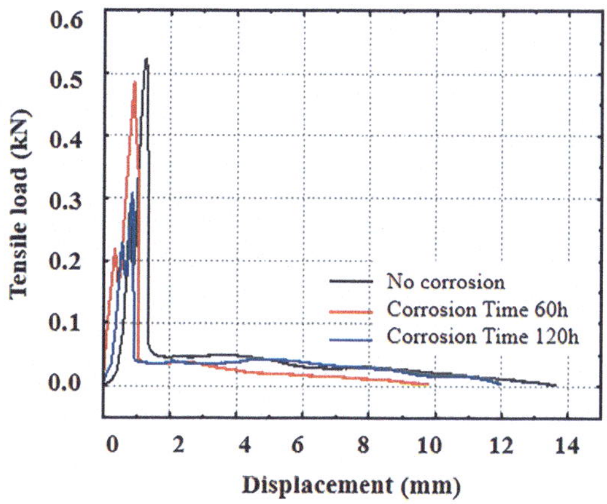

Fig. 9 Test results of steel wire-HDPE matrix after LCO_2 corrosion

Table 7 Comparison of mechanical parameters of steel wire-HDPE matrix after corrosion

Parameter	Value		
Corrosion time (h)	0	60	120
Shear strength (MPa)	18.7	14.58	9.89
Debonding initiation displacement (mm)	1.40	0.91	0.83

decreases by 1.5% compared to the one without corrosion, and the tensile strength decreases by 2.3%; When the corrosion time is 120 h, the yield strength of the steel wire decreased by 2.1% compared to the non-corroded one, and the tensile strength decreased by 2.9%. In addition, it can be observed that as the corrosion time increases, the critical fracture displacement of the steel wire shows a decreasing trend, from 7.57 mm without corrosion to 6.06 mm after 60 h of corrosion time, and then to 4.72 mm after 120 h of corrosion time.

For the steel wire-HDPE specimens, Fig. 9 shows the test results without LCO_2 corrosion, 60 h LCO_2 corrosion, and 120 h LCO_2 corrosion, and Table 7 gives calculated values. It can be seen that from the initial stage of displacement loading until the maximum tensile load is reached, the relationship between tensile load and displacement shows a linear growth trend. At this stage, the interface between the steel wire and HDPE remains intact and no debonding failure occurs; As the tensile load continues to increase, the slope of the load displacement curve decreases significantly until the tensile load reaches its maximum value; Then the tensile load displacement curve rapidly enters a sudden drop section, and the tensile load decreases sharply with the increase of displacement, which means the steel wire and HDPE undergo large-scale debonding failure. In the final stage, as the tensile displacement increases, the steel wire is pulled out of the HDPE by the fixture. It can be clearly seen from the surface of the pulled-out steel wire that there is no foreign object residue at the bonding interface between the steel wire and HDPE, indicating that the failure of the specimen is due to debonding failure at the steel wire HDPE interface, and there is no failure form such as HDPE damage.

3 Simulation Analysis of Steel Wire-HDPE Interface

Based on above experimental data, the finite element software ABAQUS is used to numerically simulate and analyze the shear strength test of the steel wire-HDPE bonding interface at −27.5 °C condition and a tensile speed of 25 mm/min. In the process of establishing a simulation analysis model, the model uses cohesive elements based on cohesive force theory to simulate the mechanical behavior of the steel wire-HDPE bonding interface. An exponential model is used to characterize the damage evolution process of the steel wire HDPE bonding interface. Then, the obtained tensile load displacement relationship curve is compared with the shear test results to verify the rationality and reliability of the simulation analysis model.

3.1 Simulation Model for the Interface of Steel Wire and HDPE

In ABAQUS finite element software, referring to the size parameters of the shear strength test specimens obtained, an interface model of steel wire and HDPE was established using the part module, as shown in Fig. 10. Set the basic mesh size to 0.5 mm, maintain uniform distribution of mesh seeds, and locally refine the circular holes and steel wire sections of the HDPE matrix, set 20 uniform seeds along the circumference. The element types of the steel wire model and the matrix model are 8-node linear reduced integral element C3D8R. For boundary conditions of the model, the constraint conditions of the substrate are applied according to the fixture clamping method in the shear strength tensile test, and full constraints are applied at the position where the upper surface of the substrate contacts the fixture. For steel wire, apply a constant tensile speed at its top ($v = 5$ mm/min).

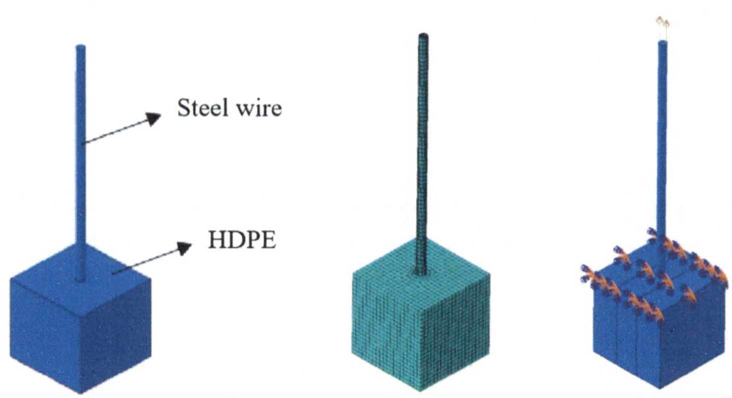

Fig. 10 Schematic diagram of tensile model for interface strength between wire and matrix

3.2 Cohesive Element Connection and Damage Condition

The connection method between the steel wire and the substrate is set as cohesive element connection based on cohesive force model. For the cohesive force model, it divides the contact interface into two different blocks, one of which is a completely separated surface; The other part is the existence of cohesive force between the two contacting surfaces, which forms a cohesive zone and presents a state where the two contacting surfaces may separate under load. In the cohesive zone, the distance between the two contact surfaces is less than the critical value of the opening displacement defined by the cohesive zone. At this point, the opening displacement of the cohesive contact surface exhibits a specific variation relationship, which is called the traction separation rule. As the applied load increases, the cohesive force increases with the opening displacement of the contact surface. When the opening displacement reaches its maximum value, the cohesive force also reaches its maximum value. At this point, structural damage occurs at the contact interface, and then the cohesive zone enters the damage evolution stage. The trend of change is that the opening displacement increases and the cohesive force begins to decrease to zero. Throughout the process, the area enclosed by the curve and the coordinate axis is the critical fracture energy required for the contact surface to fail. At this point, the cohesive force unit is completely disconnected, and the contact interface separates (Fig. 11).

The criterion for determining the transition of cohesive force units from the initial damage stage to the damage evolution stage is called the quadratic nominal stress (CSQUADSCRT) criterion, and the calculation method for the damage value D is expressed as:

$$
\begin{aligned}
D &= \left(\frac{\langle t_n \rangle}{t_{n\max}}\right)^2 + \left(\frac{t_s}{t_{s\max}}\right)^2 + \left(\frac{t_t}{t_{t\max}}\right)^2 \\
t_n &= \begin{cases} t_n, & t_n \geqslant 0 \\ 0, & t_n < 0 \end{cases}
\end{aligned}
\tag{1}
$$

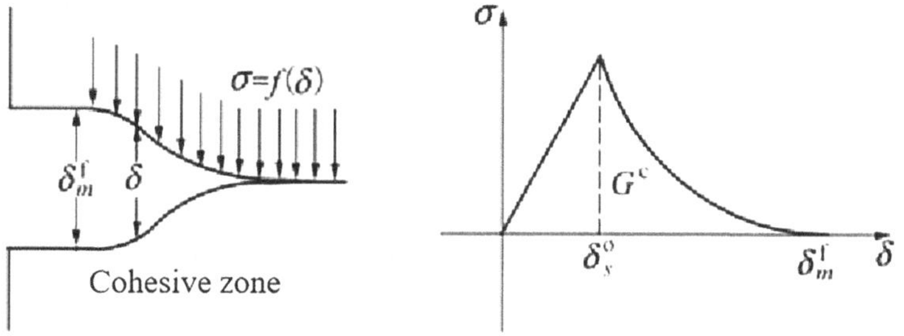

Fig. 11 Schematic diagram of tensile model for interface strength between wire and matrix

When the stress state of the cohesive element satisfies the maximum nominal stress criterion, it enters the damage evolution stage. In this stage, the damage variable D is introduced to describe the reduction of the stiffness of the cohesive element. During the material damage evolution process, D monotonically increases from 0 to 1. In order to more accurately estimate the mechanical performance of the steel wire HDPE interface in the hose, the values of normal strength and stiffness in the cohesive contact properties should be consistent with the values of tangential strength and stiffness. The relationship between the stress components of cohesive units and the damage variable can be expressed as:

$$
\begin{gathered}
t_n = \begin{cases} (1-D)\bar{t}_n, & \bar{t}_n \geq 0 \\ \bar{t}_n, & \bar{t}_n < 0 \end{cases} \\
t_s = (1-D)\bar{t}_s, \quad t_t = (1-D)\bar{t}_t
\end{gathered} \tag{2}
$$

When the hose is subjected to a large load, the steel wire will undergo plastic damage, and the ductile damage failure criterion can be used to quantitatively characterize the damage to the steel bar. For reinforced steel wire, when its equivalent plastic strain accumulates to a critical value during loading, the material is in a damaged state. The failure judgment index of the failure criterion is shown in the following formula and is represented by "DUCTCRT" in the subsequent damage cloud map.

$$
\omega_D = \int \frac{d\bar{\varepsilon}^{pl}}{\bar{\varepsilon}_D^{pl}\left(\eta, \dot{\bar{\varepsilon}}^{pl}\right)} \qquad \eta = -\sigma_m / \sigma_{\text{von-Mises}} \tag{3}
$$

In the forms of contact, the cohesive contact properties of no corrosion, 60 h corrosion and 120 h corrosion are set separately as shown in Table 8. It should be noted that, in order to more accurately predict the mechanical performance of the steel wire-HDPE interface in the hose, the values of normal strength and stiffness in the cohesive contact property are kept consistent with the values of tangential strength and stiffness.

Table 8 Cohesive contact properties of steel wire-HDPE interface

Parameter	Value		
Corrosion time (h)	0	60	120
Normal stiffness K_n (N/m^{-3})	13.57×10^9	16.07×10^9	11.99×10^9
Tangential stiffness K_s, K_t (N/m^{-3})	13.57×10^9	16.07×10^9	11.99×10^9
Tensile strength t_n (MPa)	18.97	14.58	9.89
Shear strength t_s、 t_t (MPa)	18.97	14.58	9.89
Fracture energy (kJ/m^2)	17	15	12

3.3 Validation of Interface Model

The calculated values of tensile load and displacement at the steel wire loading end were extracted in finite element software and compared with the shear test values for verification. The curve graphs are shown in Figs. 12 and 13, which intuitively show that the finite element simulation calculation results are in good agreement with the shear strength test data.

For the interface shear strength that has not undergone corrosion, the average maximum tensile load obtained through shear strength testing is 595.83 N (shear strength of 18.97 MPa), and the debonding initiation displacement of the interface is 1.40 mm. The corresponding simulation analysis results show a maximum tensile load of 537.54 N (shear strength of 17.11 MPa), with a relative error of 9.79% compared to the experimental value. The debonding initiation displacement is

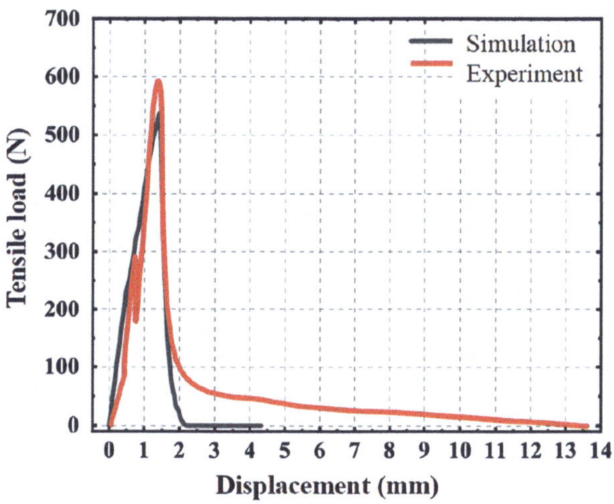

Fig. 12 Comparative analysis of interface shear strength without corrosion

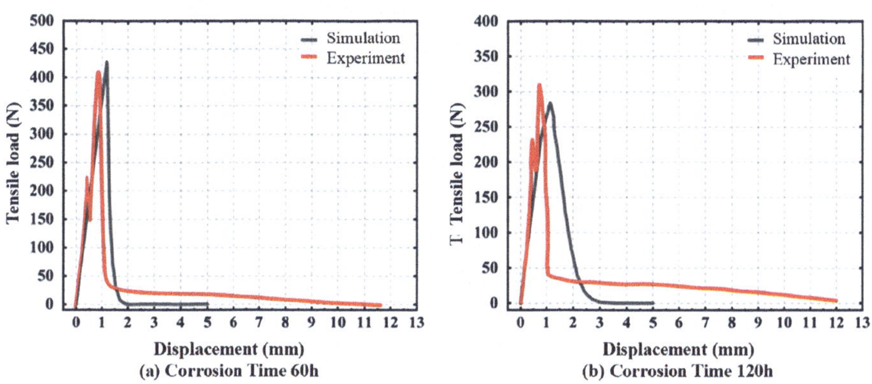

Fig. 13 Comparative analysis of interface shear strength with LCO_2 corrosion

1.41 mm, with a relative error of 0.714% compared to the experimental value; Regarding the interfacial shear strength of 60 h LCO_2 corrosion, the average maximum tensile load obtained through shear strength testing is 458.17 N (interfacial strength is 14.58 MPa), and the initial displacement of debonding at the interface is 0.91 mm. The corresponding maximum simulation analysis results consider the mechanical behavior and damage mechanism of floating LCO_2 cryogenic hose at the interface. The tensile load is 427.14 N (shear strength is 13.60 MPa), with a relative error of 6.77% compared to the experimental value, and the initial displacement of debonding is 1.08 mm, with a relative error of 15.74% compared to the experimental value.

4 Analysis of Interface Damage Mechanism Based on Finite Element Simulation

4.1 Finite Element Model

In order to analyze the influence of steel wire and steel wire-HDPE interface on the mechanical properties of floating LCO_2 cryogenic hose before and after LCO_2 corrosion, a mechanical performance analysis model for the hose is established considering the attenuation of steel wire mechanical properties and the changes in mechanical properties of the steel wire-HDPE interface, as shown in Fig. 14.

The dimensional parameters of the hose are the same as those introduced in Sect. 2.1. During the model construction process, the embedded technology is employed to combine high-density polyethylene with the reinforced steel wire, simulating their bonding characteristics through heating. In the finite element model, the element

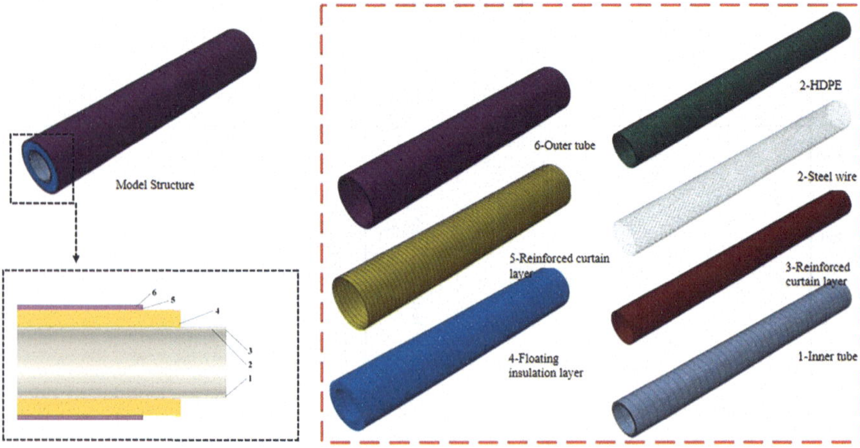

Fig. 14 Finite element model

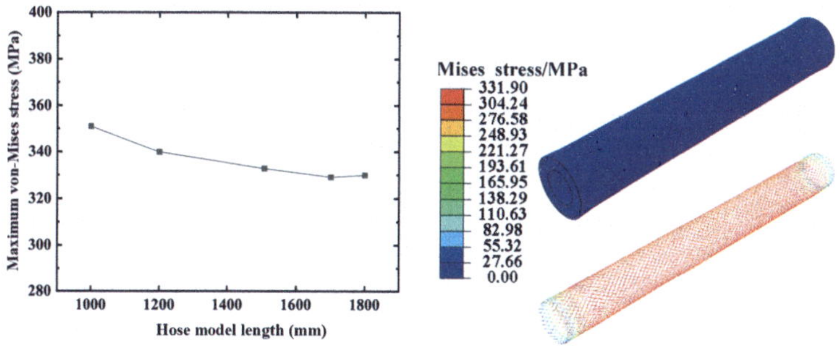

Fig. 15 Analysis results of model length independence

type for the aramid cord layer is SC8R, while the steel mesh is B31, and other areas including the inner tube, outer tube, and floating insulation layer are C3D8. At both ends of the cryogenic hose model, full fixed constraints and different load applications (such as tensile, bending, and torsional loads) are set, and internal pressure loads are added to the inner surface of the hose model. In addition, a temperature field is added to the analysis model of the hose, referencing the operating temperature of LCO_2 transport ships (−27 °C). By analysis of the impact of model length on results shown in Fig. 15, it is concluded that when the model length exceeds 1507.5 mm (twice the spacing between reinforcing wires), the maximum Mises stress of the hose is located on the reinforcing wire, and the variation tends to stabilize, with a change of less than 0.46%. Therefore, the length of the hose model is determined to be 1507.5 mm.

4.2 *Tensile Load Condition*

When considering the effect of low temperature, the details of the damage of the reinforced layer under the combined action of internal pressure and tensile load on the hose are shown in Fig. 16, and it can be clearly found that the the damage value of the reinforced layer increases with the increase of axial displacement, and the damage value of the inner pipe cord is larger than the one of the reinforced. When the axial displacement is less than 15.08 mm, the damage value of the inner pipe cord is larger than that of the reinforcing steel wire and the outer pipe cord, which indicates that at this stage, the internal pressure is the main factor leading to the damage of the hose, and it also indicates that the inner pipe cord is the key component to resist the internal pressure. With the increase of corrosion time, the ultimate tensile displacement of the hose shows a decreasing trend, the ultimate tensile displacement of the hose is 89 mm without considering the interface, 84 mm without considering the corrosion interface, 81 mm with considering the corrosion interface of the LCO_2

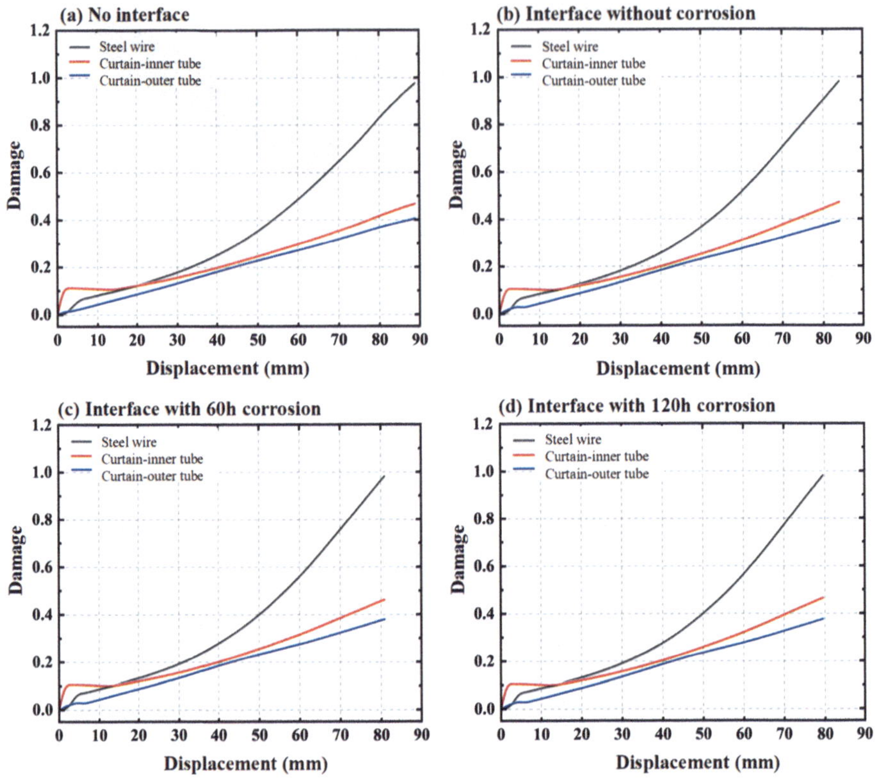

Fig. 16 Damage details of reinforced layers under low temperature + internal pressure + tensile load. (**a**) No interface; (**b**) Interface without corrosion; (**c**) Interface with 60 h corrosion; (**d**) Interface with 120 h corrosion

corrosion of 60 h, and 80 mm with considering the corrosion interface of the LCO_2 corrosion of 120 h. The maximum reduction is 11.25%.

Combining the damage details of the reinforced steel wire-HDPE interface during the internal pressure and tensile load of the hose considering the effect of low temperature, as shown in Fig. 17, it can be seen that, with the increase of axial displacement, the damage value of the reinforced steel wire-HDPE interface in different corrosion states shows the state of increasing firstly and then smooth, which results in this phenomenon for the reason that the reinforced steel wire-HDPE interface is close to debonding in the case of large load when the load is relatively large; when the axial displacement is small, the damage value of the interface increases abruptly, which is also due to the fact that the internal pressure is the main factor leading to the damage of the hose in this stage. It can also be found that with the increase of corrosion time, the damage value of the reinforced steel wire-HDPE interface increases. However, the damage value of the reinforcing steel wire-HDPE interface is much smaller than that of the reinforcing layer, indicating that when the hose subjected to the internal pressure and tensile load is considered to

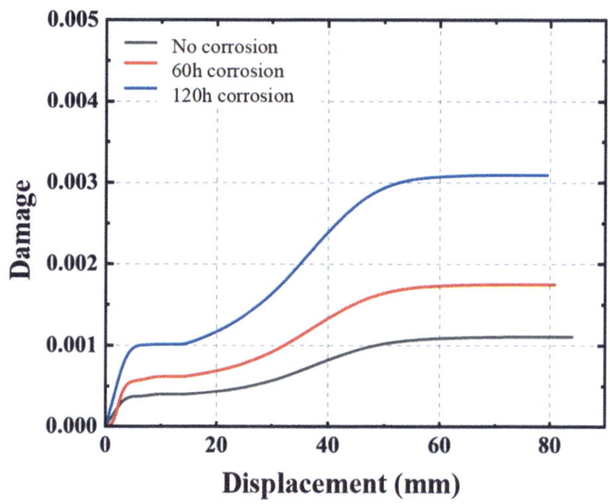

Fig. 17 Damage details of steel wire-HDPE interface under low temperature + internal pressure + tensile load

be affected by low temperature, the damage of the hose mainly occurs in the reinforcing steel wire and reinforcing cord, of which the reinforcing steel wire is the most vulnerable component to failure, and the inner tube cord is the key component to resist the internal pressure load.

4.3 Bending Load Condition

Figure 18 shows the details of damage to the reinforcing layer of a hose subjected to a combination of internal pressure and bending load when the effects of low temperature are considered. It can be found that with the decrease of bending radius, the damage value of the reinforced layer gradually increases, and the maximum damage value of the reinforced layer of the hose occurs in the outer pipe cord in the bending radius greater than 3 m, the maximum damage value of the reinforced layer occurs in the reinforced steel wire when the bending radius decreases gradually, which is because when considering the effect of low temperatures, the basic mechanical properties of the reinforced steel wire parameter changes occur and its stress response becomes larger in the same deformation, making its damage value larger. The ultimate bending radius of the interface hose under these four conditions remains basically unchanged, with the value of 2.08 m.

Considering the low temperature effect of the hose by the internal pressure and bending load under the joint action of the reinforced steel wire-HDPE interface damage details shown in Fig. 19, the interface damage gradually increases with the bending radius decrease. The interface damage in the ultimate bending state occurs a sudden increase, indicating that the possibility of the debonding of interface increases. When the hose to consider the effect of low temperature and the existing

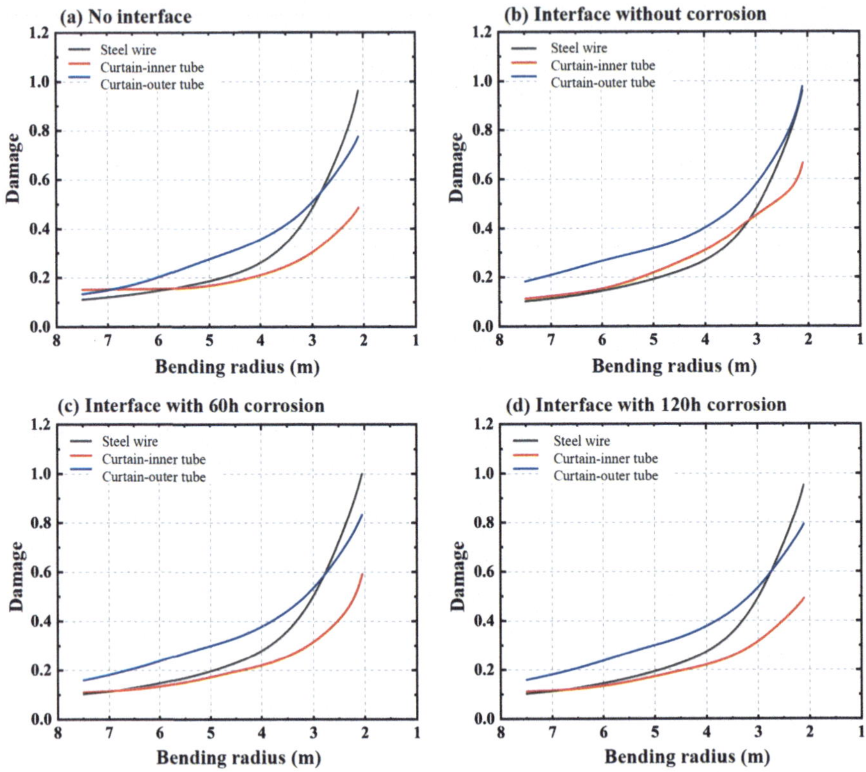

Fig. 18 Damage details of reinforced layers under low temperature + internal pressure + bending load. (**a**) No interface; (**b**) Interface without corrosion; (**c**) Interface with 60 h corrosion; (**d**) Interface with 120 h corrosion

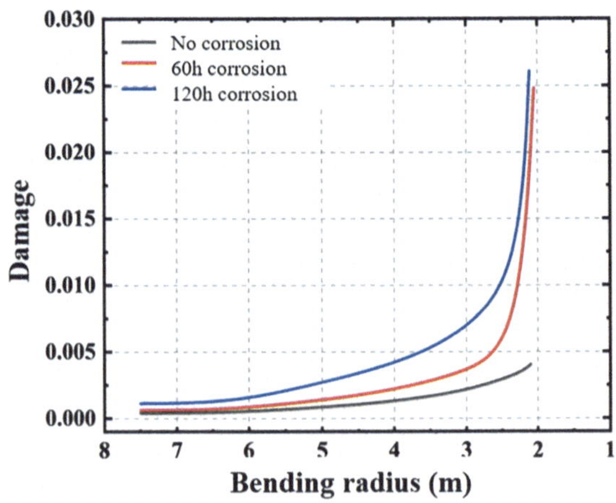

Fig. 19 Damage details of steel wire-HDPE interface under low temperature + internal pressure + bending load

of the internal pressure, the parts which most likely reach damage change with a process. When the bending radius is larger, the outer pipe cord is the most likely to damage, while the bending radius decreases, the damage part turns into the reinforce steel wire.

4.4 Torsional Load Condition

The details of the damage value of the hose affected by low temperature by the combined action of internal pressure and torsion load are shown in Fig. 20. In the small torsion angle, the damage value of the inner pipe cord increases under the action of internal pressure and is larger than that of the reinforcing steel wire and the outer pipe cord, which indicates that the internal pressure carried by the hose is the main load that causes damage in such a state, and the inner pipe cord is a key

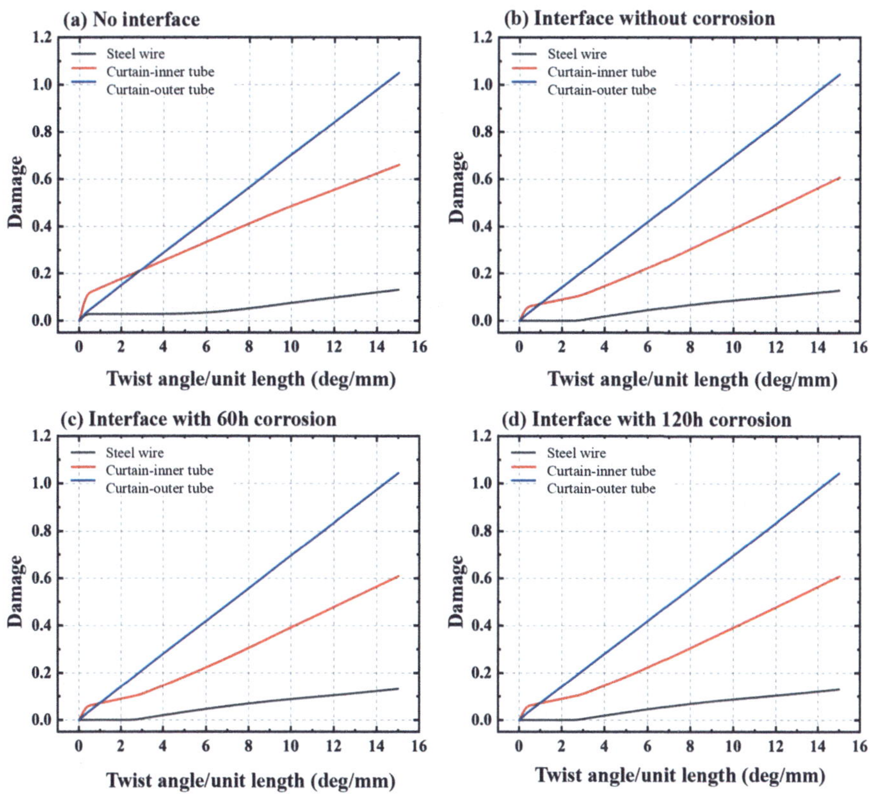

Fig. 20 Damage details of reinforced layers under low temperature + internal pressure + torsional load. (**a**) No interface; (**b**) Interface without corrosion; (**c**) Interface with 60 h corrosion; (**d**) Interface with 120 h corrosion

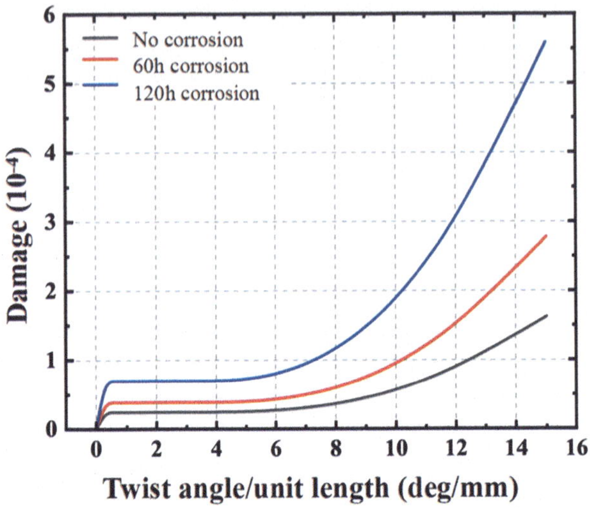

Fig. 21 Damage details of steel wire-HDPE interface under low temperature + internal pressure + torsional load

component to resist the load of internal pressure. However, as the torsion angle increases, the damage of the hose is still concentrated in the outer cord.

Figure 21 shows the details of the reinforced steel wire-HDPE interfacial damage for the combined effect of internal pressure and torsional loading when the hose is considered to be affected by low temperature. The interfacial damage values gradually increase with the increase of torsion angle and with the increase of LCO_2 corrosion time. However, the interfacial damage value is much smaller than that of the reinforcing layer, indicating that in the whole process of torsion to the ultimate torsion state of the hose, the reinforcing layer is the main part to reach damage.

5 Conclusions

In order to accurately obtain the mechanical behaviors and damage mechanisms of floating LCO_2 cryogenic hose, this paper conducts mechanical property and LCO_2 corrosion experiments on the specimens for the hose, laying the foundation for a more realistic evaluation of the changes in mechanical properties. Comparative analysis of changes in mechanical indicators before and after corrosion is made to explore the damage of combined load condition, providing reference for the safety assessment of hose assemblies after long-term service. The following conclusions can be summarized.

1. The experimental platform for LCO_2 corrosion on steel wire and steel wire-HDPE interface is established using equipment such as CO_2 high-pressure reaction kettle with low-temperature jacket and refrigeration cycle bath. The peel strength of the steel wire HDPE interface in the reinforced layer of the hose is determined

through mechanical performance testing, and the variation of the shear strength of the steel wire HDPE interface with LCO_2 corrosion time is analyzed.

2. A simulation analysis model is established for a floating LCO_2 low-temperature hose considering the enhancement of the steel wire HDPE interface. Based on the simulation analysis model and research results of the interface strength between steel wire and HDPE, the damage model of a flexible hose considering interface strength attenuation under low temperature, internal pressure, tensile, bending, and torsional loads is established.
3. The variation laws of reinforcement layer damage and reinforcement steel wire HDPE interface damage under the combined effects of internal pressure and tensile load are obtained when considering the influence of low temperature. Research has shown that the component that is most prone to failure when hoses are subjected to tensile loads is the reinforcing steel wire; The attenuation of the interface strength between steel wire and HDPE caused by LCO_2 corrosion will result in a decrease in the ultimate capacity of the hose.

References

1. Pelto MS. Impact of climate change on North Cascade Alpine glaciers, and Alpine runoff. Northwest Sci. 2008;82(1):65–75.
2. Rosenzweig C, Parry ML. Potential impact of climate change on world food supply. Nature. 1994;367(6459):133–8.
3. Hoegh-Guldberg O, Bruno JF. The impact of climate change on the world's marine ecosystems. Science. 2010;328(5985):1523–8.
4. Parton WJ, Scurlock JMO, Ojima DS, et al. Impact of climate change on grassland production and soil carbon worldwide. Glob Chang Biol. 1995;1(1):13–22.
5. Gao Q, Zhang P, Duan M, et al. Investigation on structural behavior of ring-stiffened composite offshore rubber hose under internal pressure. Appl Ocean Res. 2018;79:7–19.
6. Tonatto MLP, Tita V, Araujo RT, et al. Parametric analysis of an offloading hose underinternal pressure via computational modeling. Mar Struct. 2017;51:174–87.
7. Gao P, Gao Q, An C, et al. Analytical modeling for offshore composite rubber hose with spiral stiffeners under internal pressure. J Reinf Plast Compos. 2021;40(9–10):352–64.
8. Amaechi C, Wang F. Mathematical modelling of bonded marine hoses for single point mooring (SPM) systems, with catenary anchor leg mooring (CALM) buoy application—a review. J Mar Sci Eng. 2021;9:1179.
9. Amaechi C, Adefuye E, Oyetunji A, et al. Numerical study on plastic strain distributions and mechanical behavior of a tube under bending. Inventions. 2022;7:9.
10. Xia M, Takayanagi H, Kemmochi K. Analysis of multi-layered filament-wound composite pipes under internal pressure. Compos Struct. 2001;53(4):483–91.
11. Xia M, Takayanagi H, Kemmochi K. Bending behavior of filament-wound fiber-reinforced sandwich pipes. Compos Struct. 2002;56(2):201–10.
12. Bai Y, Wang N, Cheng P. Collapse of RTP (reinforced thermoplastic pipe) subjected to external pressure. Wuhan, China: ICPTT; 2012.
13. Bai Y, Wang Y, Cheng P. Analysis of reinforced thermoplastic pipe (RTP) under axial loads, vol. 2012. Wuhan, China: ICPTT; 2012. p. 708–24.
14. Bai Y, Xu W, Cheng P, et al. Behavior of reinforced thermoplastic pipe (RTP) under combined external pressure and tension. Ships Offshore Struct. 2014;9(4):464–74.

15. Xu Y, Bai Y, Fang P, et al. Structural analysis of fiberglass reinforced bonded flexible pipe subjected to tension. Ships Offshore Struct. 2019;14(7):777–87.
16. Wei D, An C, Wu C, et al. Torsional structural behavior of composite rubber hose for offshore applications. Appl Ocean Res. 2022;128:103333.
17. Zhang F, Zhang Y, Li Z, An C, Li L. Theoretical and FEA investigation of the mechanical behavior for offshore LCO2 floating hose. Ocean Eng. 2024;312:119107.
18. Xiang Y, Yuan Y, Zhou P, et al. Research on metal corrosion in carbon capture, utilization, and storage: progress and challenges. Chin Eng Sci. 2023;25(3):197–208. (in Chinese).
19. Feng H, Liu Z, Wu G, et al. Research progress on corrosion and material selection of CCUS offshore pipeline facilities. Mater Protect. 2024;57(6):102–12. (in Chinese).
20. Shang Y, Li X, Liang Y, et al. Research on the corrosion law of liquid CO2 on 304 stainless steel. Pet Chem Equip. 2021;24(2):91–6. (in Chinese).
21. Li Y. Study on the interface mechanical properties of cord rubber composite materials. Harbin Institute of Technology; 2009. (in Chinese).
22. Guo X, Yuan X, Zhang Z, et al. Interface failure mechanism of rubber cord composite materials. Polym Mater Sci Eng. 2022;38(11):88–95. (in Chinese).
23. He J, Xian G, Zhang YX. Numerical modelling of bond behavior between steel and CFRP laminates with a ductile adhesive. Int J Adhes Adhes. 2021;104:102753.
24. National Steel Standardization Technical Committee of China. Tensile testing of metallic materials-part 1: Room temperature test method: GB/T 228.1-2021. China Standard Press, Beijing. 2021. (in Chinese).
25. National Technical Committee of China for Standardization of Plastic Products. Steel wire mesh reinforced polyethylene composite pipeline for water supply: GB/T 32439-2015. China Standard Press, Beijing. 2015. (in Chinese).

The Signal Analysis of Lamb Wave in Steel Plates Strengthened by CFRP with Interface Debonding

1 Introduction

Due to the corrosive effects of seawater, microorganisms and their metabolites, as well as the cyclic loading from wind, waves and currents, marine steel structures are prone to corrosion and cracking. CFRP repairing and strengthening is a popular way in ship and ocean engineering structures, civil engineering, aeronautical engineering et al. [1, 2], as shown in Fig. 1, as its good mechanical properties, corrosion resistance and easy operation. For the strengthened or repaired structures, the interface debonding is easy to happen as the deformation incoordination at the interface of two materials, and the debonding is difficult to see from the outside. It is necessary to perform a non-destructive testing to evaluate and predict the interface debonding.

There are many studies on CFRP repairing technology, Chen et al. [4] proposed a method to calculate the stress intensity factor of the crack tip at the interface after repair and proved the effectiveness of CFRP sheet in repaired cracked offshore platform structures. Alaa et al. [5] studied the influence of different loading rates

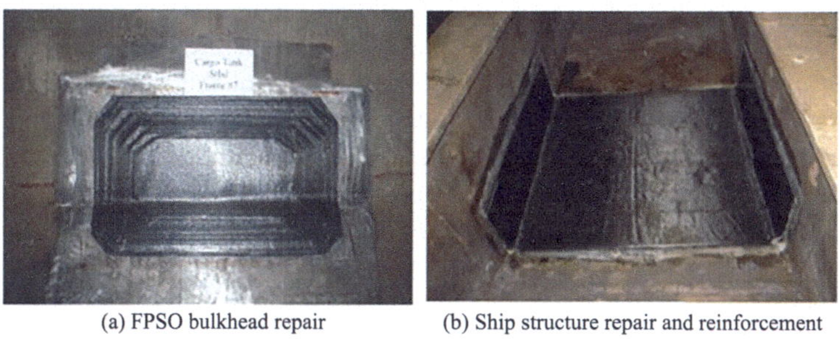

(a) FPSO bulkhead repair (b) Ship structure repair and reinforcement

Fig. 1 Application of CFRP repair technology in ocean engineering [3]. (**a**) FPSO bulkhead repair. (**b**) Ship structure repair and reinforcement

Y. Zhang, *Interfacial Mechanics of Offshore Engineering Equipment*,
https://doi.org/10.1007/978-981-95-3200-1_14

on the bonding properties by combining experiments and finite element models and analyzed the influence of different types of reinforced materials. Zhang et al. [6] used theoretical method and finite element method to establish an interface failure prediction model and studied the stress distribution of adhesives in repaired steel plates with different crack lengths. Zhang et al. [7] studied the durability of CFRP strengthening pipeline with the bending moment and seawater immersion and predicted the service life of repaired pipeline. Elchalakani et al. [8] studied the effect of corrosion penetration in the wall thickness, its extent along the pipe, the type and number of the CFRP sheets on combined flexural and bearing strength of the CFRP strengthened and rehabilitated pipes. Yu et al. [9] used finite element method to study the buckling failure of CFRP repaired marine pipeline with internal corrosion under external pressure.

Common non-destructive inspection methods include ultrasonic infrared thermal imaging, vibration testing, X-ray testing, eddy current testing, and ultrasonic guided wave testing [10, 11]. Compared with other debonding detection technologies, the detection technology based on Lamb wave has the characteristics of wide detection range, high detection efficiency and accurate detection results [12, 13]. Lamb waves was originally developed by Horace Lamb in 1916 to describe the characteristics of waves propagating in plates, and it is sensitive to the detection of typical damage in composites, such as fatigue crack, delamination, plastic fracture, and porosity [14]. Worlton [15] established the equations relating phase velocity to frequency and plate thickness, obtained the dispersion curve of aluminum plate structure, and discussed the multimodal phenomenon of Lamb wave and the distinguishing characteristics of various modes. This is early research into the propagation laws of Lamb waves in plate structures. Since the end of the twentieth century, scholars have begun to deeply study the propagation law of Lamb wave in plate structures and composite structures and to use the Lamb wave for structural damage detection. Alleyne and Cawley [16] studied the interaction between small groove damage and different modes of Lamb waves through experiments, and qualitatively evaluated the groove damage of structures. Saravanos et al. [17, 18] used piezoelectric transducers to study the feasibility of using the Lamb wave to detect delamination in composite beams. Deng et al. [19] studied the physical process of the cumulative second harmonic generated by the propagation of the first Lamb wave in the time domain under the second-order perturbation approximation. Senyurek [20] used Lamb to study the impact damage detection of the actual wing of the Boeing 737 aircraft. Luca et al. [21] used ABAQUS to simulate the guided wave propagation in the damage process of composite plates and studied the influence of different parameters such as the shape, size, and extent of the damage on the reflected wave. Metya et al. [22] studied the creep characteristics of 9Cr–1Mo iron using third harmonics generated by Lamb waves of different frequencies. Rahim et al. [23] summarized the effects of environmental and operating conditions parameters such as temperature, moisture, load, vibration, and other parameters on Lamb wave propagation.

In the research of debonding detection, many authors have investigated the propagation law of Lamb waves in debonding honeycomb sandwich structure, composite laminate and adhesive structure by experiments and finite element method. Song et al. [24] obtained the debonding image results from the probabilistic analysis method and estimated the size of the debonding in the honeycomb sandwich structure. Ng et al. [25] studied the scattering characteristics of Lamb wave in the debonding of laminates were and proved that the A0 mode Lamb wave had a potential application in debonding damage detection of composite adhesive layers through the finite element method and experiments. Sikdar et al. [26, 27] studied hidden debonding regions in honeycomb sandwich composites using monitoring technique based on Lamb wave and investigated the debonding effects on Lamb wave propagation in a bonded composite structure under variable ambient temperature conditions. Allen et al. [28] studied the debonding problem of adhesive joints by using Lamb wave and found that there was a good correlation between the debonding width and the combined harmonic energy.

A review of past literature shows that research has been carried out on Lamb wave dispersion characteristics, propagation laws, and structural damage detection. However, there are few studies on the signal analysis of the Lamb wave used for quantitative debonding assessment in the CFRP repairing or strengthening steel structure. In this paper, the experimental model and finite element model of Lamb wave signal analysis are constructed. The effects of debonding position and debonding degree are discussed, and the relationship between the signal characteristics and the debonding parameters are obtained.

2 Experimental Model

In order to study Lamb wave propagation in the CFRP strengthening steel structure, a specimen of CFRP strengthening steel plate is prepared. The specimen is composed of a steel plate, adhesive, and CFRP, as shown in Fig. 2. The geometry of the steel plate is according to ISO 8501-1-2007 [29], and the material is Q235 whose

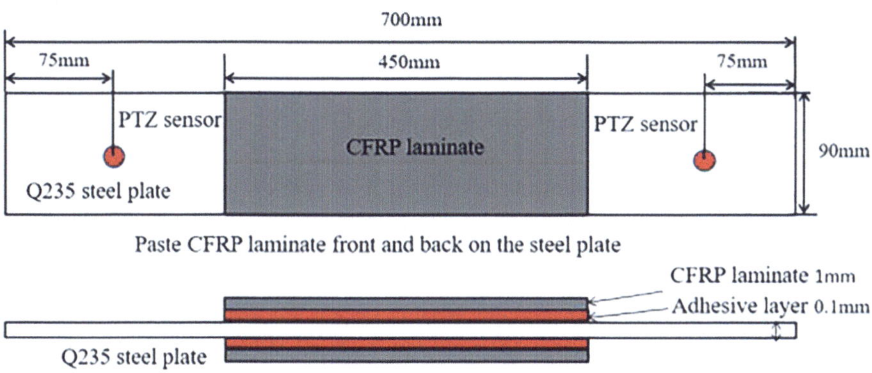

Fig. 2 Geometric dimensions of CFRP strengthening steel composite structures

Table 1 The mechanical properties of Q235 steel

Elastic modulus/GPa	Yield strength/MPa	Tensile strength/MPa	Poisson's ratio
200	297	420	0.3

Table 2 The material properties of CFRP

Density/g/cm^3	Elastic modulus/GPa	Tensile strength/MPa	Poisson's ratio	Elongation/%
1.76	240	3350	0.3	1.5%

Table 3 The adhesive properties

Elastic modulus/GPa	Shear modulus/GPa	Tensile strength/MPa	Elongation/%
2.8	22.08	33	1.85

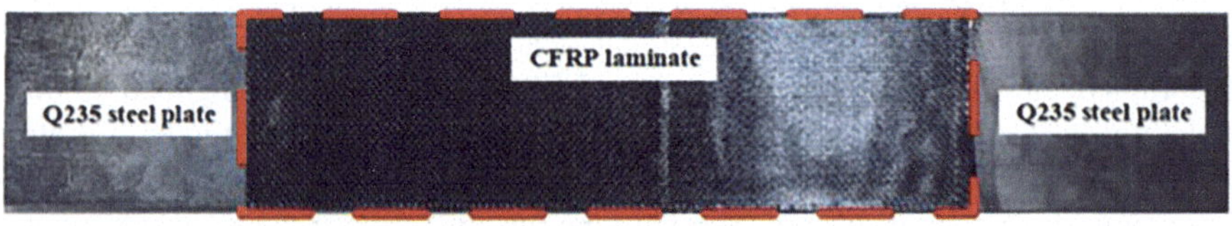

Fig. 3 The specimen of CFRP strengthening a steel plate

Table 4 The piezoelectric ceramic sheet material parameters

Elastic stiffness factor/10 GPa								
C_{11}	C_{12}	C_{13}	C_{22}	C_{23}	C_{33}	C_{44}	C_{55}	C_{66}
13.2	7.3	7.1	11.5	7.3	13.2	3	2.6	2.6
Piezoelectric strain constant/C/m^2					**Permittivity/F/m**			**Density/kg/m^3**
D_{13}	**D_{23}**	**D_{33}**	**D_{52}**	**D_{61}**	**PerX**	**PerY**	**PerZ**	ρ
−4.1	−4.1	14.1	10.5	10.5	7.124	7.124	5.841	7500

material parameters are shown in Table 1. The CFRP laminate T700 has four layers with two-way woven CFRP prepreg, and its laying angle is 90/−90/0/45°. The thickness of the bonded CFRP sheet is 1 mm, and the material parameters are shown in Table 2. Araldite 2015 is used as the adhesive between the steel plate and the CFRP sheet. The thickness of the adhesive layer is 0.1 mm, and the material parameters are shown in Table 3. The prepared structural specimen is shown in Fig. 3.

The signal testing equipment used in the experiment is from Beijing Soft Island Co., Ltd. The piezoelectric transducer (PZT) is RS-35C broadband piezoelectric ceramic sensor, and the material parameters are shown in Table 4. The equipment also includes DS5-8B four-channel signal acquisition instrument, DC28 signal amplifier (100 times magnification), low-noise signal cable, notebook computer display software. The experiment layout is shown in Fig. 4.

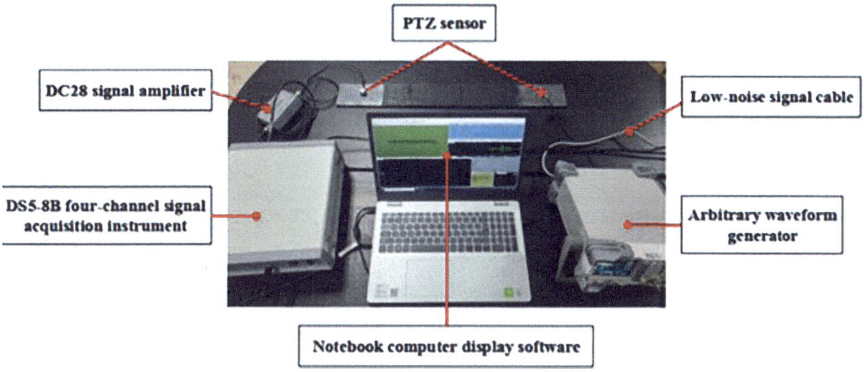

Fig. 4 The overall experimental system layout

The complexity of the CFRP strengthening steel structure determines that the neutral plane of the structure is difficult to find, so it is difficult to study the dispersion characteristic using analytical method. In this paper, the frequency range of the excitation signal is studied by using the finite element modal method. The resonance frequencies are extracted by finite element modal analysis, and the phase velocity (C_p) and group velocity (C_g) are calculated according to Eqs. (1) and (2).

$$c_p = \frac{w}{k} = \frac{L}{p} f_1 \tag{1}$$

$$c_g = \frac{\Delta w}{\Delta k} = \frac{LL'(f_1 - f_2)}{(L' - L)P} \tag{2}$$

p-the number of wavelengths

L-the length of the original model

$L^{'}$-the length of the deformed model

f_1, f_2-the resonant frequency of the original model and the deformed model

w_1, w_2-the angular frequency of the original model and the deformed model

λ-the wavelength, $\lambda = L/p$

k-the wave number, $k = 2\pi/\lambda$

According to the direction of particle motion relative to the middle of the plate, Lamb waves can be mainly classified into two main types of modes. The mode of symmetric motion of particles relative to the neutral plane of the plate is called the symmetric mode (S0, S1 …), and the mode of antisymmetric motion of particles relative to the neutral plane is called the antisymmetric mode (A0, A1 …). Each mode has a corresponding vibration pattern and intrinsic frequency. In this paper, the three most widely used modes in ultrasonic non-destructive testing, A0, S0 and A1, are selected for analysis. The dispersion curves of the CFRP strengthening steel plate obtained from the modal analysis are shown in Fig. 5. From the dispersion curve, it can be seen that the group velocity and phase velocity change relatively smoothly

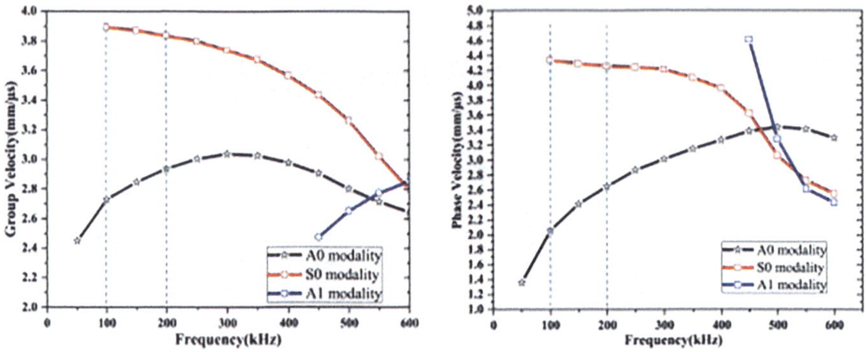

(a) The group velocity dispersion curve (b) The phase velocity dispersion curve

Fig. 5 Group and phase velocity dispersion curve. (**a**) The group velocity dispersion curve. (**b**) The phase velocity dispersion curve

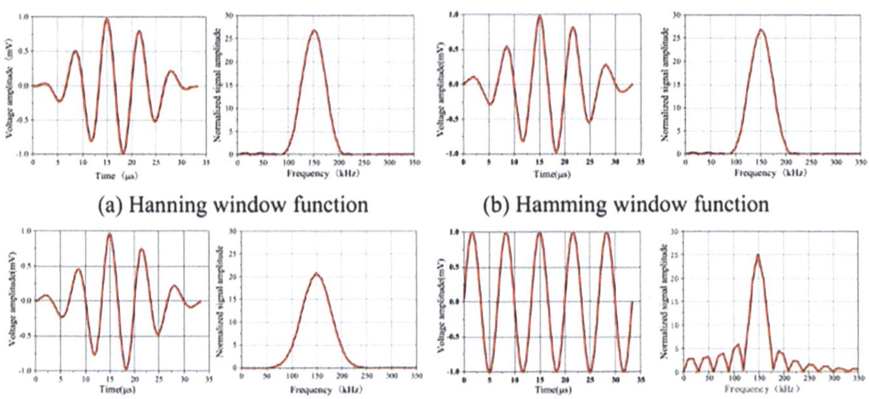

(a) Hanning window function (b) Hamming window function

(c) Gaussian window function (d) Rectangle window function

Fig. 6 Diagram of excitation signals modulated by different window functions. (**a**) Hanning window function; (**b**) Hamming window function; (**c**) Gaussian window function; (**d**) Rectangle window function

and the structure mode is relatively single in the frequency range of 100–200 kHz. In this frequency range, the separability between different modes is strong, which can achieve a better excitation effect and minimize the dispersion of Lamb wave in the process of propagation. The frequency of the excitation signal is selected as 150 kHz that is the middle value of the above frequency range.

In order to suppress the possible dispersion phenomenon of Lamb and to ensure that the generated Lamb wave has high sensitivity and low attenuation characteristics, in addition to selecting a suitable frequency, a suitable window function should be selected to modulate the excitation signal. The excitation signals modulated by different window functions (including Hanning window, Hamming window, Gaussian window and rectangle window function) are shown in Fig. 6. So as to evaluate

Table 5 Quality factors by four different window functions

Parameters	Hanning window	Hamming window	Gaussian window	Rectangle window
Center frequency (kHz)	150	150	150	150
Quality factors	3.82	3.56	3.14	Big leak

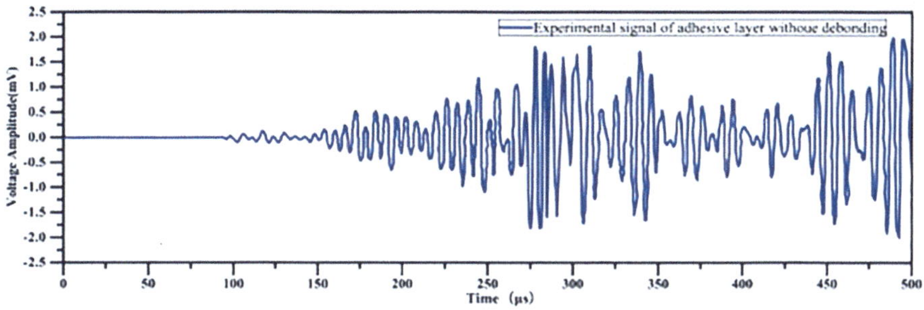

Fig. 7 Experimental signal of CFRP strengthening steel plate

the modulation effect of different window functions, the quality factor Q is introduced, the calculation formula of Q is shown in Eq. (3), and the results are shown in Table 5.

$$Q = \frac{f_0}{|f_2 - f_1|} \tag{3}$$

In the Eq. (1), f_0 is the center frequency of the excitation signal, f_1 and f_2 are the frequency where the signal strength is $1/\sqrt{2}$ times of the peak strength at the central frequency respectively, $|f_2 - f_1|$ represents the signal bandwidth.

By comparing the quality factors of different window functions in Table 5, it can be found that the Hanning window has the best modulation effect. The sinusoidal signal with frequency of 150 kHz, period of 5 and amplitude of 1 V modulated by Hanning window is selected as the excitation signal. The modulated excitation signal expression is shown in Eq. (4), where the A represents the maximum amplitude of the signal, the f_c represents the center frequency of the signal, the n represents the number of signal peaks, and the $w(t)$ represents the window function. In this paper, $A = 1, f_c = 150\text{kHz}, n = 5, w(t) = \left(1 - \cos\frac{2\pi f_c t}{n}\right)$.

$$u(t) = Aw(t)\sin(2\pi f_c t) \tag{4}$$

After the suitable excitation signal has been determined, the non-destructive testing experiment of CFRP strengthening steel plate is carried out. The time domain signal of the Lamb wave propagating obtained by the nondestructive testing experiment in 0–500 μs is extracted as shown in Fig. 7. It can be seen from the Fig. 7 that the head wave signal is received at 90 μs. The first wave signal reaches the PZT2 at 150 μs and the signal strength gradually increases after 150 μs due to the gradual superposition of the waves. The position of the PZT sensor will influence the time when wave superposition occurs. With all other experimental facilities unchanged, the greater the distance between PZT1 and PZT2, the later the time when wave superposition occurs. The arrival time of head wave and the signal energy with different damage conditions can be used to reasonably evaluate and predict the debonding of the structure.

3 FEM Analysis

The finite element model (FEM) of CFRP strengthening steel plate is constructed using ABAQUS software (Fig. 8) and the geometric dimension and material parameters of each part are consistent with the parameters in the experiment. In the interaction setting, CFRP surface and epoxy surface, epoxy surface and steel plate surface, PZT surface and steel plate surface are set as tie.

In order to truly reproduce the real situation of Lamb wave propagation in the structure, the PZT is not simplified in FEM. The excitation PZT and receiving PZT model are established respectively, and the PZT is tied to the upper surface of the steel plate. The electric potential of the lower surface of the excitation PZT and the receiving PZT is set to 0 V, and the upper surface of the PZT1 applies the excitation signal corresponding to in the experiment.

The C3D8R element is used for the steel plate and adhesive layer, SC8R element is used for CFRP, and the C3D8E piezoelectric element is used for PZT. In order to ensure the calculation accuracy, it is necessary to ensure that the number of nodes per wavelength exceeds 10 [16]. The length range of each element should satisfy the Eq. (5), where the $\lambda_{\min}$ represents the minimum wavelength of the Lamb wave in the propagation process, and L represents the mesh length.

$$L \leq \frac{\lambda_{\min}}{10} \tag{5}$$

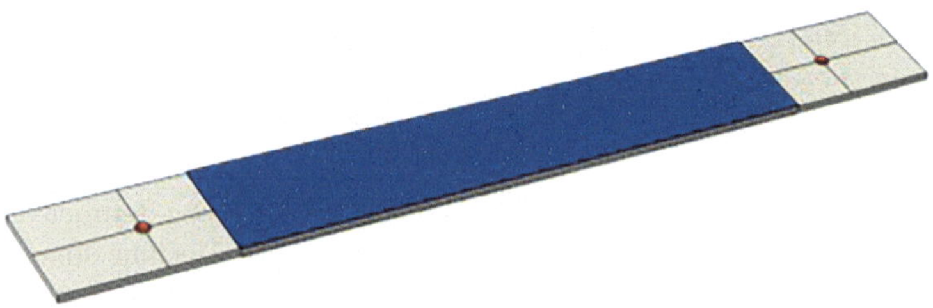

Fig. 8 The FEM model of CFRP strengthening steel plate

Fig. 9 The meshed model of CFRP strengthening steel plate

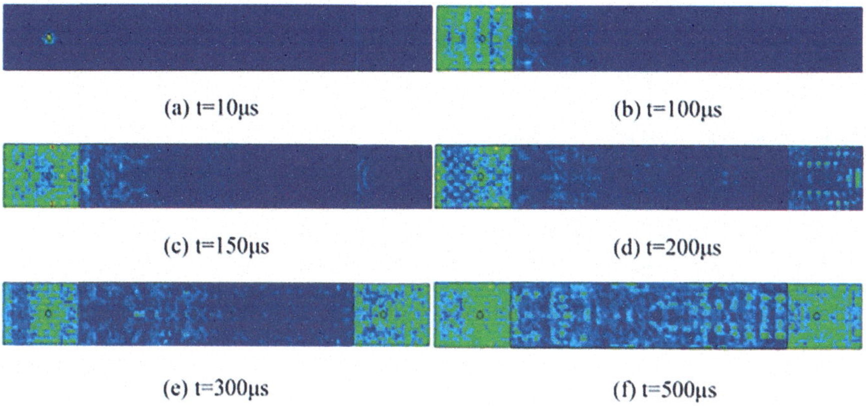

Fig. 10 The signal propagation at different times. (**a**) $t = 10$ μs; (**b**) $t = 100$ μs; (**c**) $t = 150$ μs; (**d**) $t = 200$ μs; (**e**) $t = 300$ μs; (**f**) $t = 500$ μs

According to the dispersion curve of Lamb wave in the CFRP strengthening steel structure, it can be seen that the group velocity of the A0 mode C_{A0} at the selected excitation frequency of 150 kHz is 2841.9 m/s. Therefore, the minimum wavelength of the Lamb wave A0 mode signal propagating in the composite structure can be calculated according to Eq. (6). The mesh length L should satisfy Eq. (7), and the mesh is shown in Fig. 9:

$$\lambda_{\min} = \frac{C_{A0}}{f_{\max}} = 2841.9 \div 1.5 \times 10^5 = 18.95 \ \ (\text{mm}) \tag{6}$$

$$L \leq \frac{\lambda_{\min}}{10} = 18.95 \div 10 = 1.895 \ \ (\text{mm}) \tag{7}$$

Figure 10 shows the complete wave propagation cloud diagram of Lamb wave and Fig. 11 shows the time domain signal extracted from the receiving PZT. It can be seen that the Lamb wave excited by exciting PZT propagates along longitudinally and gradually reaches the receiving PZT after 150 μs, which is consistent with the experimental results.

In order to intuitively compare the difference between the experimental results and the finite element simulation results, the voltage is extracted on the receiving PZT, and the results are compared in Table 6. The maximum error is 18.8324% at 150 μs. This error is mainly due to the uneven thickness of the adhesive layer, which

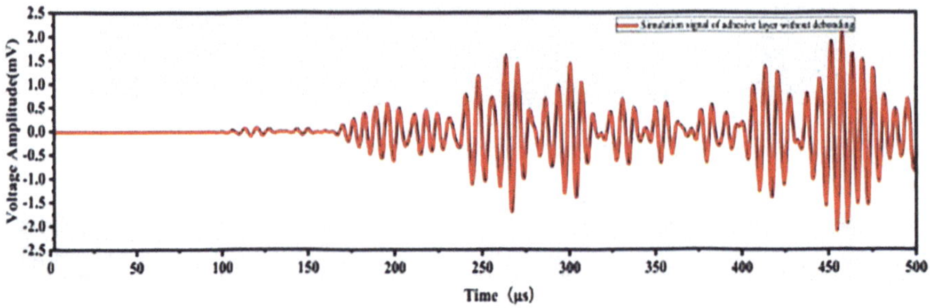

Fig. 11 Simulation signal at the receiving PZT

Table 6 Amplitude comparison between experimental signal and FEM signal

Time (μs)	The signal amplitude of finite element simulation (mV)	The signal amplitude of experimental (mV)	Relative error (%)
50	−0.0157	−0.0162	3.0864
100	−0.0245	−0.0263	6.8441
150	0.0631	0.0531	18.8324
200	−0.3133	−0.2694	16.2955
250	−0.2230	−0.2572	13.2970
300	1.3271	1.2752	4.0699
350	0.2160	0.1932	11.8012
400	0.1345	0.1412	4.7450
450	0.4938	0.5281	6.4950
500	−0.6480	−0.7188	9.8497

is caused by manually pasting CFRP and curing the adhesive layer with the industrial hot air gun. The comparison results show that the FEM can accurately simulate the Lamb propagation in the CFRP strengthening steel plate.

4 Signal Analysis for the Case with Interface Debonding

As mentioned in the above, interface debonding may happen between CFRP and steel plate. In this section, the finite element models with different debonding conditions are established, the appropriate signal processing methods are selected to analyze and process, and the fitting relationship between damage and signal is studied.

4.1 The Case of Different Debonding Positions

This section analyzes the characteristics of signals for different debonding positions. Debonding is simulated by establishing the incomplete epoxy adhesive in the finite element model, and the missing area of the epoxy adhesive model is equal to the debonding area in the simulation process. Five finite element models with different debonding positions which have consistent debonding degree (the debonding degree is 10%) are established, and the schematic diagram of different debonding positions is shown in Fig. 12. The electrical potential of the top surface of the receiving PZT is extracted, and its variation with time is shown in Fig. 13.

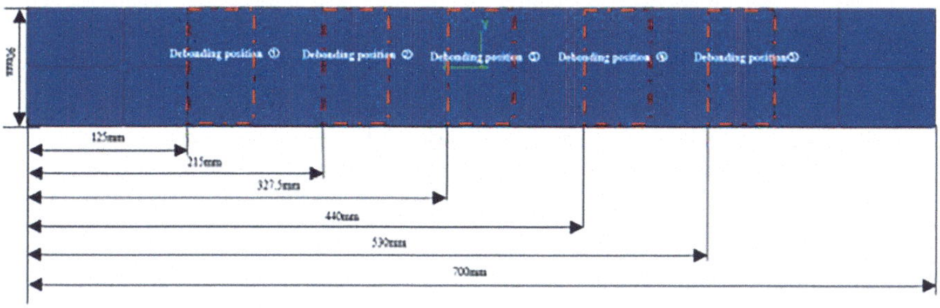

Fig. 12 Schematic diagram of different debonding positions

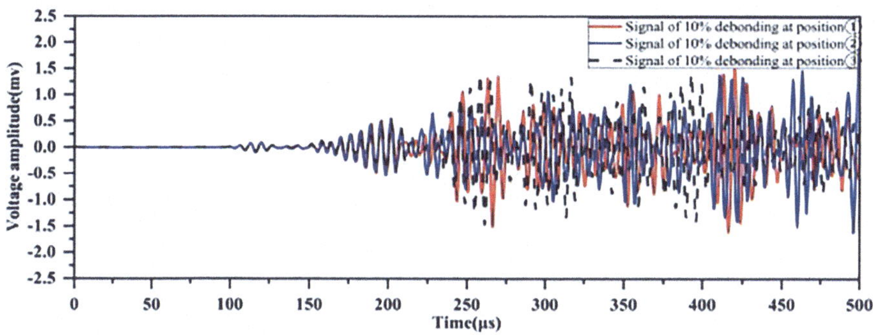

(a) Time domain signals at position 1, 2 and 3

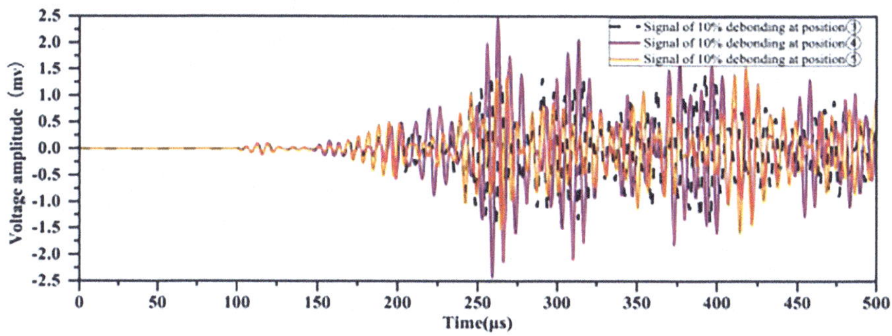

(b) Time domain signals at position 3, 4 and 5

Fig. 13 Time domain signals of different debonding positions. (**a**) Time domain signals at position 1, 2, and 3; (**b**) Time domain signals at position 3, 4, and 5

Comparison of the time domain signals from different debonding positions shows that the signals are well matched before 200 μs, making it difficult to determine the relationship between the debonding position and the signal. Appropriate signal processing methods shall be selected to process the signals. The scattered signal waveform can reflect the details of the signal itself in the time domain. In this paper, the wavelet time-frequency conversion method is selected to analyze the signals, and the time-of-flight (TOF) is introduced to find the relationship between the first arrival wave package in the received signals and the different debonding positions. This method obtains an approximate sinusoidal wave frequency by wavelet translation and scaling based on the initial signal. In order to make the signal processed by the wavelet time frequency conversion method have better smoothness, this paper uses db4 series wavelet basis functions. The time difference between the head wave signal and the excitation signal with different debonding positions are shown in Fig. 14 and Table 7. It can be clearly found that the time difference is obvious, and the time difference between debonding position ① and debonding position ⑤ is 45.820 μs. The relationship between the debonding position and the time difference is fitted by linear regression, and the fitting expression is shown in Eq. (8), where x represents the debonding position of the adhesive layer, y represents the arrival

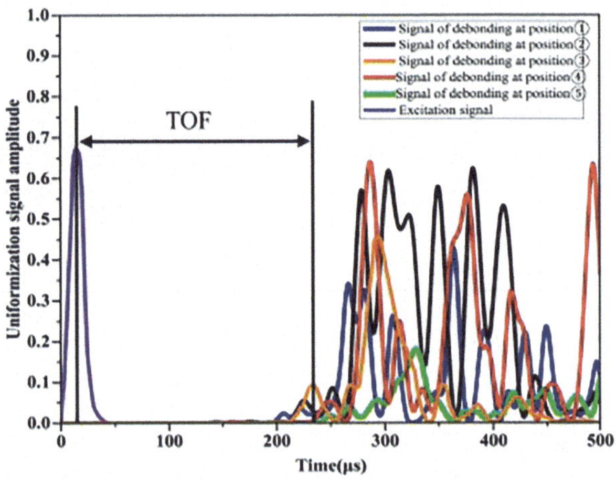

Fig. 14 The signal arriving time with different debonding positions

Table 7 The time difference with different debonding positions

Signals	Arriving time of head wave (μs)	Time difference between the head wave signal and the excitation signal (Δt/μs)
Signal of debonding position ①	206.592	191.319
Signal of debonding position ②	221.277	206.004
Signal of debonding position ③	226.774	211.503
Signal of debonding position ④	249.174	233.901
Signal of debonding position ⑤	252.412	237.139

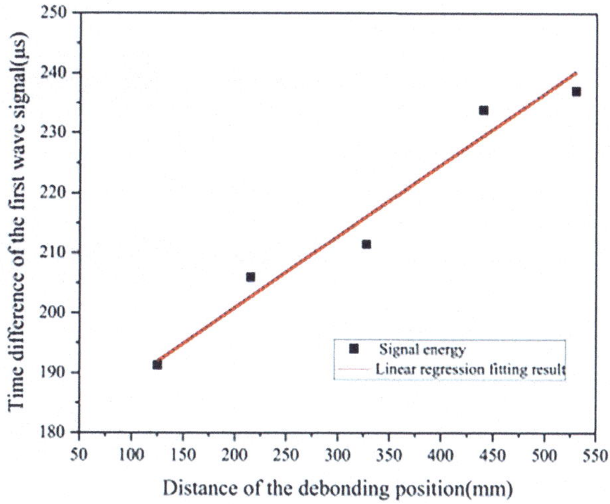

Fig. 15 The relationship between time difference and debonding position

time difference between the head wave signal and the excitation signal. The fitting accuracy parameter R^2 is 0.99809, and the fitting curve is shown in Fig. 15.

$$y = 0.12x + 117.1 \tag{8}$$

4.2 *The Case of Different Debonding Degrees*

In the previous section, the relationship between signals and different debonding positions is investigated. In this section, finite element models are set up with debonding degrees of 5, 10, 15, 20, 25, 30, 35 and 40%, where the debonding percent is the ratio of debonding area to CFRP strengthening area. The debonding position is at the center of the steel plate model, and the debonding degree is changed by changing the length of the debonding area (the debonding width is 90 mm). Figure 16 shows the comparison of time domain signals between different debonding degrees and without debonding. It can be seen that it is difficult to assess the degree of debonding intuitively from the time domain signals alone. Similarly, the arrival time difference between the head wave and the excitation signal Δt with different debonding degrees is calculated by the wavelet time-frequency conversion method, as shown in Table 8. It can be seen that Δt gradually decreases with the increase of the degrees of debonding, and the time difference between 10% debonding and 40% debonding is 19.071 μs, which means that the signal is not as separable as the signal of the different debonding position processed by this method.

The wavelet packet analysis method is selected to process the obtained signal, and the signals are decomposed by three-layer wavelet packet. Figure 17 represents the

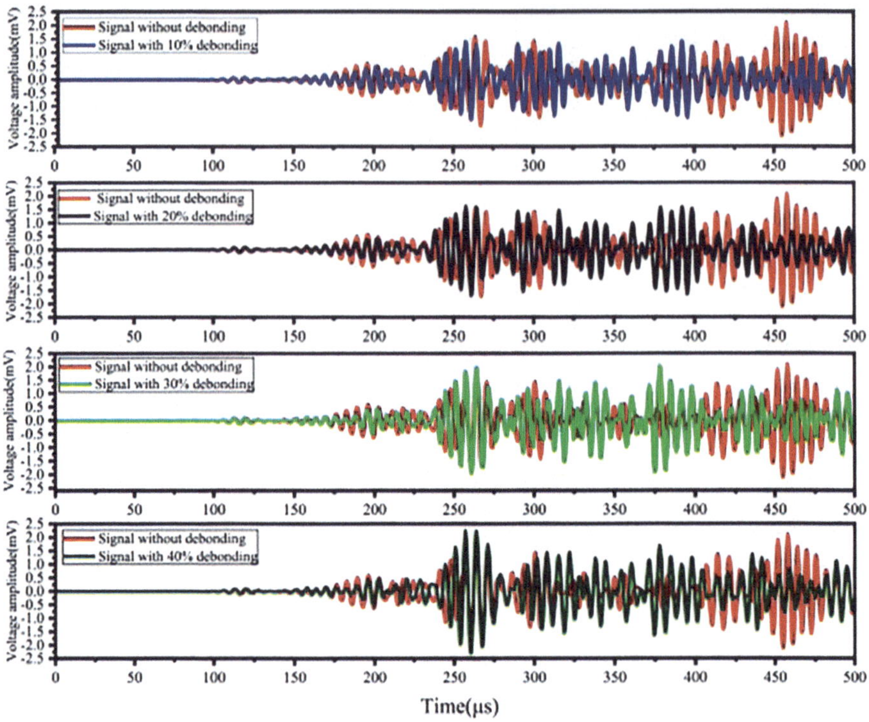

Fig. 16 The comparison of time domain signals between different debonding degrees and without debonding

Table 8 The time difference with different debonding degrees

Debonding degree	Arriving time of head wave (μs)	Time difference between head wave signal and excitation signal (Δt/μs)
10% debonding	226.774	209.625
20% debonding	216.964	199.815
30% debonding	211.872	194.723
40% debonding	206.902	190.554

change in signal energy of each decomposition channel with different debonding degrees. It can be found that the energy proportion of the first channel (energy proportion of the low-frequency signal part) can clearly characterize the relationship between the signal and the debonding degree.

The calculation results of the wavelet packet analysis method are fitted, and the fitting diagram between the debonding degrees and the energy proportion of the first channel is obtained (as shown in Fig. 18). Compared with other fitting methods, the cubic polynomial fitting results are better ($R^2 = 0.99661$), and the expression is more intuitive. The fitted expression as shown in Eq. (9), where x represents the debonding

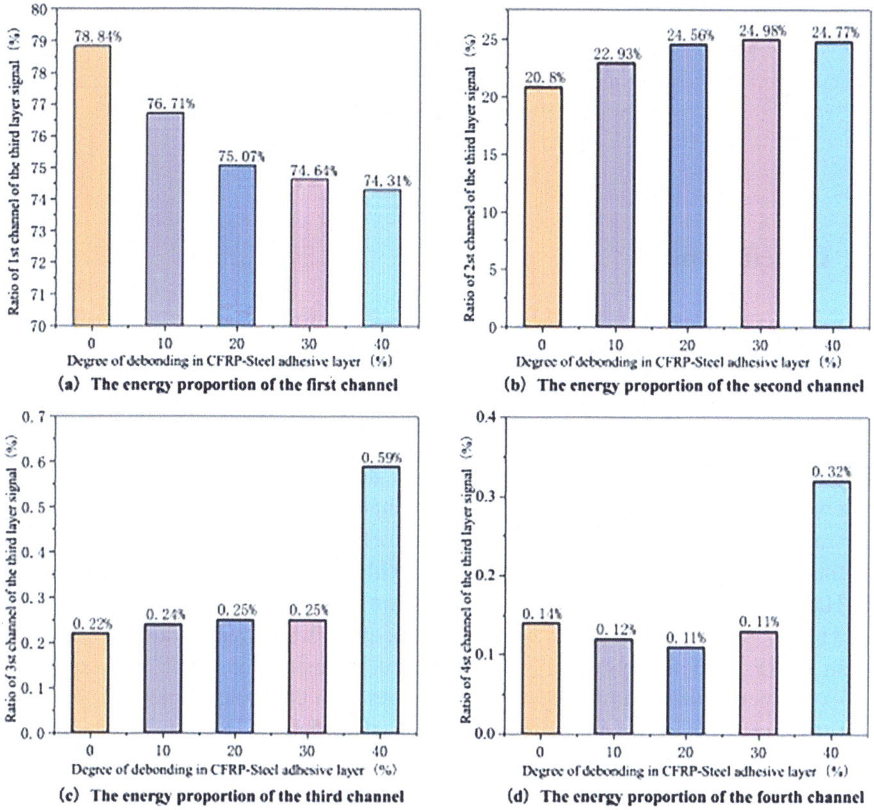

Fig. 17 The signal energy ratio of different decomposition channel. (**a**) The energy proportion of the first channel. (**b**) The energy proportion of the second channel. (**c**) The energy proportion of the third channel. (**d**) The energy proportion of the fourth channel

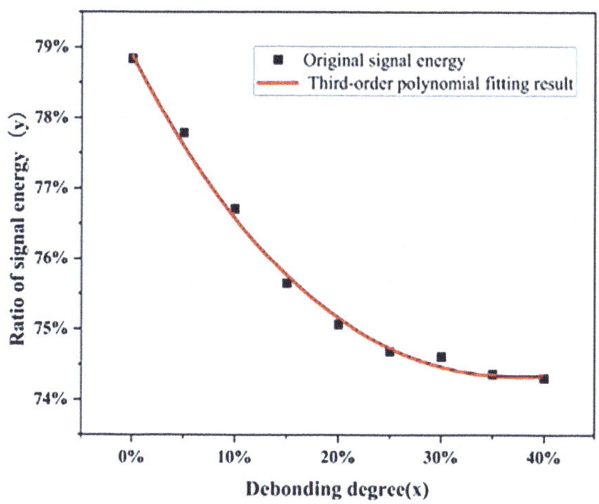

Fig. 18 The relationship between ratio of signal energy and debonding degree

degree of the adhesive layer, y represents the ratio of the first channel signal energy to the total signal energy.

$$y = 0.79 - 0.28x + 0.54x^2 - 0.33x^3 \tag{9}$$

5 Conclusions

In this paper, the interface debonding of steel plate strengthened by CFRP is studied. The signal of Lamb wave in steel plates strengthened by CFRP with interface debonding is analyzed by experimental method and finite element method. The main work and related conclusions are as follows:

1. The nondestructive testing experimental device based on Lamb waves is constructed. According to the dispersion characteristic of Lamb wave in CFRP strengthening steel plate and the modulation effect of different window functions, the sinusoidal signal with frequency of 150 kHz and period of five modulated by Hanning window is selected as the excitation signal.
2. The electromechanical coupling finite element model of Lamb wave propagation in CFRP strengthening steel plate is constructed using ABAQUS software. The experimental time domain signal is compared with the finite element results, and the correctness of the finite element model is verified.
3. The finite element models with different debonding conditions are established. The signal waves are analyzed by the wavelet time-frequency conversion method and wavelet packet analysis method, and the fitting expressions between the signal characteristics and the debonding degrees and positions are obtained. The fitting expressions are obtained under the condition of the same size of steel plate specimen, the same type of adhesive and the rectangular debonding shape. The fitting expression for the debonding position is obtained by analyzing the signals of 10% debonding with different debonding positions, and the expression for the debonding degree is obtained by analyzing the signals of debonding positions at the center of the steel plate with different debonding degrees.

The experimental and modeling methods, signal analysis methods and fitting expressions in this paper can provide reference for the debonding detection of CFRP strengthening steel structure. The method proposed in this paper is applicable to the individual detection of debonding positions and debonding degrees and cannot achieve the simultaneous detection of both. Follow-up research can be conducted by combining big data and artificial intelligence algorithms to achieve comprehensive detection of debonding positions, debonding degrees and debonding shapes for different CFRP strengthening structures.

References

1. Zeng JJ, Gao WY, Liu F. Interfacial behavior and debonding failures of full-scale CFRP-strengthened h-section steel beams. Compos Struct. 2018;201(1):540–52.
2. Gholami M, Sam ARM, Yatim JM, et al. A review on steel/CFRP strengthening systems focusing environmental performance. Constr Build Mater. 2013;47:301–10.
3. Turton TJ, Dalzel-Job J, Livingstone F. Oil platforms, destroyers and frigates—case studies of QinetiQ's marine composite patch repairs. Compos Part A Appl Sci Manuf. 2005;36(8): 1066–72.
4. Chen TH, Chen GM, Lin H, et al. Simulation study of CFRP based repair effect for cracked tubular of offshore platforms. China Pet Mach. 2008;36(10):1–4. +50+96.
5. Al-Mosawe A, Al-Mahaidi R, Zhao XL. Bond behaviour between CFRP laminates and steel members under different loading rates. Compos Struct. 2016;148:236–51.
6. Zhang Y, Zhang K, Zhao H, et al. Stress analysis of adhesive in a cracked steel plate repaired with CFRP. J Constr Steel Res. 2018;145:210–7.
7. Zhang Y, Liu Z, Xin J, et al. The attenuation mechanism of CFRP repaired corroded marine pipelines based on experiments and FEM. Thin-Walled Struct. 2021;169:108469.
8. Elchalakani M, Karrech A, Basarir H, et al. CFRP strengthening and rehabilitation of corroded steel pipelines under direct indentation. Thin-Walled Struct. 2017;119:510–21.
9. Yu J, Xu W, Yu Y, et al. CFRP strengthening and rehabilitation of inner corroded steel pipelines under external pressure. J Mar Sci Eng. 2022;10(5):589.
10. Renshaw JB, Lhota JR, Muthu N, et al. Thermographic inspection of pipes, tanks, and containment liners. AIP Conf Proc. 2015;1650(1):20–5.
11. Pan SS, Zhao XF, Zhang Z. Autoexcitation-based accelerometer array for interface separation detection of concrete-filled steel tubular arch bridge. Appl Mech Mater. 2014;578:995–9.
12. Yelve NP, Mitra M, Mujumdar PM. Detection of delamination in composite laminates using lamb wave based nonlinear method. Compos Struct. 2017;159:257–66.
13. Wang R, Wu Q, Yu F, et al. Nonlinear ultrasonic detection for evaluating fatigue crack in metal plate. Struct Health Monit. 2018;18(3):869–81.
14. Yang S, Lee PH, Cheng JW. Evaluation of general corrosion on pipes using the guided wave technique. J Chin Inst Eng. 2016;39(4):418–28.
15. Worlton DC. Experimental confirmation of lamb waves at megacycle frequencies. J Appl Phys. 1961;32(6):967–71.
16. Alleyne DN, Cawley P. Optimization of lamb wave inspection techniques. NDT & E Int. 1992;25(1):11–22.
17. Sarvanos D, Birman V, Hopkins D. Detection of delaminations in composite beams using piezoelectric sensors. Adaptive Structures Forum; 1994. p. 1754.
18. Saravanos DA, Heyliger PR. Coupled layerwise analysis of composite beams with embedded piezoelectric sensors and actuators. J Intell Mater Syst Struct. 1995;6(3):350–63.
19. Deng M, Xiang Y, Liu L. Time-domain analysis and experimental examination of cumulative second-harmonic generation by primary lamb wave propagation. J Appl Phys. 2011;109(11): 113525.
20. Senyurek VY. Detection of cuts and impact damage at the aircraft wing slat by using lamb wave method. Measurement. 2015;67:10–23.
21. De Luca A, Sharif-Khodaei Z, Aliabadi MH, et al. Numerical simulation of the lamb wave propagation in impacted CFRP laminate. Procedia Eng. 2016;167:109–15.
22. Metya A, Tarafder S, Balasubramaniam K. Nonlinear lamb wave mixing for assessing localized deformation during creep. NDT E Int. 2018;98:89–94.
23. Gorgin R, Luo Y, Wu Z. Environmental and operational conditions effects on lamb wave based structural health monitoring systems: a review. Ultrasonics. 2022;105:106114.
24. Song F, Huang GL, Hu GK. Online debonding detection in honeycomb sandwich structures using multi-frequency guided waves. In: International conference on smart materials and nanotechnology in engineering, Weihai, China, vol. 7493; 2009. p. 74931K.

25. Ng CT, Veidt M. Scattering characteristics of lamb waves from debondings at structural features in composite laminates. J Acoust Soc Am. 2012;132(1):115–23.
26. Sikdar S, Ostachowicz W. Ultrasonic lamb wave-based debonding monitoring of advanced honeycomb sandwich composite structures. Strain. 2018;55(1):e12302.
27. Sikdar S, Fiborek P, Kudela P, et al. Effects of debonding on lamb wave propagation in a bonded composite structure under variable temperature conditions. Smart Mater Struct. 2018;28 (1):015021.
28. Allen JCP, Ng CT. Debonding detection at adhesive joints using nonlinear lamb waves mixing. NDT E Int. 2022;125:102552.
29. ISO 8501-1. Preparation of steel substrates before application of paints and related products visual assessment of surface cleanliness. Switzerland: International Organization for Standardization; 2007.

The manufacturer's authorised representative in the EU is Springer Nature Customer Service Centre GmbH, Europaplatz 3, 69115 Heidelberg, Germany. If you have any concerns regarding our products, please contact ProductSafety@springernature.com

Printed and bound by CPI Group (UK) Ltd, Croydon, CR0 4YY
21/09/2026
02220008-0009